Transform one phone line int[o]

Voice Over DSL

Understanding and implementing VoDSL

CMP Books

www.cmpbooks.com
An imprint of CMP Media Inc.
Converging Communications Group

Voice over DSL

copyright © 2002 Richard Grigonis

All rights reserved under International and Pan-American Copyright conventions, including the right to reproduce this book or portions thereof in any form whatsoever.

Published by CMP Books
An imprint of CMP Media LLC.
12 West 21 Street
New York, NY 10010

ISBN Number 1-57820-106-3

July 2002

First Edition

For individual orders, and for information on special discounts for quantity orders, please contact:

CMP Books
6600 Silacci Way
Gilroy, CA 95020
Tel: 1-800-500-6875 or 408-848-3854
Fax: 408-848-5784
Web: www.cmpbooks.com
Email: cmp@rushorder.com

Distributed to the book trade in the U.S. by

Publishers Group West
1700 Fourth St., Berkeley, CA 94710

Design by: Saúl Roldán
Cover by: Danel Roldán

Manufactured in the United States of America

Contents

Preface ix

Acknowledgments xi

Chapter 1: The Evolving Telephone Network 1
  From Telegraphy to Telephony to Voice over DSL (VoDSL) 1
  Sending Voice Through a Wire
    — The Public Switched Telephone Network (PSTN) 3
    Who Really "Invented" the Telephone? 4
    Starting from the Customer's Premises. 6
      The Subscriber Network Interface (SNI) 9
      Customer Premises Wiring Topology 10
      Twisted Pair Cabling 11
  Powering the PSTN 15
  The Local Loop / Last Mile 17
    The Local Loop's Topology 18
    The Serving Areas Concept 20
    Loading Coils 25
    CDOs, RTs and DLCs 25
    Time Division Multiplexing and the T-Carrier 27
  Getting the Signal Back to the CO 28
    Pair Gain 29
    Line Concentrators 30
    DLCs and Pair Gain are Bad for DSL 30
  The Central Office 31
    The Cable Vault 31
    The Main Distribution Frame 32
    The Switch 34
  The 1984 Divestiture 42

Chapter 2: **Data-over-Voice, Voice-over-Data** 45

Facsimile 46
Interoffice Signaling 47
Circuit-Switched Networks versus Packet-Switched Networks 48
    Circuit-Switched Networks 48
    Packet-Switched Networks 49
        Connectionless Service 50
        Connection-Oriented Service 51
        Fast Packet Services 51
        ATM's Architecture 57
        IP and the Internet 67
The IP / ATM Debate 71
Summary 72

Chapter 3: **DSL Explained** 73

DSL's Story 74
    Hertz vs. Bits 75
    The Birth of DSL 76
        ADSL 78
        The DSL Modem 80
        The IAD 82
        Regulatory Boosts 82
The Different Flavors of DSL 83
    ADSL: Its Advantages and Disadvantages 84
        The Distance Equation 85
        Splitters and Filters 86
        ADSL Lite 87
        Why Asymmetric? 88
        VoDSL and ADSL 88
        VDSL 89
    Symmetrical DSL 89
        IDSL 90
        SDSL 90
        HDSL2 91
        HDSL4 93
        G.SHDSL aka SHDSL — The "New" Standard 93

Clarifying How DSL Works 98
    Encoding and Upper-layer Protocol Stacks 99
        The DSLAM 101
        The DSL "Modem" or ATU-R 103
    Conditioning the Local Loop for Optimal DSL Service 103
        Load Coils 104
        Bridged Taps 106
        Digital Loop Carriers 107
        Remote DSLAMs 108
        Line Cards 110
        Remote Access Multiplexer (RAM) 111
        Attenuation 112
        Crosstalk 113
    POTS Equipment 115
In Summary 115

Chapter 4: VoDSL - The Basics 117
  The VoDSL Upgrade 118
    QoS 119
    Limitations in the DSL Technology 120
    Deploying the IAD — Complexity vs. Ease of Use 120
        Scalable IADs 121
        Fixed-configuration IADs 121
        Flexible-configuration IADs 122
        The Current Trend 122
    The Softswitch 123
    The Biggest Headache 124
  VoDSL's Ancestors can Retire Now 124
    ISDN 125
    T-1/E-1 125
    Analog 126
  How VoDSL Works 126
    The DSL Technology 126
        SDSL 126
        G.SHDSL 127
        Other Options 128

Power Considerations 128
The IAD Connection 128
    Encoding and Compression Techniques 129
The PSTN's Role 130
Decisions, Decisions 132
The Value Proposition 132
    The Service Provider's Point of View 132
    The Customer's Point of View 133
In Summary 134

Chapter 5: Four Roads to VoDSL 137

Overview of the Four VoDSLs 138
    "Digital" Isn't always Digital 142
Broadband Loop Emulation Service (BLES) 146
    ATM or Frame Relay for Loop Emulation? 150
    ATM-based VoDSL 151
        AAL2 for Voice Packets over Loop Emulation-based VoDSL 154
        AAL5 for IP Data Packets 169
Voice over Multiservice Broadband Networks (VoMBN) 172
    ATM vs. IP 173
    The Particulars of VoMBN 184
        Transporting a Voice Packet over VoMBN using IP and ATM 184
        VoMBN and the "Next Generation" Network 187
Voice-over-Ethernet-over-DSL 199
Channelized VoDSL (CVoDSL) 202
In Summary 204

Chapter 6: Customer Premise Equipment 205

The IAD - The Enabler 206
    Differing IAD Architectures 207
        Value-Added Services 210
        VPNs 211
    Meet the Vendors 213
    Interoperability 215
    Getting to Know Your IAD 215
    IAD Management 216

Soft IADs	217
Choosing an IAD	218
The Future	218
Wrap-up	218
Residential Access Devices	219
Gaming Consoles as IADs	220
Network Interface Devices and More	220
Receiving Devices	221
Analog Instruments	221
Faxes	221
Digital and IP Instruments	222
Video Conferencing Devices	223
Videophones	223
In Summary	225

Chapter 7: **Considerations When Choosing a Service Provider** 227

Service Provider Types	227
"Deconstructing" the Service Provider and its Capabilities	228
Transportation Mode	230
ATM	230
IP	230
Network Infrastructure	231
Next Gen VoDSL Solutions	233
Technical Aspects of the Provider's VoDSL	236
Interoperability	236
Scalability	237
Flexibility	237
Compression	238
Reliability and Quality	239
Fax and Modem Support	243
Network Management	251
Bureaucratic and Other Issues	251
Some Service Providers Use Outsourcing	252
In Quest of the Full Digital Loop (FDL)	253
In Summary	257

Chapter 8: VoDSL - A Compelling Business Case 259
VoDSL Call Centers 260
"In-Building" DSL and Building Local Exchange Carriers (BLECs) 261
   Alphabet City: MTUs, MDUs and MHUs 262
   The Building Local Exchange Carrier (BLEC) 263
      BLEC-specific Equipment and the Features They Offer Customers ... 266
      The Evaluation Process 272
      Can a BLEC "Outgrow" DSL? 275
   If you can't beat 'em... 281
Virtual Private Networks (VPNs) 282
   Voice over VPNs 286
   Centrex DSL is not VoDSL 290
   A VPN in Every Business and Home? 291
ASPs 291
In Summary 294

Chapter 9: Why VoDSL Will Succeed 295
   Benefits of VoDSL 296
   Advantages for the Business Community 297
   Advantages to the Service Provider 299
      ILECs 299
      Service Providers Other than ILECs 300
   Economics 302
      The Value Proposition 302
   Attractive Bundled Services 303
   Caveat Emptor — Let the Buyer Beware. 305
   The Prospects are Bright 306
      The Softswitch's Pivotal Role 307
      Applications 308
   In Summary 309

Chapter 10: The Future 311
VoDSL Transformed by Evolving Technology 311
   The Future Dominance of G.SHDSL 311
   "Next-Gen" Networks and the Rise of the Applications Server 313
   Managed VPNs 318

    What's Next 324
    Beyond G.SHDSL: VDSL and Ethernet Hybrids 326
VoDSL Usage Transformed by Government Legislation 328
In Summary 334

Appendix A: **The Open Systems Interconnection (OSI) Model** 335

Appendix B: **A Plethora of Protocols** 341

Glossary 355

Index of Figures 417

PREFACE

The idea for a practical guide to Voice over DSL (VoDSL) was conceived during discussions between the author and the CMP Book division during 2001. It was originally to be a "simple little book" explaining VoDSL's technology and business opportunities.

As things turned out, the writing and publication of this opus took six months longer to complete than I had anticipated; the the events of September 11, 2001 compelled me to drop everything I was doing in order to write the book, *Disaster Survival Guide for Business Communications Networks* (CMP Books, 2002).

Then, when I finally returned to continue work on *Voice over DSL*, I discovered that there was much more to the world of VoDSL than meets the eye. There are various competing approaches to this technology, all of which will eventually meet up with the much-anticipated "next generation" network that one day shall comprise most of America's — if not the world's — voice and data communications infrastructure. Whether we ever actually get to live in the utopian visions conjured up by the present generation of experts and technologists, remains in part up to the Byzantine workings of the United States Congress and the Federal Communications Commission.

Yes, the "simple little book" I was going to write quickly grew into a tome more expansive than anyone could have anticipated. But so much the better for the reader, for information on VoDSL has either been scant or expensive to obtain. Aside from a couple of high-priced reports, this is the first commercial book ever published solely on the topic of VoDSL.

Preface

The advantages of implementing VoDSL in a small or medium-sized business are significant, noteworthy — and any other similar adjective you can think of!

Whether you eventually decide to start a call center, do some videoconferencing, or set up a virtual private network for teleworkers, I hope you'll enjoy your exploration of the many incredible options that VoDSL can provide.

Richard Grigonis
New York City
June 2002
RGrigonis@CMP.com

ACKNOWLEDGMENTS

I want to thank CMP Books and its director, Matt Kelsey, for having the vision to see the exciting possibilities afforded by VoDSL, for giving me "the green light" and for supplying all the necessary support that brought this project to a successful conclusion.

Assistance was afforded me by everyone, but first and foremost I must single out the extraordinary work done by my editor, Janice Reynolds, herself the distinguished author of books that include *A Practical Guide to DSL* (CMP Books, 2000). Without her boundless energy and enthusiasm, this book would probably have taken an additional six months to complete.

Many thanks should also go to veteran art director Saul Roldán, who has taken my text and graphics files and has fashioned them into a marvelous work, a stunning example of typography and graphic design.

> *It is my warm and world embracing Christmas hope and aspiration that all of us — the high, the low, the rich, the poor, the admired, the despised, the loved, the civilised, the savage may eventually be gathered together in a heaven of everlasting rest and peace and bliss — except the inventor of the telephone.*

— **Mark Twain, 1890.**

CHAPTER ONE

The Evolving Telephone Network

Voice over Digital Subscriber Line (VoDSL) is but the latest chapter in a long saga of how a huge, somewhat primitive telephone system has become a modern digital communications system capable of delivering integrated voice and data services to anyone who rents a telephone line.

From Telegraphy to Telephony to Voice over DSL (VoDSL)

Once upon a time, in the mid-19th Century, the only rapid long-range electrical telecommunications system was the telegraph network, nearly all of which was owned by the Western Union Telegraph Company. Telegraph lines needed lots of expensive conducting iron wire. Moreover, telegraphy itself was inefficient. One telegraphic message occupied a whole line during its transmission. Western Union needed a way to squeeze many simultaneous telegraphic signals onto the same wire.

In the early to mid-1870s inventors such as Elisha Gray (1835-1901) and Alexander Graham Bell (1847-1922) initially worked on what was called a "harmonic telegraph," which had the ability to send several telegraphic signals over the same line simultaneously in either direction. Thomas Edison (1847-1931) had already patented a simple multiplex telegraph in 1870, and in 1875 Jean Maurice Emile Baudot (1845-1903) of France increased the transmission speed by a factor of four via a difficult-to-use multiplexing system. Even so, it was thought that 30 or 40 simultaneous signals could be made to occupy the same telegraph line. So important was this idea that Western Union had set up a $1 million prize to anyone who could multiply the capacity of the telegraph network.

Telegraphy is a naturally digital medium. The dots and dashes of Morse code are easy to detect since they are massive pulses that either appear in the receiving electromechanical "sounder" devices that make the dot and dash sounds, or they don't. Normally, trying to send more telegraphic messages on the same line would cause the dots and dashes to overlap, thus making the messages unintelligible. The electro-harmonic or electro acoustic multiplex telegraph, however, consisted of a series of different sized steel reeds. Each reed was vibrated by an electromagnet and generated a unique audible tone. The current vibrating one reed, when passed over a telegraph line, would set in motion at the other end a reed exactly corresponding to the first in rate of vibration, and cause it to yield the same note, while a reed tuned to a different note would be unaffected. By interrupting each vibrating reed in Morse code style, the corresponding reeds at the other end of the telegraph line would be interrupted also, causing clearly recognizable audial breaks in the singing tone emitted by the vibrator. Hence, the "twin" reeds vibrated at specific frequencies in sympathy with each other, establishing separate channels.

Inventors soon realized that if many such frequencies could be sent in this way, and if the human voice consists of various frequencies, then perhaps a telephone could be developed.

Around mid-1875, Alexander Graham Bell became sidetracked from his work on a harmonic telegraph when he found he could transmit sound. (His harmonic telegraph was also based on multiple steel armatures or "reeds" of different specific lengths at each end of a line.) One day (June 2, 1875) when his assistant Thomas Watson tested a reed by plucking it, another reed in a receiver vibrated even though the battery current was switched off. Bell realized that there was still enough residual magnetism in the iron cores of the transmitter/receivers so that manually vibrating one reed would set up an alternating current (or "undulating current" as they called it in those days) that vibrated another reed in another device at a distance. Furthermore, Bell pressed his ear against one reed and was amazed to discover that he could hear faint sounds of the other reeds being plucked by Watson in the other room. Bell now understood that a single reed relay could receive complex sounds, such as speech. This line of research ultimately led to a working telephone. Elisha Gray later claimed he discovered a similar phenomenon back in 1867, but unlike Bell, Gray never documented his sources.

Thus, attempts to send several digital telegraph signals through a wire led to the telephone, although it's greatly unlike the telegraph in that the subtle, smooth electrical waveforms it transmits closely resemble (or are an analog of) the acoustic sound waveforms of speech. The telephone converts or transduces varying sound waves into nearly identical varying electrical currents.

Today, in a situation mirroring the past, new communications technology is able to send more telephone conversations (and data) simultaneously over the legacy copper wires that lead to your home or business. While this new technology is called Voice over Digital Subscriber Line (VoDSL), there is, ironically, nothing digital about it! As we shall see, DSL uses *analog* frequencies higher than those used for ordinary telephone conversations, so it doesn't interfere with the existing phone system. As we shall also see, DSL can squeeze additional voice channels and data onto phone lines using a technique called *Discrete Multi-Tone (DMT)*, which works by separating incoming and outgoing information into many small frequency bands that are used as separate channels. Each unique channel has associated with it modem circuitry that deals with that channel (or "subcarrier") alone.

You might therefore call VoDSL "the harmonic telephone"!

Some of the links with our technological past are expensive to break. Because of the vast distances involved in telecommunications, great pains have always been taken to inexpensively upgrade the legacy infrastructure. Just as many telegraph lines laid in the mid-19th Century were converted into telephone lines in the 1870s and 1880s, today many telephone lines are being stripped of extraneous coils and extra hanging bits of wire, all in anticipation of their conversion to "DSL lines."

The old copper wires, it seems, can still be taught a few new tricks.

Sending Voice Through a Wire — The Public Switched Telephone Network (PSTN)

When pressed to explain how the telephone system works, just about everybody outside of the communications industry offers the same vague explanation: A phone is connected via a wire (actually a set of wires in a cable) to a wall jack, or outlet. The wires then go to some box, though a wall, out to a "telephone pole" and then all the wires in the area converge on a Central Office, where a big, room-filling thing called a "switch" can route a phone call long distances to other big switches, one of which will route the call to the phone of the person you're calling.

That's not too bad a start for describing what we refer to as the PSTN, or Public Switched Telephone Network (in Europe and some other places they call it the GSTN, or General Switched Telephone Network). The PSTN furnishes its subscribers with POTS, the Plain Old Telephone Service. Everything new in the way of communications either replaces, is added to, is built upon, or must interface in some way with this original network infrastructure, construction of which began in the U.S. in 1876 following the "invention" of the telephone by Alexander Graham Bell (1847-1922) and which continues to this day throughout the world.

Who Really "Invented" the Telephone?

I put the word invented in quotation marks since in recent years revisionist historians, while acknowledging that Bell made the first authentic demonstration of the transmission of speech by telephone ("authentic" in modern parlance meaning that it was recognized for what it was immediately, received tremendous publicity and acclaim, and led to perhaps the greatest single business success of all time) are also fueling a growing controversy as to whether other inventors of the mid-19th Century actually developed enough key components of the telephone to be considered co-inventors, or perhaps even had already designed workable devices capable of transmitting speech.

In 1860, after eight years of research, a brilliant though sickly and impoverished 26-year-old teacher in Germany named Johann Philipp Reis (1834-1874) built the first in a series of different telephonic devices, each of which he called *das telephon*.

At some point in 1860 or early 1861, Reis believed his device could transmit intelligible speech. Supposedly, the first words sent were "*Ein Pferd frisst keinen Gurkensalat*" which translates into English as "A horse does not eat a cucumber salad." (Not exactly as dramatic as Bell's "Mr. Watson — come here — I want to see you," but it served its purpose.) There were problems with Reis' phone, however, and he died before he could perfect his invention.

The next figure to appear on the scene, Antonio Meucci (1808-1889), may in fact be the real inventor of the telephone. Born in Florence, Italy, he moved to Cuba where he did some initial experiments on a telephonic device, which he called a *telegrafo parlante* ("speaking telegraph"). Meucci actually built such a device in Havana in 1849, when Alexander Graham Bell was just a two-year-old living in Scotland.

Meucci moved to Staten Island, New York, on May 1, 1850 and continued experimenting. Meucci's telephone would be completely transformed into a sophisticated electromagnetic device as it went through 26 design changes over the following decades. He would later call it the *telettrofono* or "telectrophone."

Meucci left documentation indicating that he had perfected what are today called "loading coils," as well as a way to eliminate the so-called "side-tone" effect which resulted in the speaker hearing an echo of his own voice. Anti-sidetone circuits would not be heard of again until 1918, when nine patents relating to them were registered by George Ashley Campbell (1870-1954).

Though impoverished, Meucci and three Italian partners established the Telettrofono Company on December 12, 1871. But to file a U.S. patent cost $250 in the year 1871 — the equivalent of about $3,600 in the year 2001 — so he instead took out a less expensive $10 caveat ($143 in 2001 currency) which is a one-year notice of intent to file a patent (these caveats or "notices of invention" were a 19th Century version of what we would today call a provisional patent). Entitled "Sound Telegraph," this caveat (No. 3335) was filed on December 28, 1871 and was renewed on December 9, 1872 and December 15, 1873. But Meucci, now ill and living on public assistance, allowed the caveat to lapse in December 1874. Ten dollars might have changed history.

Meucci's "day in court" against Alexander Graham Bell and his company consisted in part of a relatively swift patent infringement trial (*American Bell Telephone Co. et al. vs. Globe Telephone Co. et al., 1885-1887*). Meucci's own testimony explained in every detail his experiments, and there were affidavits sworn by various witnesses between 1880 and 1885 which substantiated Meucci's ability to send voice through a wire more than 20 years prior to Bell's first patent of 1876.

To "dispose" of the affidavits of all witnesses who actually talked through Meucci's telephones, Bell's attorneys employed a completely outrageous trick that they had used in a previous telephone patent infringement case. The trick involved the claim that Meucci's phone at that time was not electrical, but mechanical — that it was the equivalent of the "two cans and a string" toy used by children!

That, along with some other strange legal shenanigans led Judge William Wallace to rule against Meucci on July 19, 1887.

Meucci's great modern vindicator is Professor Basilio Catania, former head of Italy's Central Telephone Research Laboratories and recipient of the 1988 Eurotelecom and 1991 Marconi Prizes.

Since his retirement in 1989, Catania has gone about the arduous task of revising the history of telephony; predominately through a series of articles, speeches, but also, through his ongoing work on a four volume biography of Meucci, *Antonio Meucci - The Inventor and His Times (Antonio Meucci — L'Inventore e il suo Tempo)*. Two volumes (in Italian), which can be purchased via the Internet at www.levrotto-bella.net, were available at the time this book went to press.

Most notable was his eye-opening lecture, "Antonio Meucci, Inventor of the Telephone: Unearthing the Legal and Scientific Proofs," delivered October 10, 2000 at New York University's home for Italian studies, the Casa Italiana Zerilli-Marimo.

Our next victim of history is Elisha Gray (1835-1901). Gray is famous for supposedly arriving at the U.S. Patent Office "two hours" after Alexander Graham Bell's associates delivered Bell's patent application. There is much more to the Gray-Bell controversy, but it's too complex to discuss here in its entirety. One should read A. Edward Evenson's brilliantly researched book, *The Telephone Patent Conspiracy of 1876* (Jefferson, North Carolina: McFarland & Company, Inc., 2000), to get the full details.

Starting from the Customer's Premises. . .

Regardless of who invented the telephone, for many people it quickly became a household necessity. Farmhouses without running water soon had them. Indeed, the telephone even made the skyscraper possible — without it, could you imagine thousands of messenger boys running up and down all of those stairs? And yet for nearly a century after its beginnings, the substrate of the PSTN infrastructure (the metallic "highway" for voice telephone calls) was a humble pair of copper wires — albeit millions of miles of it.

But in the 19th Century the "pair" of wires was actually just one wire. The oldest telephone circuits were "earth return" circuits, which consisted of a single transmission wire from the phone company to each phone, with a metal rod stuck into the ground at each end — the electrical circuit was thus "completed" through the earth between the rods. The telegraph industry had used this design for decades — it saved on wire. However, recall that telegraphy is really a robust "digital" medium and is more immune to electromagnetic noise than a subtle analog system such as the telephone. If the soil was too dry or too wet the circuit would lose the earthed ground. Such a "Magneto Exchange" was partially powered by a local acid-filled wet cell battery inconveniently housed in the subscriber's phone and which had to be replaced by the telephone company every three to six months. "Partially powered" by the battery means that the caller had to turn a crank attached to a generator-like magneto to infuse a large voltage surge in the line to ring up the central office, a surge which might also be heard in adjacent telephone lines.

Also, the development of alternating current (AC) power by Nikola Tesla (1856-1943) and the Westinghouse Electric Corporation in the 1880s meant that phone callers would have to listen to a 60 cycle hum accompanying their conversations. Single lines were thus quite noisy, picking up electrical emissions from power lines, adjacent telephone lines, telegraph lines, streetcars, and electrical machinery. The generally unendurable cacophony consisted of hissing, screeching, bubbling, clicking, and many other sounds that made conversation difficult to hear. Some users even claimed they could hear the low-pitched reverberations of thunderstorms 30 or

Chapter One

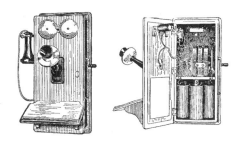

Fig. 1.1. Magneto set telephone from late 1890s to early 1900s. Note crank attached to magneto and cylindrical "wet cell" batteries. (From *Hawkins Electrical Guide* published in 1917.)

Fig. 1.2. 1890s Western Electric Common Battery Telephone. No more crank or magneto!

more miles away. The telephone lines running east and west, cutting across the Earth's magnetic field, were noisier than the lines running north and south. The line noise was worse at night than during the day, and reached a crescendo at the Witching Hour of Midnight.

On July 19, 1881 Bell received a patent for using a second wire to achieve a stable ground by providing an insulated return path (back to the phone company) for the signal current, an arrangement now called a *voice circuit*. Over a 20-year period the fledging one-wire "grounded" system was replaced with the noise-eliminating, two-wire "metallic" system. Around the same time, the electrical battery for talking and signaling was relocated to the central office under the "central battery" system developed by G.L. Anders of London in 1882 — a system which also eliminated the phone's crank and magneto. In 1893, the first central office exchange with a common battery began operation in Lexington, Massachusetts; and, in 1888, a standard "common battery" system was developed by Dr. Hammond V. Hayes of the U.S. (Hayes, the first trained scientist to be hired by the newly-created AT&T, came from MIT in November 1885. He was appointed the Director of the Mechanical Department in Boston, which did AT&T's research and development work and was the genesis of Bell Laboratories.)

Iron and steel wires then in use were substituted with more electrically conductive ones made of copper. Copper wire had not been used previously because it was too weak. However, while building a private telephone system for the Ansonia Brass and Copper Company in 1877, amateur telegrapher and early telephone innovator Thomas B. Doolittle (1839-1921) invented a "hard drawn" copper wire manufacturing process that improved the tensile strength of copper wire considerably. Doolittle's process was later used to furnish the wire needed for the nationwide long-distance network, which began on March 27, 1884 with the opening of the Boston-New York line.

All conventional wired phones are now connected to a phone company's central office via two copper wires, known for historical reasons as "tip" and "ring." The names derive from a switchboard operator's cordboard plug. The tip wire was connected to the tip of the plug, and the ring wire was connected to the slip ring around the jack. A third conductor on some jacks was called the sleeve. Traditionally, the tip wire is a common electrical ground and is colored green, while the ring wire is red.

Four-wire home phone wiring was standard for many years in North America, making possible two separate lines per location. This wire consists of one inner pair of red and green wires (line 1), and another pair of yellow and black wires (for a possible line 2). If a third pair of wires is added to the bundle, they will be blue and white. Additional wire pairs will have a tip wire in a "secondary color" (white, red, black, yellow, violet) overlaid by stripes in a "primary color" (blue, orange, green, brown or slate), while the ring wire will be in a primary color but with secondary color stripes.

Unless your phone is already mounted on the wall, the wire pair starts out leading from your conventional phone (or computer modem) to a wall jack. During this first stretch the wires are simply side-by-side in a thin, inexpensive form called silver satin.

At the wall jack, things start to get interesting.

Wall jacks can have up to five types of hole arrangements or *pinout*, depending on installation (for example, high-speed digital, conventional voice analog, or Local Area Network for computers). There are an equal number of possible modular cable plugs that fit into these pinouts, but for telephony the two most common plugs are the RJ-11 and the RJ-14 (computers plug into data networks and Digital Subscriber Lines using another kind of plug, the RJ-45; the RJ-45 has eight pins while the RJ-11 is a six conductor jack usually wired for four conductors and so has four pins). A single standard analog telephone is almost always plugged into a wall jack via the RJ-11 plug which uses only two of the wires in a four (or more) strand wire. The RJ-14 uses four wires and can handle two lines, or two-line phones.

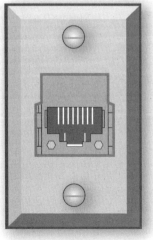

Fig. 1.3. Wall Jack.

Your home or business may have several wall jacks or "phone jacks." There are two kinds of jacks: Surface mounted and flush mounted. Surface mounted jacks are attached to the baseboard, mopboard or "washboard" as they say in Britain (the molding that conceals the joint between your interior wall and the floor). Surface mounted jacks protrude from the wall, while flush mounted jacks are level with the wall surface.

The Subscriber Network Interface (SNI)

The wires from the wall jacks must run back to a place in the building where the line from the phone company enters your home or business.

Whereas stringing phone cable through the walls of older homes is a somewhat informal affair, the design of building for commercial use, such as business offices allow for *risers*, or space reserved to house cabling and other building services. The cabling can include "backbone" communications cabling, which is the part of a premises distribution system that includes a main cable route and structure for supporting the cable.

Whether it be a home or office, the cabling ultimately terminates at the entry point where phone service enters the building. This place is called the *distribution point*. You may have more than one distribution point in your home or office. The wires must be connected here in some fashion. How they do so depends on what kind of *Standard* or *Subscriber Network Interface (SNI)* has been installed by the phone company at or near the distribution point.

New telephony installations at the customer premises include a *network interface box* or *customer user interface* as the SNI. For a house, a 6-inch x 4-inch box sports modular test jacks that allow you to determine if a line is "live" by plugging a phone into one, as well as a terminal strip into which you plug your building's communications-related *internal wiring*.

Older residences, multi-tenant units and businesses with up to 50 phones probably have a larger flat rectangular box located on the side of the building's exterior, inside of which is a *punchdown block* instead of a network interface box. A punchdown block is also known by other names such as a punch-down block, a connecting block, a quick connect block, a terminating block, or a "cross connect block" (since it connects

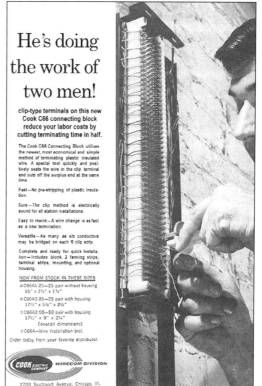

Fig. 1.4. Type 66 punchdown block, shown in an old advertisement.

one group of wires to another). The most commonly used punchdown block was Western Electric's 66-type block, followed by Nortel's Bix block. A punchdown block has *Insulation Displacement Connections (IDCs)* that enable you to easily grab a wire or cable and immediately "punch it down" or force it into the connecting block's insulation displacement connector, piercing the cable jacket to make a connection and thus eliminating the need to peel off the cable's plastic shielding before attachment.

The older Type 66 block is generally a square piece of plastic, usually white, with rows of metal pins sticking through it. You connect wires to it by "punching" them down onto the pins. Each pin has a y-shaped opening on the end, and as you punch the wire into this y shape, the wire is mashed between the sides of the y orifice. This forms a strong mechanical and electrical connection to whatever is hooked up to the pin on the other side of the Type 66 block. Typically, the pin's other side is factory-wired to a connector on one side of the block. In other cases, the block's rows of pins are connected together horizontally, which means that the installer must "punch down" something on either side of the pins to make a connection.

Having a punchdown block also means that an installer (perhaps you) will need a punchdown tool to connect your inside wiring to the SNI. Although most punchdown tools permit installation of wires on Type 66 blocks, other IDC blocks demand unique punchdown dies for punchdown tools, such as those for the 110 Connector, the 45-degree angled Krone, and Nortel's Bix.

Customer Premises Wiring Topology

The cabling between the wall jacks and SNI can take on a certain shape or *topology*. The two most common topologies are the star or "homerun" configuration and the series, or "loop" method.

The star or "homerun" configuration is now the most common wiring technique. This configuration attaches each phone jack to a copper wire pair that runs directly to the SNI or in-building phone system.

In the old days, phone jacks were connected in a series, or "loop" method, where one wire pair makes a big loop, connecting all of the phone jacks in sequence. There are still homes in North America where one multi-wire cable (up to four pairs or more) loops throughout all the rooms and the cable is opened at each phone location so that the wall jack can tap into one of the wire pairs. Many building constructors used this technique to cut costs. This method isn't used any more because although it uses less wire than the star or "homerun" pattern, any damage to any link in the chain will stop all phone access from the premises, which also makes this configuration harder to troubleshoot. Series or "loop" connections also increase line interference and can't really be used with high-speed data communications.

If you're using a small in-building phone system, such as a *Private Branch eXchange (PBX)* or *key system,* the POTS lines will probably connect to the system using RJ-11 plugs, and the system will then connect to the outside line(s).

With either of these two wiring topologies, the wires that connect the phone jacks to the network interface is much different than the silver satin wire that connects the phones to the wall jacks.

Most older, non-twisted pair cabling consists of four wires (two pairs) in a single, flat, untwisted sleeve or jacket. The individual wires have solid color insulation (green and red, yellow and black — but only the green and red wires are normally used on a single-line system). It's a bad idea to use this kind of flat-line cordage to carry two separate phone lines since its straight side-by-side wires generally suffer from "mutual induction" or "crosstalk." When you talk into a telephone's carbon microphone, your speech changes the voltage traveling through one pair of wires, and this sets up a changing magnetic field that induces a current in adjacent wire pairs. This, in turn, leads to an annoying situation where the conversation on your line intrudes or "crosses over" into another line. To help prevent crosstalk in the U.S., the Federal Communications Commission (FCC) limits the amount of power that phone companies can use to send signals over the network. And this cap on signal strength limits the data throughput of even 56 Kbps modems to a maximum of 53 Kbps!

Twisted Pair Cabling

It was Alexander Graham Bell who found that having the two wires in each pair twist around each other in a gentle spiral (like a spiral staircase or the strands of a DNA molecule) reduced crosstalk and other interference by confining each wire pair's fluctuating magnetic field. (Bell was granted his patent for this on July 19, 1881.) Also, each pair has a different *twist pitch* (so the pairs appear twisted to each other), and the pairs are each displaced randomly relative to the other pairs in the cable jacket.

All cabling systems strive for perfect "balance." A balanced line has two conductors, with equal currents moving in opposite directions, thus canceling each other out. An unbalanced line has just one conductor; the current in it returns through a common ground or earth path.

When a twisted wire pair approximates perfect "balance," the two wire conductors of each pair become geometric duplicates of each other and the interference-induced currents on the cable equalize and are subtracted out when an electronic receiver detects them. Good pair balance in a cable also helps to decrease unwanted radio emissions — currents induced on unbalanced cabling act just like tiny loop antennas, radiating away signal energy with a magnitude dependent upon the degree of mismatch between the wire pair's conductors. Perfectly balanced cable has infinite noise immunity and doesn't radiate any electromagnetic waves.

ElectroMagnetic Compatibility (EMC) describes the ability of cable to minimize radiated energy levels (emissions) and resist noise interference from outside sources (immunity). Signals traveling through badly designed or manufactured cabling suffer from excessive *attenuation*, also called "loss" or "roll off." As signals of any sort travel, they become weaker, and attenuation increases. Distance restrictions (i.e., loop reach) for any DSL is a function of "insertion loss" (attenuation loss, measured in decibels or dB) which varies in direct proportion to the electrical primary constants of the telephone loop, such as resistance, capacitance, inductance, and operating frequency. At higher frequencies, the insertion loss / attenuation becomes greater so all parties need to agree at which "standard" frequency the attenuation loss is measured. For analog voice, the average loop has an attenuation loss of 4 dB at 1 kHz. For most DSL loops, the attenuation loss should be no more than 35 dB at 200 kHz.

In any case, most multi-line phone systems and in-building computer networks now use twisted pair copper wires. Modern twisted pair wiring consists of matched pairs of wires, normally two or three pairs in a single jacket for telephony applications, or four pairs for high-speed data networking applications and *Local Area Networks (LANs)*. Of each wire pair in the cable, one wire has a colored insulation with a white stripe, and its partner is white with a matching colored stripe (i.e. white with blue stripe and blue with white stripe).

UTP versus STP

Twisted pair copper cables can be *Unshielded Twisted Pair (UTP)* or *Shielded Twisted Pair (STP)*. As its name implies, shielded twisted pair cabling encases the signal-carrying wires in a conducting shield that also acts as an electrical ground. Don't confuse STP with a third category of cabling we haven't yet discussed, called *coaxial cable*, or "coax." Coax is the kind of cable that plugs into your cable TV box. It consists of one center wire surrounded by insulation, which is in turn surrounded by a foil or mesh shield, and all of this is sheathed in an outer insulated jacket.

As for STP cabling, it first appeared many years ago not for telephony, but for computers communicating with each other via IBM token ring networks.

One might think that since the more expensive STP cable is physically encased in a shield, all outside interference and internal wave emissions are therefore blocked.

Actually, this is not the case.

Being a conductor, a well-grounded shield can act as an antenna, receiving stray electromagnetic wave or magnetic field noise and converting it into a current that flows along the shield. Current flow can also appear in the form of a *ground loop*, where the ground voltage at each end of the cable differs. In both cases, the current in the shield induces an equal and opposite current flowing in the twisted pairs. As long as the two currents are symmetrical, they cancel each other out and no noise appears at the signal receiver. However, any shield damage or imperfection can bring about an asymmetry between these currents, resulting in noise. Indeed, any electrical imbalance, either in terms of resistance or capacitance, will cause a current equal in value to the difference in the balance between the conductors to flow to ground and "noise-up" the circuit. STP cable lives up to its reputation for preventing radiation or blocking interference only as long as the entire end-to-end link is in pristine shape — perfectly shielded and properly grounded.

Some STP cables use a heavy, thick, braided shield. These are hard to install, since one cannot sharply bend them around corners or obstacles. Other STP cables use a much thinner outer foil shield. These cables, called *Screened Twisted Pair (ScTP)* cables or *Foil Twisted Pair (FTP)* cables, are thinner, lighter and less expensive than braided STP cable, but they're perhaps even more difficult to install — if you momentarily exceed the cable's rated minimum bending radius or maximum pulling tension force you could instantly tear the shield.

In the case of UTP cabling, eliminating the shield reduces the cost, size, and installation time of the cable and connectors and eliminates the possibility of ground loops. A shield that loses its ground at one end will no longer prevent magnetic field interference, and if a cable (and therefore the shield) is too long it no longer even acts as a ground. Also, interference in UTP cabling is not as big a problem as one might assume, since contemporary UTP cable is carefully designed and manufactured to rely on sophisticated balancing and filtering techniques.

The twists in the cable pairs when exposed to noise (external electromagnetic interference) allow current to be induced equally on both conductors, and so the noise is canceled out at the receiver. With properly designed and manufactured UTP cable, the desired results are easier to achieve and maintain than the grounding and shielding continuity of an STP cable. Indeed, extensive tests have rated some UTP cabling as better than their more expensive, harder to install and more difficult to maintain STP brethren.

Because of its immense popularity, UTP cable has undergone much evolution, with specialized versions of it having been developed to serve a variety of applications.

These cables may differ in the number of cable pair twists per inch, or in the individual wire sheaths or the entire cable jacket. Since the quality of these various types of UTP cables ranges from telephone grade wire to high bandwidth cabling, it was natural that standards organizations began to rate the many kinds of cable and group them into categories.

For example, Underwriters Laboratories (UL) and the Electronic Industry Association/Telecommunication Industry Association (EIA/TIA) have each formulated five grades to classify UTP cabling. The EIA/TIA rates UTP cable from Category 1 to Category 5, while UL uses the term "Level" instead of "Category" and also uses Roman numerals (I, II, III, IV, V). The divisions are roughly equivalent, particularly grades 3 to 5. Also, the International Standards Organization (ISO) and the International Electrotechnical Commission (IEC) have developed the ISO Generic Cabling Standard, referred to as IS 11801, which is based in part on EIA 568. Like EIA 568, IS 11801 uses a 3 to 5 rating system to qualify cabling performance; but EIA uses the term "Class" instead of "Category." Classes C and D corresponding to Category 3 and 5 respectively. Yet another organization, CENELEC, the European Committee for Electrotechnical Standardization, now officially recognized as the European Standards Organization in its field by the European Commission, also has a roughly equivalent standard, EN 50173. The European Standard for Generic Cabling generally references CENELEC rather than the ISO\IEC specifications for cable. Both shielded and unshielded cables are presently defined by EN 50173.

For the purposes of this book we'll be using the classifications of the EIA/TIA-568 standard, which covers telecommunications UTP cabling in modern commercial buildings. This standard also specifies the electrical and physical requirements for STP, coaxial cables, and optical fiber cables.

The basic requirements for UTP include the following:
- Four individually twisted pairs per cable.
- Each pair has an impedance of 100 Ohms +/- 15% (when measured at frequencies of 1 to 16 MHz).
- 24 American Wire Gauge (AWG) wire is generally used for the individual wires in each pair. Under the AWG standard, the thickness (gauge) of the wire increases as the AWG number decreases. The gauge is calculated so that the next largest diameter always has a cross-sectional area that is 26% greater. A change of 6 AWG numbers results in a doubling or halving of the wire's diameter, while a change of 3 AWG numbers gives a doubling or halving of the cross-sectional area. Under AWG, 24 AWG wire is 0.0201 inch or 0.511 mm in diameter. Under the EIA/TIA standards, one can optionally use 22 gauge (0.0253 inch or 0.643 mm diameter) copper wires

instead (24 gauge cable has approximately 25% more signal loss for any measured length than the thicker 22 gauge cable).

Additionally, EIA/TIA-568 specifies the wire sheath color-coding, overall cable diameter, and other electrical characteristics, such as the maximum allowable crosstalk and attenuation.

The nice thing about UTP is that it can be used for just about anything. A high quality Category 5 cable installed in your premises means that you can send through it anything from analog telephone (POTS) service to computer networking and high bandwidth data transmissions. The same type of UTP cable (and connectors) can be used for an entire building, unlike STP.

Of course, bad installation practices and substandard connectivity components can degrade the quality and capacity of a cabling system. If an installer bends a cable too much or untwists a wire pair more than 0.5 inch when terminating it onto a connector, crosstalk and noise appears. All components (such as data jacks, panels, and patch cables that connect a PC to a wall outlet) must be certified for Category 5 installations and to work together to make an end-to-end Category 5 ("CAT-5") installation.

Powering the PSTN

Power for the PSTN in the U.S. starts out as -48 *Volts of Direct Current (VDC)*. In many countries "safe low voltage" is considered to be 50 VDC and below (though some countries use voltages in the 36 VDC to 60 VDC range). Early central offices were powered by the same kind of lead acid batteries used in automobiles, and by string-

Five Categories of UTP

The EIA/TIA's five UTP categories are described below:
Category 1: Pre-1983 traditional UTP telephone wiring, no data transmission.
Category 2: Telephone and low-speed data transmission up to 4 Mbps (LocalTalk).
Category 3: 10 Mbps Ethernet, ISDN BRI. The minimum standard for new installations. Maximum cable segment length of 100 meters (328 feet).
Category 4: Data up to 20 Mbps (Token Ring).
Category 5: Data up to 100+ Mbps (OC3 ATM, SONet, 100BaseTX Fast Ethernet). Most popular for new installations. Maximum cable segment length of 100 meters (328 feet). The very best enhanced version (Category 5e) can handle 155 Mbps ATM.

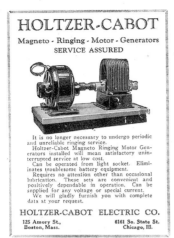

Fig. 1.5. In the late 1800s and early 1900s, the central office made "ring voltage" by using special Ringing Voltage Generators. A regular electric motor (powered by the -48 VDC used in the central office) would drive a magneto similar to those that were cranked by hand on the phones themselves. The magneto would thus supply the ringing voltage. Differently geared generators would produce ring voltage at a special number of cycles per second to handle party line service, where the bells would have to ring at a particular frequency. In extreme cases as many as 30 or more customers shared a party line, each having a distinctive ring (such as two long rings and two short ones). Many ears were sure to be listening in on any given call and few secrets were kept in any community.

ing four 12 Volt car batteries in series you get 4 x -12 VDC or -48 VDC, otherwise known as the "talk battery." The original alleged reason for using negative voltage instead of positive was that negative voltage was somehow safer. In actuality, when a phone line gets wet, having a wire at negative potential forces the wire's metal ions to go from the ground to the wire. Positive voltage would do the opposite, causing rapid corrosion of the wire. Indeed, since outside plant cables come in 500-foot reels, the many splice points along the cable can be a source of corrosion, and so a sealing current is applied to the cable to prevent oxidation (rust).

Conversely, various small and large positive voltages are used in modern electronic circuitry. Computerized customer premises telephony equipment, such as PBXs, may use something like +24 VDC.

Although the voltage may start out from the central office at -48 VDC, the electricity travels through resistors and various circuits as needed, sometimes reducing the voltage by increasing electrical resistance on the wire (electrical resistance is measured in Ohms).

When your telephone is in an "on-hook" (unused) state the "Tip" wire is almost at 0 VDC, while the "Ring" wire is about -48 VDC with respect to earth ground. When you activate the phone and go "off hook," the tip is now about -20 VDC and the ring becomes less negative, changing from -48 VDC to about -28 VDC. Thus, during normal phone operation there is about an 8 VDC voltage difference between the two wires going into the telephone. If you could measure the DC resistance produced by your phone, it would be in the 200 to 300 ohm range (telephone regulations typically specify that the total DC resistance must be less than 400 ohms, though the large digital switches in the central office can deal with up to 1,500 ohms). Current flowing through the phone ranges from 20 to 50 milliAmps (mA), though it's generally around 23 milliAmps in the U.S.

Aside from a lightning strike, the only instance where a telephone line in your home or business could ever possibly give you a shock is when you're about to receive a call and "ringing voltage" appears on the line. To get your phone to ring, the central office sends a "ringing voltage" of from +40 to +150 VDC (usually +90 VDC) as a pulsating positive direct current that pulsates at between 20 to 40 cycles per second, but usually at 20 cycles in the U.S. and 25 in Europe. Unlike alternating current (AC), this voltage doesn't cycle between minus and plus; instead, it pulses from zero to +90 VDC. In the case of ordinary subscriber lines, these positive pulses come in through the (otherwise negative) red ring wire and return to the central office via the green tip wire, which is always near "ground."

The Local Loop / Last Mile

When you pick up your phone and it goes off-hook, a connection across the tip and ring wires is made and a path for the 23 mA or so of DC current is established. By completing the circuit you've created a loop of flowing electricity up to about 18,000 feet in length, all the way to the phone company's central office.

The name *local loop* or *subscriber loop* was soon applied to the physical connection leading from just outside a subscriber's premise (generally from the *demarcation point* or "demarc" on the subscriber's building, which is usually something like a junction or connector box) to the central office. The local loop includes the *outside plant (OSP)*, which is all of the buried and aerial telephone cabling, poles, conduit and outside equipment that not only connects the subscriber to the central office, but central offices to each other.

From the phone company's point of view, the cabling leading to the subscriber is "the last mile" (or so) of the physical connection — from the subscriber's point of view it is, of course, the first mile! The term "the last mile" grew in popularity when high bandwidth data access services began to be overlaid on the local loop – "the last mile" suggests some kind of final, difficult and exhausting task. Indeed, efforts to bring high bandwidth digital data access and voice services (such as DSL and Voice

The PSTN's Outside Plant

As we've said, the outside plant (OSP) consists of most of the local loop: Pole lines, cabling, conduit, aerial and drop wires, range extension devices, and even manhole covers. In his best-selling *Newton's Telecom Dictionary*, Harry Newton writes that "Microwave towers, antennas, and cable-system repeaters traditionally are not considered outside plant." For the purposes of this book, however, we'll include anything other than the big switches themselves.

over DSL) over old twisted pair wires designed for low-bandwidth analog voice communications has been an ornery undertaking.

The Local Loop's Topology

Back in the 1870s when inventors Alexander Graham Bell and Thomas Watson were testing the first telephones, one telephone was connected to another via a dedicated cable called a "tieline." If, say, eight telephones were connected to each other in this way, each user on this dense meshed network of wires would have to serve as his or her own switching system, plugging the phone into the wire leading to the phone of the person to be called. The end of the wire of the person to be called would need some sort of device (bell, light bulb, etc.) to indicate that that particular wire should be plugged into the receiving phone.

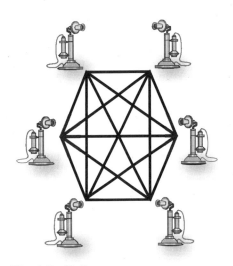

Fig. 1.6. Tielines.

This primitive kind of tieline network is complicated and wastes wire. Obviously, a more efficient network would consist of just one switching entity, with each phone now requiring only a single pair of wires to connect to the switch at the network's center, known as a Central Office or "exchange." Years later the definition of the word "exchange" would be expanded to include an entire city or local calling area, and the local phone company would be referred to as the *Local Exchange Carrier* or LEC.

In 1877 a primitive exchange (actually just an extended party line) connected the Capitol Avenue Drug Store in Hartford, Connecticut with several local doctors. The first genuine local exchange carrier, or central office, was the New Haven District Telephone Company established in New Haven, Connecticut on January 28, 1878. The exchange initially served 21 subscribers on eight party lines but by the following month the system had to be expanded to serve 50 subscribers. Subscribers used Bell's "Butterstamp" phone, which for the first time combined the receiver and the transmitter into one hand-held unit (a pushbutton was used to alert the operator). The price for the service was $1.50 a month, or about $26 in 2001 U.S. dollars.

Other exchanges quickly appeared. The second exchange opened for business in Meriden, Connecticut, three days later. San Francisco started up theirs on February 17, 1878, followed by Albany, New York on March 18th with many more appearing on the

Fig. 1.7. George Willard Coy (1836-1915), a Civil War veteran whose left arm had been shattered by a shell fragment at Petersburg, designed the world's first switchboard for the first commercial local exchange carrier, the New Haven District Telephone Company, situated at 219 (now 741) Chapel Street in New Haven, Connecticut (the company later became Southern New England Telephone). It went into service on January 28, 1878. In the replica seen here, each metal arm in the board's center connects to a group of eight subscriber telephones via a series of contacts arrayed in a circle. Eight arms across the bottom of the board connect to one of eight telephone lines. An "annunciator box" at the upper right informed the operator which of the eight lines was in use. Just below the annunciator is a cylindrical device developed by Thomas A. Watson, Alexander Graham Bell's associate, which alerted subscribers of an incoming call. The board made strange cackling noises and became known as "Watson's Buzzer" or, more locally, "Coy's Chicken." (The replica shown here belongs to the New Haven Colony Historical Society.)

Fig. 1.8. The world's first telephone exchange building in New Haven, Connecticut. This photo was taken in 1967 by Robert Fulton III as part of a Historic American Buildings Survey conducted by the U.S. National Park Service.

scene in rapid succession. The 1880 U.S. census showed that there were 148 telephone systems, 48,414 subscribers and 54,319 telephones then in existence.

The first telephone exchange outside of the United States opened in Hamilton, Ontario, Canada on July 15, 1878. The next exchange to appear outside the U.S. was in London, England, in 1879.

The Serving Areas Concept

The cabling that converges on every LEC's central office resembles a tree, with homes and businesses at the tips of the branches and the trunk representing the central office. This standard reference design for loop distribution was set up in 1986 by Bellcore (now Telcordia) for U.S. urban and suburban areas. It's called the *Serving Areas Concept (SAC)* plan. Under the SAC plan, each city is divided into one or more *Wire Centers (WCs)* — each handled by a local central office switch. A typical WC handles 41,000 subscriber lines grouped into about 10 *Carrier Serving Areas (CSAs)* each with an average size of 12 square miles (31 square kilometers), the actual size depending on the size and population of the city.

Even back in 1986, Bellcore must have realized that the local copper loops would eventually be digitized in some way. It recommended "cleaning up" the lines, removing extraneous devices such as coils, hanging wires (bridge taps) and analog carriers, and called for a so-called single-gauge loop design standard (CSA standard 22.2) to replace Bell Lab's traditional dual gauge design standard, the *Resistance Loop Design* (also known as the Supervisory Limit or Loss Design Chart) — a crude architecture that simply limited the maximum loop DC resistance on a straight copper pair to 1,500 Ohms — and suggested a 9 dB maximum signal loss over the loop at 1 KHz (under modern CSA rules, the average loop has 600 Ohms DC resistance and a 4 dB loss at 1 KHz).

This was the only real resistance design rule; the Resistance Loop Design allowed for different wire gauges on the same line, as well as all the other items that would later prove troublesome for DSL installation. When the CSA rules came into effect, most telephone companies did remove much of the paraphernalia sitting on their lines, but few changed their dual-gauge loop design standard to Bellcore's single gauge CSA standard. As a result, many embedded dual gauge loops exist in today's outside plants.

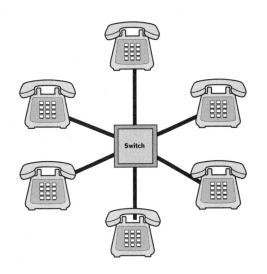

Fig. 1.9. The centralized switch configuration is more effective since it enables "virtual" tielines to be forged as needed for each call. The PSTN's switching system has become automatic and computerized over the years, but the topology and composition of a great deal of the local loop (those twisted pairs of copper wires) is very much the same as it has been for decades.

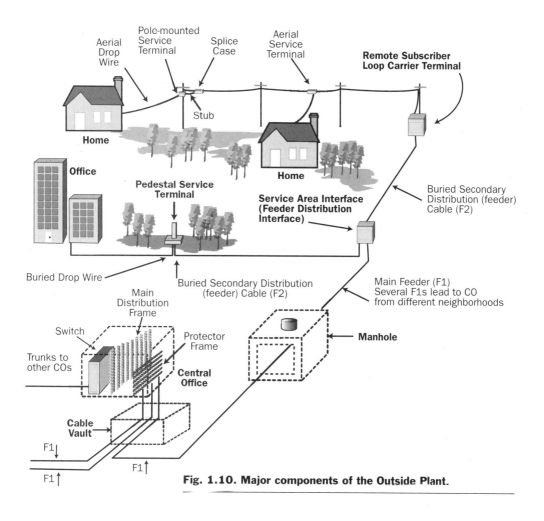

Fig. 1.10. Major components of the Outside Plant.

Since equipment vendors tend to design their DSL equipment to meet the Bellcore CSA 22.2 standard, there can sometimes be a discrepancy between the equipment specifications and the actual characteristics of copper loops. For example, most equipment specifications call for a 22 AWG (0.644 mm wire diameter) loop reach of 15,000 feet (4572 meters), 24 AWG (0.511 mm diameter) loop reach of 12,000 feet (3657.6 meters), and 26 AWG (0.404 mm diameter) loop reach of 9000 feet (2743.2 meters), all of which meet the CSA standard. Connecting such new DSL equipment to older dual gauge loops means that the actual reach of the DSL equipment will differ from its specifications. As a result, the service provider or installer must actually test the copper pair leading to the customer to ensure that the proper bandwidth can be delivered.

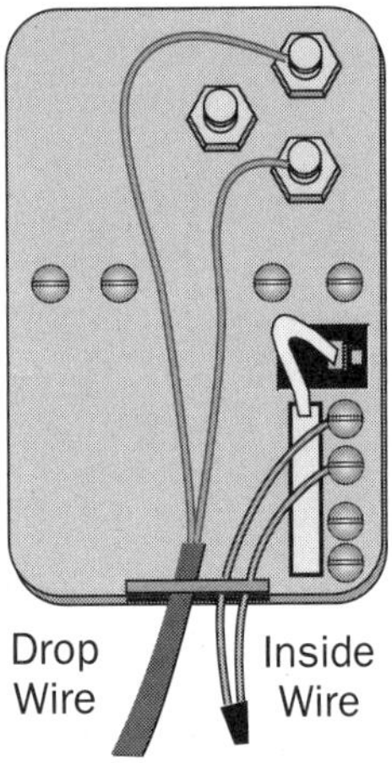

Telephone wiring inside your home or business first connects to the telco's wire at a fused "house protector" or "station protector" at the subscriber network interface's demarcation point. This is where your wiring ends and the phone company's outside plant begins. It's usually a gray box with modular telephone jacks inside arranged so that the LEC can start to troubleshoot service problems without disturbing anyone in the building. In offices a type 66 termination block may be used at the building's network interface, and the wire from the outside will be plugged into the block using a 25 pair Amphenol-type plug known as an RJ-21X. This allows for the phone company technicians to do a "quick-disconnect" of the whole building from the network for troubleshooting purposes.

From your (the subscriber's) demarcation point, an aerial or buried twisted wire pair called a *drop wire* goes to a *service terminal block* (a device similar to a punchdown block that's used to connect one group of wires to another). Service terminals can be a small grayish green box sitting on a concrete pedestal anchored on the ground, appropriately called *pedestals* (the smallest of these looks like a waist-high, square pole). Terminals such as pedestals can also be situated in buildings, underground, or on a pole.

Fig. 1.11. Station Protector.

Fig. 1.12. By the late 1880s it occurred to the phone company that the growing number of distribution and feeder cables was becoming unmanageable and should be collected into binders and buried in the ground. For example, along West Street, New York, every pole was a 90 foot-high Norway pine, each with 30 cross-arm boards upon which was hung 300 wires.

The subscriber's twisted wire pair is connected (along with those of four or five neighboring subscribers) to binding posts on a terminal block within the service terminal enclosure. The terminal enclosure's wire must now connect with the larger aerial (or underground) cable that services the neighborhood, called the *distribution cable*.

If the service terminal and the distribution cable are on a pole, the terminal might be in a Western Electric "Ready Access" enclosure, which is a small black rectangular enclosure, about 15 inches (38 centimeters) long x 3 inches (7.6 cm) wide x 5 inches (12.7 cm) that's connected directly to the aerial distribution cable. Up to 12 drop lines from nearby customer locations can enter the box and can be attached to the binding post terminal block. The enclosure is in-line with the distribution cable and the cable passes through it, connecting to the block and gathering up the wire pair connections.

Larger capacity service terminals capable of handling about 30 wires include the Western Electric 53A4 and N-type Pole Mount Cable Terminals, which are slightly larger than the ready access boxes and are mounted on a pole just below the aerial distribution cable. Their terminal blocks connect to the distribution cable via a short wire (called a *stub*) that runs from the back of the enclosure to a splice case on the distribution cable.

The total paraphernalia that makes up the subscriber drop wire-to-distribution cable connection is the subset of the outside plant called the *local exchange plant*.

The distribution cable and its supporting structures and equipment comprise the part of the outside plant known as the *distribution plant*. Distribution cable can bundle 50 to over 100 wire pairs, such as the old 100-pair air-core Plastic Insulated Conductor (PIC) cable. Since distribution cable is secondary feeder cable, it's called F2. F2 cables from many neighborhoods converge on the nearest *Serving Area Interface (SAI)* cabinet, also known as a *Feeder Distribution Interface (FDI)*. Just before the F2 cables connect with the SAI, they go underground and travel through conduit. There are so many of them and they weigh so much it would be impossible to string them above ground from pole to pole.

The SAI is often called a B-Box, cross-connect box, pedestal, Bridging Head or *Bridged Access Point (AP)*. When you see one it's usually enclosed in a big grayish-green cabinet (such as a Western Electric SAI 55 or 22/E terminal case), sitting on the ground or up on a pole. Inside, the SAI is revealed to be simply a kind of giant terminal block where the wire pairs in each distribution cable terminate on one or more panels while a larger feeder cable terminates on adjacent panels. The wire pairs can terminate on screw binding posts or punchdown block-like "quick-connect" connectors. Jumpers are run on the opposite sides of the panels to cross connect the pairs from the distribution to the feeder cables, linking from 400 to 3600 wire pairs (the inside of the SAI's frame can typically tilt forward to allow access to the other side of the panels). Some SAIs have room for expansion and so have "floater pairs" that aren't yet hooked up to subscriber lines.

These larger feeder cables emerging from the SAI may be a *branch feeder*, a cable that routes pairs from the SAI to another cross-connect with a larger F1 or *main feeder*, or (as is usually the case) the cable emerging from the SAI may be the main feeder itself. Main feeder cables are the principal players in the subset of the outside plant called (obviously) the *feeder plant*. Main feeders can contain from 600 to 3600 pairs. Since F1 cables are so important, the central office force-feeds dry air (3% humidity) into the cable interior using a mechanical air compressor and dryer. The air is fed into the cable at a pressure of about 10 pounds per square inch (PSI) or 0.70307 kg per square centimeter to prevent water from entering the cables.

Leaks inevitably appear in these pressurized cables, which means that the pressure in the cable diminishes as the length increases. Most telephone companies set a minimum pipe endpoint pressure standard of 7.5 PSI (0.527302 kg per square cm). If the pipe pressure falls below this standard, the cable's protection against moisture is in jeopardy.

To keep the pressure up until the leaks can be fixed, air pressure in the system is often kept to acceptable levels by using an air pipe that follows the cable route and introduces pressure at various fixed points, generally by connecting to a manifold which distributes air to the cables in the utility hole.

In areas of high population density (metropolitan areas) more than one F1 cable must terminate at the SAI in order to handle voice and data traffic generated in a single Serving Area. F1 cables are incredibly heavy so they're always routed underground in plastic pipe conduit (such as *innerduct*, a corrugated plastic tube about two inches in diameter, or the even stronger Schedule 80 PVC pipe) or steel ducting covered with concrete and backfill, or (in the old days or in metropolitan areas) large concrete tunnels with the cables set on racks on the walls. This cabling is accessible only via manholes or vaults, ideally spaced at anywhere from 750 to 6000 foot (228 to 1829 meter) intervals. Extra empty conduit is usually also buried for service expansion or replacement purposes.

Normally the big F1 cables head straight underground to the phone company's central office, but there are usually some additional factors that complicate matters, making it difficult to retrofit advanced DSL access services to the system, as we shall see.

If the customers / subscribers are situated far away in rural areas, their network may be subject to the *Rural Area Network Design (RAND)* plan; such a rural serving area may cover 130 square miles (336.7 square kilometers). Under such a system the main feeder F1 cables branch out not only into F2's, but the F2's branch out into F3's, and those in turn may branch out into F4's and even F5's before drop wires finally split off of the cable to the customer. At each junction point there's an SAI-like box called a *Rural Area Interface (RAI)*, where 50 or more pairs are connected in cables exceeding 30,000 feet (9144 meters) in length.

Loading Coils

Once you start working with copper wires longer than 18,000 feet (5486 meters), you run into voice transmission problems. The *capacitance*, or capacity of a pair of wires to store an electrical charge, increases as the wire pair grows in length, which ruins the quality of voice transmissions. Too many high frequencies can jumble the system.

To reduce the capacitance on lines in excess of 18,000 feet, a series of *loading coils* (or "load coils") are installed on the line. These are small inductor coils wound on a powdered iron core. Loading coils are housed in weatherproof cases mounted on poles (sometimes in service terminal enclosures) or in manholes. The two most popular models of loading coils are the H-88 and the D-66.

For a wire pair longer than 18,000 feet, an H-88 coil is first spliced into the cable 3000 feet (914.4 meters) from the central office; additional coils are spaced at 6000 foot (1828.8 meter) intervals, give or take 120 feet (36.57 meters). Thus, H-88 coils would appear at 3000 feet (914.4 meters), 9000 feet (2743.2 meters), 15,000 feet (4572 meters), and so on, from the central office.

D-66 coils are not quite as inductive as H-88's. Cable pairs working beyond 18,000 feet would have the first loading coil installed at 2250 feet (685.8 meters) from the central office, with additional coils spaced at 4500 foot (1371.6 meter) intervals thereafter. Thus, D-66 coils would be situated at 2250 feet (685.8 meters), 6750 feet (2057.4 meters), 11,250 feet (3429 meters), and so on.

Accidentally placing load coils at other than the specified intervals degrades voice grade transmission, cutting off the frequencies above 1700 Hz. This also interferes with modems and can cause Caller ID (which uses frequencies between 1200 Hz and 2200 Hz) to fail.

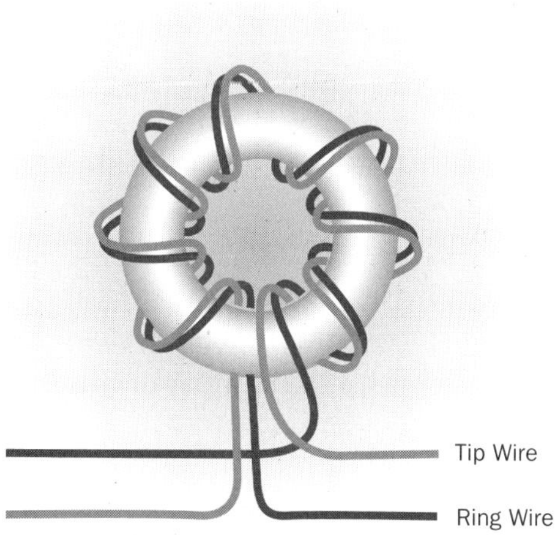

Even when they're working correctly, loading coils can introduce phase delays and cut off frequencies above the voice band of 3400 Hz, which means they must be removed from cable pairs that are used to carry high frequency, high bandwidth "digital" services such as ISDN, T-Carrier and DSL.

Fig. 1.13. Toroidal Loading Coil.

CDOs, RTs and DLCs

Phone companies soon realized that it made better technological and maintenance sense to eliminate loading coils, replacing them with different kinds of devices that would prevent any stretch of the loop to exceed 18,000 feet in length. In the late 1920s direct dialing was starting to become more popular than operator-completed calls, which made it possible for distant (e.g. rural or "outstate") customers to be served by a small, unattended "community switch," better known as a *Community Dial Office (CDO)*, controlled remotely by the nearest central office. In the 1960s, CDOs were usually connected by an analog carrier or T-1 to the central office. The original CDOs used simple Strowger-type step-by-step switches and were considered to be a kind of end office ranking just below a central office equipped with a Class 5 switch. The CDOs couldn't generate their own dial tone, nor did any trunks sprout from them to the core network — long distance service still had to go out through the central office. One of the few things that a CDO *could* do was generate "lifeline" power to keep the local phones up and running. Also, some CDOs could handle local calls without first going to a central office, though most older ones would route a call out to the central office and then back through the CDO. Once the connection was set up the call could be dropped back to the local switch, freeing up the channel to the central office. Prior to the breakup of AT&T in 1984, the final 1983 census of the Bell System indicated that there were 3700 CDOs in the U.S., more than any other kind of switch.

Since many customers were quite distant from central offices, so were the CDOs. This meant that the service still needed extra cable, repeaters or line extenders, scores of additional poles and increased maintenance. And if the CDO served remote payphone lines, even more equipment was needed on top of what was required for routine phone service

CDOs were plagued with technical and maintenance problems, and so around the late 1960s the Bell System began upgrading them. If a rural community had grown and there was sufficient traffic, a CDO could be retrofitted to become a *package office* which used a Bell standard No. 5 Crossbar switch. This was considered expensive "overkill" for some areas, so, starting in the late 1960s, CDOs were replaced by a device now called a *Digital Loop Carrier (DLC)*, which AT&T originally called *Subscriber Loop Multiplex (SLM)*, *Subscriber Loop Carrier (SLC)* or *Subscriber Line (or Loop) Carrier Circuit (SLCC)*, pronounced "slick". The first "slick" appeared in 1971.

The DLC consists of two subsystems: a cabinet called a *Remote Terminal (RT)*, a new intermediate point where the local loops can terminate, and the *Central Office Termination (COT)* which is the central office end of the circuit.

A DLC takes a bunch of analog POTS cable pairs, converts them to 64 Kbps digital signal channels, and then "multiplexes" them, allowing them to share one high-speed line back to the central office. The RT contains both the DLC's demultiplexing electronics as well as a little SAI-type terminal block for connecting the pairs to distribution cables on the side leading toward the subscribers. Since this digital "feeder" back to the CO is not a regular multipair cable but a single digital link, this arrangement is also known as a "carrier-derived feeder."

Not all DLCs use digital lines all the way back to the central office. Some older switches only connect to analog lines, so the CO will convert the digital signal from the DLC back to analog so that it can actually connect to the switch. The switch then converts the analog signal back to digital, since ISPs (AOL, Earthnet, etc.) must have a high bandwidth digital connection to the telephone network.

This causes problems when owners of 56 Kbps analog modems try to connect to the Internet. The 56 Kbps modem design is based upon the assumption that the whole call path is digital except for the subscriber loop. The additional analog-to-digital conversion introduces quantization noise that forces V.90 56 Kbps modems to drop back to V.34 operation at a theoretical maximum of 33.6 Kbps.

This is also why 56 Kbps modems have problems in hotels or offices where the calls must first travel through an in-building Private Branch Exchange phone system. A PBX system also converts analog calls to digital signals so that they can be electronically switched or stored by the system.

Time Division Multiplexing and the T-Carrier

The most popular way to digitally multiplex a bunch of signals is *Time Division Multiplexing (TDM)*. Each channel is sampled rapidly in a rotating sequence and gets its turn to send information during its own "time slot" or periodic time segment when it is free to transmit. A separate "pipe" is used for incoming multiplexed signals, which must be demultiplexed and the individual channels reconstructed.

The remote terminal contains the DLC's multiplexing and demultiplexing circuitry (special line cards) along with a small SAI-type terminal block for connecting the pairs to distribution cables coming from the subscriber premises. Because the feeder is not a multipair cable but a digital link, this configuration is also known as a "carrier-derived feeder."

Until recently the most popular kind of digital link was the *Trunk* or *T-Carrier* system. The lowest capacity version of this is the Trunk Level 1 or T-1, which can combine 24 voice channels into a serial bit stream over an ordinary four-wire copper cable. Japan's version is the J-1. Europe and the rest of the world has a 30 channel version called the E-1.

A whole hierarchy of digital signaling for greater and greater capacity exists for these systems:

North America (T-Carrier)		
T-1	1.544 Mbps	24 voice channels
T-1C	3.152 Mbps	48 voice channels
T-2	6.312 Mbps	96 voice channels
T-3	44.736 Mbps	672 voice channels
T-4	274.176 Mbps	4032 voice channels
Japan (J-Carrier)		
J-1	1.544 Mbps	24 voice channels
J-2	6.312 Mbps	96 voice channels
J-3	32.064 Mbps	480 voice channels
J-4	97.728 Mbps	1440 voice channels
J-5	397.000 Mbps	5760 voice channels
Europe (E-Carrier)		
E-1	2.048 Mbps	30 voice channels
E-2	8.448 Mbps	120 voice channels
E-3	34.368 Mbps	480 voice channels
E-4	139.264 Mbps	1920 voice channels
E-5	565.148 Mbps	7680 voice channels
E-6	2200.000 Mbps	30,720 voice channels

Most DLCs use Trunk Level 1s. Low bandwidth digital transmissions such as a T-1 can travel over copper wires back to the CO, but higher bandwidth transmissions can easily fit on one of the new *fiber optic* cables. Indeed, practically all new feeder cables are no longer made of copper, but of incredibly thin fiber optic materials instead. A fiber no thicker than a human hair can carry a tremendous number of calls via a light beam; it has far greater capacity than any copper cable. AT&T installed the first transcontinental fiber optic line in 1986.

Getting the Signal Back to the CO

If the T-1/E-1 or T-2/E-2 lines leading to the central office are copper, then signal repeaters will be used for long lines. The repeaters are placed at the same positions on the line where one would expect loading coils so that pre-existing manholes can be used to service them. Although digital signals need signal repeaters, digital signals retain their informational integrity over long distances, unlike periodically amplified analog signals, which collect additional noise with each successive amplification. Some extremely remote locations cannot be served with a copper or fiber line at all and so their local loops are instead converted to microwave beams that are transmitted from tower to tower until they reach the central office. This is called the *Wireless Local Loop (WLL)*. Even more remote locations link to the outside world via satellite.

Pair Gain

Extending the distance a phone service subscriber can be from a central office is wonderful, but the fact that a DLC can consolidate many copper wire pairs into a single copper cable or light conducting fiber is actually more important. One doesn't have to live 100 miles from a central office to experience this technology. Even phone customers living a short distance from a central office may find that a large apartment complex is being constructed near them, and since the local phone network feeder cables are already "maxed-out," extra capacity will need to be provided. This can be achieved via the miracle of *pair gain* (also termed "pair-gain" or "pairgain"), which is the general term for bestowing increased capacity on a pair of wires since usually, through the use of a digital encoding technology, the necessary capacity can be provided — thus they "gained wire pairs back" for telco reuse. Pair gain allows the local loop's burgeoning traffic to continue to fit over existing F1 cable.

The other meaning of the term "pair gain" is the actual number of the subscriber lines minus the number of trunk wire pairs. For example, 24 subscriber lines can connect to the central office and get full service with two T-1 cables (separate transmit and receive wire pairs). So this arrangement has a pair gain of 24 minus 2 or 22.

DLCs with pair gain generally have what's called "concentration" abilities, which frees up even more trunks back to the central office. Since the laws of probability suggest that not every subscriber will be making a call at the same time, there need not be a 1-to-1 ratio between the subscriber lines and the trunk channels. Thus, call setups can be supported where many more subscriber lines can be served than there are digital channels back to the host switch.

One early SLC, the SLC-40, had no concentration. The 40 analog lines of the 40 subscribers it served were multiplexed and transmitted as 40 channels on T-1 trunks. Around 1970, the first device to appear with pair gain and concentration capability was the *Subscriber Loop Multiplex (SLM),* which handled 80 subscribers. It could switch the active users from among these subscribers into 24 lines (it was presumed that there would never be more than 24 of the 80 subscribers making a call at the same time) and these were multiplexed and transmitted to the central office switch as a T-1 trunk. At the central office, "deconcentration" or "expansion" was performed by switching from the concentrated traffic on the shared 24 lines back to the original 80 lines by means of a 24 x 80 crossbar matrix switch.

Popular DLC / SLC / SLCC equipment still in operation includes the Western Electric SLC-5, SLC-8, SLC-40, SLC-96, SLC-2000 and GTE's MXU, ranging in size from eight to hundreds of lines. For many years, the leading pair gain workhorse for large suburban tracts and sprouting apartment complexes was the Subscriber Line Carrier 96, or SLC-96. An SLC-96 can concentrate 96 analog phone lines onto a single four-wire digital circuit, a T-2.

Line Concentrators

A *line concentrator* resembles a DLC except that it's bigger and exclusively handles concentrated service, terminating multiple T-Carrier trunks to large number of subscriber lines. It specializes in saving as many pairs as possible, far more than does an ordinary DLC / SLC. For example, a subscriber carrier may have a ratio of one or two subscriber lines per digital trunk channel, while a remote line concentrator tries to get away with assigning 600 or more users to only about five 24 channel T-1 lines (5:1 ratio). Concentrators can efficiently support concentration ratios up to 9-to-1 while maintaining required grade of service to residential subscribers. I've even heard of a line concentrator made by MIC Electronics Limited in India (www.micsoft.com) that can be used at concentration ratios ranging from 1:1 to 120:1 with no decrease in the grade of service.

Concentrators also may include circuitry known as *intra-calling*, which enables users whose lines terminate within the concentrator to connect to one another without forwarding calls using channels leading to and from the central office. Intra-calling features can be achieved with just one channel to the central office serving to manage the connection. If the carrier line fails, most concentrators can still provide service between local users in the serving area.

DLCs and Pair Gain are Bad for DSL

The problem with DLCs and pair gain devices is that by digitizing many voice signals and making them fit into a single communications "pipe," each subscriber is limited to a voice-grade circuit with a bandwidth of 64 Kbps. This completely prevents the provisioning of high bandwidth DSL technologies, some of which can transmit up to 8 Mbps at frequencies of 1.1 MHz, unless the DSL signal can be intercepted in the remote terminal before it gets to the DLC and be converted into a format compatible with the DLC, or if the telco can retrofit the DLCs with new line cards that provide DSL compatibility. New models of subscriber carrier systems, of course, support various services.

Not all telcos have DLCs on their local loops. Some use a serving wire center (SWC) where there may be a switching system that can do intra-calling.

The Central Office

Ultimately of course, just as all roads lead to Rome (or used to), all F1 cables lead to a central office of a telco's local exchange. To a home or business subscriber this marks the end of the "last mile" and the beginning of the worldwide telephone system. From the perspective of the telco, however, such a central office is often called an "end office" since it's the part of the network closest to the consumer.

The Cable Vault

Feeder cables enter the central office in a large underground room called a *cable vault*. Each feeder may contain up to 3600 pairs of wires, and a typical suburb can be served with ten or so feeders. In urban areas the whole vault may admit up to 120,000 or more copper wire pairs, coaxial cables and optical fibers.

Aside from the main cable feeds, the cable vault also contains equipment to pressurize air in the cables, an *Uninterruptible Power Supply (UPS)*, and one or more diesel generators in case the UPS fails.

Also located at the building entrance point or in the cable vault itself is some sort of *Line Protection Unit (LPU)* to earth-ground lightning-induced electrical surges traveling along the cable sheaths. Lightning is essentially a really big spark that averages 30 million volts and 25,000 amperes in current (though the largest recorded lightning strike was about 100 million volts and 280,000 amperes) — it takes about 1 microsecond for the current to reach its peak value, and about 50 microseconds to decay to half that value. In an average year, lightning strikes the United States about 20 million times and kills more people than tornadoes — between

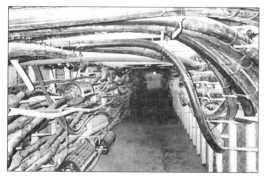

Fig. 1.14. Example of a cable vault. Mostly optical fiber is entering the building.

90 and 200 Americans each year.

As you'd expect, lightning is bad news for the PSTN's whole outside plant. All exterior equipment, and equipment connected to aerial and buried copper cables can be damaged. Aerial cables are especially vulnerable to lightning strikes. Lightning's enormous energy pulse momentarily raises the potential of ground at that strike point and then travels along the copper pairs to the central office where it finds a lower ground potential.

At the central office cable vault (or in a well-grounded protection frame), primary lightning protection often consists of line protection units taking the form of wall mounted hybrid surge arrestor devices using a three element failsafe gas tube arrestor and a solid state Metal Oxide Varistor (MOV) / resistor / capacitor combination to provide high surge current protection and transient suppression.

In some rural and traditionally troublesome areas such as the Gulf Coast of Florida, lightning strikes aerial cable so frequently that an extra measure of safety is applied at the cable vault. A short 25 or 50-foot length of thin AWG 24 or 26 gauge cable is spliced into a thicker 22-gauge cable. This "fuse cable" will vaporize like a fuse when a lightning surge travels up the line, breaking the circuit and stopping the electrical bolt in its tracks. Under normal conditions, however, this little segment of thin wire adds additional attenuation to signals on the line, which interferes with high frequency communications, e.g. T-1 and DSL.

The Main Distribution Frame

In any case, the cables emerge from holes in the cable vault wall and are routed along a support frame, then are directed up through the ceiling to the frame room. It is here that the cables meet the *Main Distribution Frame (MDF)*.

Fig. 1.15. A Main Distribution Frame (MDF) at the central office.

The frame is where the cable separates to individual pairs and attaches to connectors. The frame not only links the switching systems in the exchange to the subscriber cable network but also links trunks from one exchange to the others. The frame generally runs the length of the building, from floor to ceiling, and can be up to about five feet thick.

The frame has two sides: a vertical side and a horizontal side.

Cables from the outside attach to the vertical side, or *Vertical Distribution Frame (VDF)*, where also might be found protector fuses. The horizontal side or *Horizontal Distribution Frame (HDF)* houses another set of connectors from which multi-conductor cables run to the actual switching equipment.

Two wire twisted pair "jumper wires" run through the frame to attach the connectors on the VDF to those on the HDF. In this way any piece of equipment can be connected to any incoming cable pair. Wire-wrapping, soldering, or screw terminals on the frame are now avoided to assure the high quality signaling. For example, the ExchangeMAX Structured Cabling System for the Central Office from Lucent Technologies uses what they call the Z-IDC Interconnect System. This is an MDF block that uses *Insulation Displacement Connectors (IDCs)*. The double-sided IDC terminals can accept either one or two wires simultaneously, allowing for bridging and back-up operations. It also has hinged mounting brackets that allows cable terminations and protectors on the rear of the block, providing easy and unobstructed cross-connect and test access on the front. Also, this two-way access allows for cross-connection and testing without having to remove any of the cable protectors.

OKI Electric Industry, Ltd. took a different approach with their Smart-MDF. With Smart-MDF, the ongoing re-wiring or "jumpering" of subscriber lines on the frame doesn't have to be done manually onsite by experienced personnel. This innovative new automatic main distribution frame can alter its wiring connections with a simple key operation from a remote terminal, thus reducing operating costs and maintenance work. Smart-MDF automatically keeps all of its wiring records, eliminating yet another need for manual labor.

More and more main distribution frames must handle optical fibers instead of copper cables. Just as in the case of copper lines, an optical main distributor is used to distribute the fibers between the incoming fiber optic cable and the transmission system. Cables coming from the *Optical Line Terminations (OLTs)* are connected to the fibers of the access network.

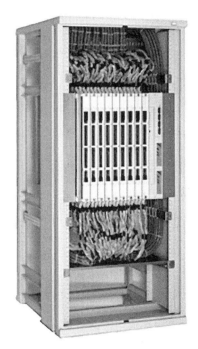

Fig. 1.16. The Smart-MDF from OKI can rewire itself under the control of a remote terminal.

The Switch

From the horizontal side of the frame, lines connect to the all-important switch. A switch is a large mechanical, electrical or electronic device that opens or closes circuits, completes or breaks an electrical path, or selects paths or circuits.

The first switching mechanism was the manually operated switchboard. The switchboard operator sat in front of a "cordboard" consisting of rows of "jack sockets" or "jacks," with each jack connected to one of the wire pairs entering the office. Above each jack were two small light bulbs. When a phone service subscriber wanted to call someone in the serving area, they took the telephone off the hook and turned a hand-cranked magneto generator, which generated extra electricity on the wire and lit a light above their circuit position on the switchboard. Later on, when the "common battery" systems appeared, a large battery at the central office supplied current through a resistor to each wire pair so that the caller didn't have to turn the crank — just picking up the handset would cause the hook switch to complete the circuit and let current flow through wires between the customer premises and the central office, which lit up the bulb above that person's designated jack on the switchboard.

Sometimes the caller also yelled "Ahoy!" or "Hello, Central!" into the phone to get the operator's attention before she actually had a chance to answer.

Early switchboards had no numbers since there were so few subscribers. The operator was expected to memorize the names of the subscribers and everybody asked for each other by name. When the National Bell Telephone Company founded Atlanta, Georgia's first local exchange in 1879, it consisted of a single switchboard which handled about 25 lines. Most of the lines were shared party lines with two or three subscribers per line, so there were about 60 subscribers. The whole "Atlanta Telephonic Exchange" was housed in a single room on

Fig. 1.17. A switchboard (indeed a whole central office!) from the late 1870s.

the top floor of the Kimball House, a hotel on the corner of Wall and Pryor Streets

Note: Unlike the central offices of today, which have underground cable vaults, the earliest COs in most cities were on a building's top floor so that aerial wires could more easily be strung out of the building to awaiting telephone poles.

The operator, seeing the light bulb lit up, plugged a cord from his/her handset with a "jack plug" or "plug" at the end of it into the jack at that position and asked who the person wanted to speak to (from about 1895 on, the Bell System adopted the standard phrase "Number, Please!" as the appropriate way to answer a switchboard). If the jack for the desired line already had a cord plugged into it then it was a "busy" line. If the jack was empty, the operator would send a ring signal to the receiving party and wait for the party to pick up the phone. Once the receiving party picked up, the operator plugged the caller's cord into the receiving party's jack, and thus "cut through" the call. A second light above the position showed when the conversation was over and the cords could be unplugged from both ends and returned to their normal inactive positions.

The Multiple or Semi-automatic Switchboard

By 1879 there were about 45,000 phone lines in the U.S. The increase in subscribers also meant that a single person sitting in front of switchboard couldn't handle the voice traffic. In 1887, less than ten years later, the number had more than quadrupled to nearly 200,000 phones in the U.S.; this phenomenal growth led to the invention of the "multiple switchboard." This type of switchboard repeated each subscriber's line jack in front of each operator, allowing any single operator to connect calls to any subscriber in the office.

But subscribers to phone service continued to explode in number, and many wanted to make calls outside of the exchange's serving area. Central offices were soon connected to each other with "trunks." Whereas subscriber loops or "lines" are dedicated access circuits connecting a home or business to the CO, a "trunk" is a shared connection between COs. The first connection between the local exchange carriers of two major cities was the link established between New York and Boston in 1883. A long distance line between London and Paris opened in 1891. In large urban areas, more than one CO was found to be necessary; these also had to be connected to each other.

In the early 1900s, the semi-automated switchboard technology reached its zenith with the Divided Multiple Switchboard using "Straightforward Operation," developed by Milo Gifford Kellogg. Using this system, a caller gave the "A" operator the name of

Fig. 1.18. A central office in 1884. Note the giant manual switchboard.

Fig. 1.19. A later switchboard from *circa* 1930. Many local calls by this time are handled by automatic electromechanical systems and only long distance calls would be handled by an operator.

the office and the number to connect to. The "A" operator then plugged into a trunk to the office desired and passed the number to the "B" operator over the trunk itself. The "B" operator was automatically connected to the calling "A" operator via an "automatic listening" function. There was also an automatic ringing feature, which meant that when the "B" Operator plugged into a customer line to ring it, the equipment rang the line immediately, returning the customary "burrr" ringing sound until the called party answered. If the called line was busy, the "B" operator plugged the cord into a special jack that returned the "buzz-buzz" sound of the busy signal.

This human operator-mediated system was too slow and complicated. Switching had to become automatic.

The Step-by-Step Switch

Almon B. Strowger (1839-1902) was a mortician in Topeka, Kansas (some say El Dorado, Kansas). A rival undertaker was getting more business and Strowger was convinced the switchboard operators at the local manual telephone exchange were being paid to divert his calls to the competitor or to report a busy signal. Other versions of the story have the competing mortician in a romantic relationship with an operator, or the competing mortician's wife or daughter working as a switchboard operator.

Strowger moved to Kansas City, Missouri, in an effort to escape his imagined enemies, but Strowger grew even more paranoid. As *Kansas City Star* reporter Joseph Popper wrote, "After his move to Kansas City, Strowger's torment seemed to become more intense. He was a frequent and unwelcome visitor at the phone company's office, shouting that his phone wasn't working and that the switchboard operators were falsely giving his customers a busy signal."

Whether it was all real or a case of paranoia, didn't matter — Strowger was determined to do away with operators. In 1887, Strowger began to develop what he called "a girl-less, cuss-less switching device."

The first working model of the Strowger automatic switching system appeared in

Number Please

The number of subscribers served by each exchange exploded. So much so that telephone directories of numbers were soon published. It seems only appropriate that the world's first telephone exchange, the New Haven District Telephone Company in New Haven, Connecticut, on February 21, 1878, published the first telephone directory (150 copies), which consisted of a single sheet of paper listing 50 names. That first phone directory was almost immediately followed by a publication of a directory of telephone subscribers by the Boston Telephone Dispatch Company.

The fast growth of subscribers also quickly led to the adoption of telephone numbers to replace names at switching offices. Telephone numbers began to accompany subscriber names in 1879 — the idea of a Lowell, Massachusetts physician named Moses Greeley Parker. Parker feared that a local measles epidemic would cause Lowell's four exchange operators to become bedridden and that the replacement operators wouldn't know what names were associated with the more than 200 jacks on the switchboards. Parker came up with the idea of a two-letter and five-digit alphanumeric system for identifying telephone service subscribers.

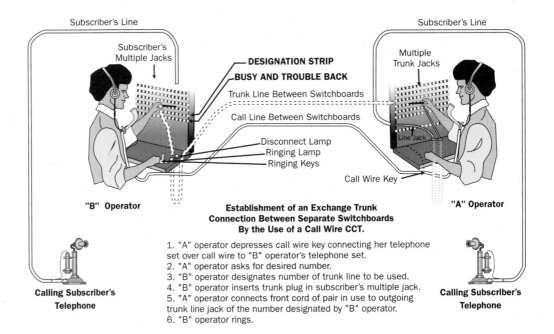

Fig. 1.20. Establishment of an Exchange Trunk Connection, pre-1915. "Long distance" calls required the use of two switchboards. This diagram, redrawn from a Connecticut phone company manual from the late 1890s, shows Operator "A" who answered incoming "outside" calls, and Operator "B" who then made the connection to local customers. Locally generated calls would first go to Operator "B" and then to "A". A manual ringing system, used by the operators to identify which circuit to use, was replaced by an automatic process in 1915.

1888. A year later, a more sophisticated commercial system was developed with his nephew, Walter S. Strowger. At some point during this time he also hired a team of electromechanical technicians to fully perfect the system.

Here's how it worked: By pressing numbered buttons on a telephone (yes, Strowger had the first push-button phone), signals traveled to a "selector" or moving wiper with an electrical contact on the end. The signals from the telephone instructed a series of electromagnets and ratchets to move the selector up and around inside of a "switch-cylinder," which was a bank of terminals arranged in a cylindrical arc 10 rows high and 10 columns wide. The selector could make electrical contact with any of the 100 terminals. The terminals could lead to either another switching system or else directly to the party being called. Thus a connection was made via a "step-by-step" switching technique where each digit dialed caused successive series of mechanical switches to bring the connection one step closer to the person called.

Although Strowger had promised his original research and development team a share of the profits of the invention, he lied, patenting what he called the "collar box" on his own (U.S. Patent No. 447918, March 10, 1891).

Strowger's invention came to the attention of Joseph Harris, a salesman traveling from Chicago who was searching for some novel idea he could promote at the World's Columbian Exposition scheduled for 1893. Harris convinced Strowger to move to Chicago where he and a friend, Moses A. Meyer, founded a new corporation, the Strowger Automatic Telephone Exchange, on October 30, 1891. With much fanfare, Strowger soon installed the first automatic exchange of 99 lines in La Porte, Indiana in 1892.

Another important innovation appeared in 1896 at the Milwaukee, Wisconsin city hall, where the new Strowger telephones were installed. The staff discovered that the buttons used for signaling numbers to the exchange had been replaced by a finger-wheel dial, a component that's still used on many telephones today.

But the entire world did not immediately adopt automatic exchanges; AT&T at first had no interest whatsoever in dial phones or automatic switching systems. Strowger then went into the building trade for a short while and retired for reasons of health, selling his patents to his associates for $1,800 in 1896 and his share in the company for $10,000 in 1898.

In 1901 Joseph Harris persuaded a group of investors to refinance and reorganize the company under the name Automatic Electric Co., known for many years simply as "AE." AE purchased the rights to sell the Strowger Automatic Telephone Exchange's selector equipment. The two companies consolidated in 1908. By the mid-1920s, AE was licensing about 80 percent of the automatic telephone equipment in the world and was the second largest telecom equipment manufacturer in the United States after Western Electric. Automatic Electric became GTE Communications, Inc. and exists today as AG Communications Systems (www.agcs.com).

Almon Strowger, a man driven by wanderlust (or perhaps just suspicions over his neighbors) moved to St. Petersburg, Florida and bought a small hotel where he died in 1902 at the age of 63.

Shortly thereafter, AT&T directed its Engineering Department to develop a 10,000 line machine switching exchange. They examined the workings of the Strowger system as well as a new (and better) rotary switch system developed by the Lorimer brothers of Canada. Nothing much happened until telephone operators went on strike during World War I. AT&T then developed a sudden interest in Strowger's technology, buying rights to it in 1916 for $2.5 million. The Bell System's first dial phones and large automatic switching exchange appeared in Norfolk, Virginia in November 8, 1919. It was a Strowger step-by-step system installed by the Automatic Electric Company.

Electromechanical Switches

Although step-by-step (or SXS) switching was reliable and very popular among smaller communities, independent exchanges and exchanges outside of the U.S. (step-by-step usage peaked in 1973 at 24,440,000 installed lines in the U.S.), the system had difficulty handing larger cities, and demanded considerable preventative maintenance. The step-by-step was therefore slowly replaced by more efficient electromechanical switches. In 1921, after eight years of research the Bell System introduced the first panel switch, which, unfortunately, presented more problems than innovative features. The use of panel switching reached its zenith in 1958 with only 3,838,000 lines installed.

In the mid-1930s the brilliant Swedish engineer Gotthilf Ansgarius Betulander developed what was soon became known as the crossbar switch. AT&T first sent a team to Sweden to look at it, then acquired the rights to it and improved on it. After a trial run that had started in October of 1937, the first No. 1 crossbar came online (or "was cut into service" as they used to say in those days) at the Troy Avenue central office in Brooklyn, New York on February 13, 1938. Many crossbars were installed in medium to large cities, peaking in 1978 as they handled over 28 million Bell System lines.

Another simple and reliable switch used in rural America (as well as Mexico and South America) was the X-Y, manufactured from the 1950s through the 1970s by Stromberg-Carlson of Rochester, New York. The company since 1982 has been part of Comdial Corporation (www.comdial.com).

Most switches from the crossbar onwards used "common control switching" which separates the computer "intelligence" of figuring out the routing of a call from the actual physical contacts that complete the connection.

Transistorized, electronic "intelligence" in a switch came with the Electronic Switching Systems (ESSs) of the mid-1960s: Western Electric's No. 1 ESS, later known as the No. 1A ESS (the first 1ESS switch went into service in May 1965 in Succasunna, New Jersey) and GTE's No. 1 EAX. These early switches, although having transistorized control functions, still had an electromechanical way of actually making the wire connections (albeit a sophisticated one). The contact component was now a two-state reed switch (a "ferreed", later called a "remreed"), a tiny sliver of alloy encased in a glass capsule that was in turn surrounded by an electromagnet. Apply voltage to the electromagnet and its resulting magnetic field would close the contact, making a connection. The Western Electric switch reversed the magnetic field to open the contacts, while GTE's contacts would pop back open once the magnetic field was shut off.

Digital Switches

The logical extrapolation of all this was the development of a totally electronic, digital switch — actually a specialized digital computer running around 22 million lines of software code. The first of these was the Western Electric No. 4 ESS for long distance lines that first appeared in 1976; a modern example is the immensely popular Western Electric (now Lucent) No. 5 ESS for local exchanges, which first saw service in Seneca, Illinois in 1982 and is still sold today. All-digital switches became possible from the 1960s onward; switches stopped sending analog signals to each other over trunks and began communicating via digital techniques such as the T-Carrier hierarchy mentioned earlier.

The PSTN Hierarchy

From the 1920s right up until its breakup by the courts in 1984, AT&T maintained a hierarchy of telephone switches that resembled a corporate personnel pecking-order chart:

Regional center: These connected 10 sectional toll centers that represented 10 regions of the country. Seven of these still exist in the U.S. and two are in Canada. One would also find international telecom gateways at this level.

Sectional center: These handled interstate calls in a geographic region, such as the Southwest or Northeast. There were 67 of these.

Primary center: These connected Class 4 switches for intrastate toll calling, and connected independent telcos and Bell Operating Companies (such as the old New Jersey Bell). There used to be around 200 of these.

Toll center: Also called a *tandem toll center*, these were the first access point into the AT&T long-distance network. They contained tandem switches that could function as a Class 5 switch if necessary. There used to be 1,500 of them.

End office: This is the *local exchange office* or *Central Office (CO)* that serves a Carrier Serving Area (CSA). All of the subscribers' "last mile" analog lines as well as digital ones from businesses and other Class 5 switches, lead to these offices. Calls between offices are routed over interoffice or tandem trunks while long distance calls are routed up to toll centers via toll trunks.

This was not a strict hierarchy. Every kind of Class switch, aside from being connected to higher and lower Class switches, were also connected to other switches of its same Class. If a call from one Class 5 switch to another Class 5 switch was blocked by too much network traffic, then the call could be routed up through a Class 4, 3, 2 or even a Class 1 switch.

Some businesses have T-1 lines connecting directly to the switch in the central office, but for many homes and smaller-sized businesses, the regular analog phone signal travels to the switch in the CO (or a digital remote switch in rural areas) where it's converted into a digital signal, multiplexed with many other phone calls into a single big digital "pipe" sent from one switch to another across the country, then, near the destination, the local switch demultiplexes the signal, converts it back to an analog signal, and finally sends it across "the last mile" of the local exchange's outside plant to the receiving subscriber's telephone.

The 1984 Divestiture

A somewhat ungainly PSTN hierarchy (see previous sidebar) was used by AT&T and the Bell System until January 1, 1984. On that date, bowing to anti-monopoly pressure by the Justice Department and the courts, the Bell System was divested into seven Regional Bell Operating Companies (RBOCs) or "Baby Bell" companies (providing local service) and a more diminutive AT&T to handle long-distance service. The small independent phone companies weren't affected very much.

RBOCs could carry local calls end-to-end on their local facilities, but had to "hand off" long distance calls to a long-distance carrier such as AT&T, MCI or Sprint. Any FCC-approved long distance carrier could offer long distance service in any local service area if they had a switching office close enough.

With so many companies now competing and intruding into each other's serving areas, the courts and the Department of Justice had to define what a "local call" was. They divided the United States into 161 (now 196) geographical *Local Access and Transport Areas (LATAs)*. LATAs neatly specify the area within which the RBOCs can offer service, and they were used to determine how the assets of the former Bell System were to be divided between the RBOCs and AT&T. The original LATAs tended to be identified by three-digit area codes: 1xx and 2xx designated NYNEX (now Verizon) LATAs; 3xx was Ameritech, etc. All local companies such as RBOCs and independents that handled calls within each LATA were now known as *Local Exchange Carriers (LECs)*.

If a call had to travel from one LATA to another, the LEC had to hand it off to one of the long distance companies, which were now called *Interexchange Carriers (IXCs or IECs)*. To accomplish this, the IXC had to establish and maintain a switching office within each LATA to be served, called a *Point of Presence (POP)*.

After the divestiture of AT&T in 1984, switches became more powerful and the network flattened out. The LEC (RBOC or independent) in each LATA has several local exchange offices (still called central offices) to which may be attached some rural remote switches. The local exchanges all connect to a tandem switch in a toll office via trunks that can be twisted pair, coaxial cable, microwave towers, but now more and more, they are fiber optic cables.

If a call arrives at a local exchange and can't be forwarded directly to another local exchange, it moves up to the toll office and becomes a more expensive toll call. Since trunks from all the local exchanges are collected in the distribution frame of the toll office, this makes it easy for long distance IXCs to connect their POPs to provide long distance (inter-LATA) service by just connecting to the toll office rather than each individual local exchange office. The POP connects to the long haul network of IXC switches. These powerful switches are now generally connected by fiber optic cables and for many years have communicated via the *Synchronous Optical NETwork (SONET)* a standard for high-bandwidth data transmission over fiber optic networks (the international term for it is *Synchronous Digital Hierarchy, or SDH)*.

CHAPTER TWO

Data-over-Voice, Voice-over-Data

In the previous chapter we saw that the traditional PSTN (telephone system) is made up of four basic components:

Subscribers: Customer Premises Equipment(CPE) that attaches to the network.

Local Loop: The "last mile" of twisted pair copper wire stretching between the subscriber and the network. It is the last remaining purely "analog" piece of an otherwise fully digital network.

Exchanges: The local and tandem (toll) switching centers in the network. The local exchanges convert analog phone calls from the subscriber into a digital format for the larger network, and *vice versa*.

Trunks: The copper, fiber optic, or microwave connections between exchanges, which combine multiple voice circuits into large digital "pipes" via synchronous Time Division Multiplexing (TDM).

Prior to automatic switching, the PSTN conveyed voice calls exclusively. Since 80 percent of the energy of the human voice falls within the frequency range of 300 to 3400 Hz, the whole telephone system was originally engineered to handle this small analog bandwidth. This was done even though the twisted pair copper wires of the "last mile" of the local loop could easily handle frequencies reaching up to 1.1 mHz or more.

Voice enjoyed nearly exclusive use of the phone system until it became evident that data would have to be sent through the telephone network too.

Facsimile

Since the days of the telegraph, businesses have wanted to send images of documents to their customers, while news services have wanted to broadcast photo images to newspapers around the world.

Facsimile is the general term for the transmission of photographs, drawings, maps, and written or printed words by electrical signals. Just as the human voice can be encoded into electrical analog waves or digital pulses, so an image can be "scanned" or broken up into small dots or lines and transmitted as electrical signals by wire or radio to a distant receiver, then reproduced on paper or film in the proper order to reconstitute the original image.

Facsimile was invented in 1842 by the Scotsman Alexander Bain. We think of "fax machines" as sending mostly documents, but prior to the 1980s most facsimile traffic consisted of "wire photos" sent by services such as Associated Press (AP) and United Press International (UPI) to subscribing newspapers. Although the principle had been demonstrated in 1882 in England and facsimile experiments with devices having semiconductor photoelectric cells were performed in England in 1902 and Germany in 1904, it wasn't until 1924 that the first commercial wire photo was sent in the U.S. (from Cleveland to New York). Furthermore, it wasn't until January 1, 1935 that regular wire photo service commenced with AP transmitting a photo of an airplane crash.

Note: Until recently, data signals were traveling over circuit paths normally occupied by voice. Indeed, most simple fax machines work the same way today: You put a document in a fax machine, input a telephone number of another fax machine, the document is scanned, you press the SEND button, and a conventional voice channel then connects the two machines, over which strange sounds convey an encoded version of the document.

Although facsimile technology seems rather simple today, the scanning and printing technology was not well developed until the late 1980s. Technically, things have come a long way since the Visual Sciences KD-111 Remotecopier of 1968, which weighed 100 pounds and featured automatic start and record, dry copying, and had as its chief selling feature the fact that it was "odorless."

Indeed, today we can use tiny devices called *modems* (short for *modulator-demodulators*) used in conjunction with desktop computers to send data over voice lines. Modems can act as fax machines (hard copy requires a printer to be attached to the computer), they can send document files or images to electronic bulletin boards, or directly to another modem via file transfer protocols such as XMODEM and KERMIT.

Interoffice Signaling

When you pick up a telephone handset and dial a number, the signals sent from the phone to the switch in the local central office travel on the same line used by your voice. You've no doubt heard these strange sounding "touchtones" or "pulses." Similarly, during much of the 20th Century, switches in the phone network also used strange sounding tones to signal each other over the interoffice trunks in order to set up, route and "tear down" (disconnect) calls. These dual frequency combinations were called *multifrequency or MF*, which sounded like touchtones but were actually slightly different. The telephone network needed such signals to deal with call-related signaling messages, billing information, maintenance test and routing / flow control signals.

Prior to 1976, these interoffice signals traveled over the same path as voice calls. This *in-band signaling* technique was a bad idea, allowing the first generation of hackers — called "phone phreaks" — to gain illegal, free access to the world telephone network merely by building an electronic "blue box" or "M-Fer" (for "multi-frequency") that could generate tones capable of disabling the metering functions of the toll circuit. In 1971, when the makers of Cap'n Crunch breakfast cereal included a toy whistle prize in every box as a promotional item, a phone phreak named John Draper discovered that the toy whistle just happened to produce a perfect 2600-cycle tone that could be used to make free phone calls.

Needless to say, the telcos soon re-engineered their network to use a form of *in-channel out-of-band signaling*, which uses tones higher than the highest 3400 Hz tone allowed in the local loop. Whereas interoffice toll trunks allow for higher frequencies to be transmitted, coils on the standard local loop filter out these tones, making it difficult to defraud the system. Eventually, the telcos decided to use *common channel signaling*, which involves building a dedicated digital signaling network separate from the voice network, but running parallel to it. The latest version of this is called *Common Channel Signaling System 7 (CSS#7 or SS7)*. SS7 has its own special switches called *Signal Transfer Points (STPs)*. These connect to *Signal Control Points (SCPs)* that contain a central database for a particular geographical region served, which is used for keeping track of such things as billing, credit card authorization, 800 number conversion tables, virtual network subscriber listings, and other special services (as you will come to see SS7 plays an important role in VoDSL service). SS7 providers interconnect their STPs and SCPs using their own network of T-1/E-1 circuits, or in some cases the same physical wires or optical fibers can be used as those for voice paths so long as the SS7 signals and voice signals occupy separate channels.

Thus, the phone companies were now burdened with two networks: a voice network and a signaling network, and that was just the beginning of things to come.

Circuit-Switched Networks versus Packet-Switched Networks

As more and more networks dissociated from the PSTN were built to handle data, a different network architecture — packet-switched — was found to have some advantages over the voice-centric circuit-switched networks. Today, telecommunications providers around the globe are struggling to create a solution that can deliver an array of communications services over a single, integrated network with the same quality, speed and reliability their users expect.

Circuit-Switched Networks

Circuit-switched networks, such as the traditional telephone network, have a number of inefficient characteristics. First, before you can communicate with the called party, you must establish a real-time electrical circuit between your telephone (or fax machine or modem) and the receiving party's device. A 3.4 kHz analog voice call is digitized into a 64 Kbps channel for a long distance connection, but after you establish this connection, the circuit — and its reserved bandwidth — remains open and dedicated to you, even if you decide to just sit there and not send any voice or data. This inefficient use of bandwidth has been likened to reserving an entire lane of a highway from New York to Los Angeles for a single automobile trip. Circuit-switched transfers (whether voice or data) are generally *full-duplex*, which means that both endpoints can send and receive simultaneously. When the call is finished, a circuit disconnect is initiated by one of the two communicating endpoints.

When the cost and time involved in providing each party with its own pair of copper wires from end to end became too great, new technologies like synchronous *Time Division Multiplex (TDM)* were created. Because the inherent bandwidth provided by a pair of copper wires far exceeded the needs of a single voice telephone channel, TDM allows more than one call to be sent over a single pair of wires.

TDM uses a single bus, or wire, to allow multiple calls to share the same transmission path. These calls are gated by a sophisticated clocking system to give the path entirely to one call for a set period of time, then to the next call, and so on. This is called a "synchronous" system because of the need to accurately synchronize each call's designated "time slot" to a standard time clock — obviously, if the timing is off, each of the 24 calls in their respective timeslots will be sent to the wrong called parties.

While TDM allows the carriers to achieve a much higher capacity using their existing copper networks, it suffers from a certain amount of inefficiency, mainly because each call periodically occupies the network path for the duration of its timeslot regardless of whether it has anything to transmit on it or not. If 24 calls share a common path, each call gets 1/24 of the "pipe's" capacity whether it needs it or not. Some sophisticated systems, however, can detect inactivity and will "steal" bandwidth when possible.

Now that you've been schooled in the development and operation of a state-of-the-art, legacy, predominantly circuit-switched *Public Switched Telephone Network (PSTN)*, note that a totally different kind of network is becoming popular for transmitting both voice and data — the packet-switched network.

Packet-Switched Networks

Unlike circuit-switched networks, packet-switched data-centric networks are quite efficient. They were designed to take advantage of the bursty nature of the typical data stream.

Examples of packet-switched networks include the Internet *(a Wide Area Network, or WAN)*, as well as such technologies as *Frame Relay, Asynchronous Transfer Mode (ATM)*, and the old *X.25* network (invented in 1964). At the customer premises one can find another packet network, *Ethernet*, which is how computers in a business communicate with each other over a *Local Area Network* or *LAN*. Packets differ from the voice channels of the "conventional" PSTN, which is a circuit-switched network.

Packet-switched networks combine the packet streams for numerous users onto a single facility enabling the network to use a greater amount of the available bandwidth. When a stream of data to be sent between two endpoints is first received it's broken up by a computer at the sender's end into small binary bundles called *packets, blocks, cells* or *frames,* depending on the transmission protocol that defines whether the packet can vary in size (up to 4 MB in some cases), or is fixed. The packet consists of the *payload* or segment of data to be transmitted along with a *header* that gives the packet a unique identity (so that the packet can be placed

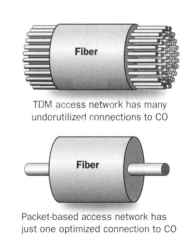

TDM access network has many underutilized connections to CO

Packet-based access network has just one optimized connection to CO

Fig. 2.1. TDM versus packet-switched access networks.

in the proper order with other packets when the transmitted data is reconstructed at the receiving end) and has control information such as the network address of the destination or target device, the network address of the sending device, the length of the packet, and other information. The packet may also have a *trailer* that has some error correction bits that can be used to determine if the packet has been garbled during transmission.

Each packet is merged into a data stream and travels separately across the network to the recipient. Depending on data traffic conditions, no two packets necessarily have to take the same route to the recipient (we'll examine this idea in more detail in a moment).

Just as a voice call is connected through the PSTN via switches, so is each data packet's header read by a *router* or *switch* that determines the route and, in particular, to what adjacent network point a data packet should be sent to further its travels toward its destination. A router is generally used as a "gateway" at the interface of two networks — for example, between a business' LAN and an external network. Switches are more efficient than routers, and are used to select a path for sending a data packet to its next destination over the massive "backbone" of the Internet, which is where the heaviest data traffic occurs between cities or nations.

There are two types of services provided by packet-switched networks — connectionless service and connection-oriented service.

Connectionless Service

With connectionless service networks, since each packet contains a complete destination address, the individual packets take whatever path is available to end up at the destination address. The packets are free agents, so they can be rerouted as the network sees fit. Connectionless service is a "best effort" service and thus uses lower level protocols (e.g. IP). "Best effort" means that the mechanism forwarding the packets is "stateless." That is, the device that is forwarding the data doesn't keep track of the data that arrived before the current packet. The packets are forwarded as long as there is bandwidth available to transport the packet. If there is no bandwidth available, the packets are simply discarded. The Internet today is a good example of best-effort service.

Thus, these connectionless, best effort networks make no delivery guarantee, although every effort is made to deliver the packets to their destination. However, connectionless service networks such as the Internet do have the ability to route around network traffic logjams (situations in which there are too many packets present in the network, leading to performance degradation) and other kinds of trouble spots, which at least helps to ensure the survivability of the packets, if not their timely arrival at the proper destinations. If the packets must absolutely, positively, get to their destination via a connectionless service, then a higher layer protocol that is capable of recovering from errors (such as HDLC, IPX or TCP) must be brought into play.

Connection-Oriented Service

A connection-oriented service (also known as virtual circuit service) is packet-based but comes closer to emulating a circuit-switched private line. This type of service sends all packets from the same source along the same route, ensuring sequential delivery at the destination point. Connection-oriented service networks (such as the X.25 network) are "rules-based" in that a multi-stage process is used to transmit packets across the network. The first stage is the call setup packet, which contains addressing fields that provide both the full source and destination addresses that are used to establish the virtual circuits (VCs). Virtual circuits can be either permanent or switched (temporary). Permanent virtual circuits (PVCs) are generally used for the most often used data transfers, whereas switched virtual circuits (SVCs) are used for sporadic data transfers.

With most of these kinds of services, when the first packet of a transmission arrives at the first switch along the route, the switch buffers the packet and examines it for bit errors — if an error is found the switch sends a message back to the device that sent the packet asking for it to send another copy. If this copy of the packet is found to be error free, the original packet is discarded.

The destination address of the packet is then read and the switch selects the best route for the packet to travel to its destination. The switch then records the packet and its route into its routing table. This process is repeated at each switch the packet encounters in its travels across the network to its ultimate destination (except for the error checking routine — if there is an error detected in the packet traversing its domain the switch just goes back one step to the last switch the packet passed through and asks for another copy). Since the first packet has left a clear trail for the related packets to follow, the subsequent packets don't need a full address, just a virtual circuit identifier, and the switches do the rest.

Fast Packet Services

Unlike circuit-switched networks, which were built from the ground up to provide QoS, the packet-switched network operator faces the problem of building a robust, properly engineered network capable of providing transport services having the same transmission quality as the traditional voice network. In an attempt to achieve this, these operators must implement various QoS protocols and architectures to guarantee the quality of the services they offer to their customers. Connectionless and connection-oriented service networks alone can't provide a high level of quality without technical enhancement; this is where fast packet services can play a major role.

Fast packet networks can greatly reduce the overhead required to send data packets over a network by making a number of assumptions about the network. First, that the network is based on a fiber (or "fibre") infrastructure. The second assumption is that the network is made up of modern, robust switch fabrics. Third, it's assumed that there is a low error rate. Finally, fast-packet services assume that the destination receiving device is intelligent enough to make an informed decision about the correctness of the data.

With such assumptions in place, when a switch receives a packet it can simultaneously parse it for errors, examine the destination address information and select a route (no buffering). If, by some chance, an error is detected, it discards the bad packet and assumes that the end receiving device has enough intelligence to detect a missing packet and to take the proper correction procedure.

While no network is totally error-free, tolerably few errors appear in the fast packet service networks so that the benefits of such streamlined networks far outweigh any inefficiency that is incurred when an error actually does occur.

Fast Packet Services include the following:

Fiber Distributed Data Interface (FDDI) Network Services (FNS): A high-speed data service that connects subscriber LANs in metropolitan areas over a shared 100 Mbps fiber backbone. FNS is an extension of the American National Standards Institute's (ANSI) FDDI LAN standard to create a 100 Mbps FDDI backbone with the ability to function as a shared public network. Users connect to the FNS backbone at native 10 Mbps Ethernet (802.3) and 16 Mbps Token Ring (802.5) LAN speeds. FNS is a very cost effective way to inter-connect buildings at native LAN speed, since there's no need to purchase and maintain special routing equipment. FNS allows virtual private network (VPN) domains to be created via special software to ensure privacy, so packets within one customer's FNS domain will won't accidentally appear in another's FNS domain.

Full Fiber Distributed Data Interface (FDDI): A high-speed data service that provides customers with dedicated FDDI networks. Customers requiring a full 100 Mbps to aggregate traffic from multiple LAN traffic can use Full FDDI Service. Full FDDI requires some minimal Customer Premises Equipment (CPE) to connect to the network. The service offers high reliability as all transmissions generally ride over all-fiber

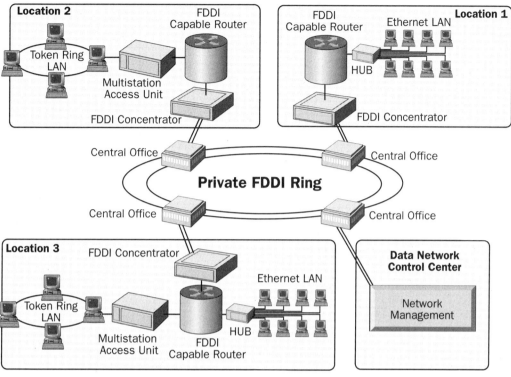

Fig. 2.2. A Central Office-based FDDI Network.

networks. Verizon's full FDDI service, for example, comes with 24x7 network management from Verizon's Data Service Center using the Simple Network Management Protocol (SNMP). A full FDDI service running on a dual counter-rotating ring configuration (two fibers) connecting the nodes at the customer premise provides fault tolerance capabilities. The network interface at the customer premises (which need not be placed in a wire closet) is a single-attached connection on fiber that carries one light frequency (single mode fiber) or many frequencies (multimode fiber).

Frame Relay (FR): A connection-oriented packet switching service similar in principle to X.25. Frame Relay relies on the High level Data Link Control (HDLC) protocol. HDLC is a link layer protocol (which is Layer 2 of the OSI Reference Model described in Appendix A) used for point-to-point and point-to-multipoint communications. HDLC encapsulates packet data in a "frame" (a sort of super-packet) which can be up to 4 kilobytes in length. Synchronous frames of packet data are routed to different destinations depending on the frame's header information. There's also a "trailer" which includes various control information and which can be used for such

> **Note:** Aside from its use in Frame Relay, other variants of HDLC include LAP-B (Link Access Procedure-Balanced), LAP-D (Link Access Procedure-Data channel), PPP (Point-To-Point Protocol), and IBM's SDLC (Synchronous Data Link Control).

things as simple error correction. Each frame starts and ends with a Flag character (7E Hex). The first two bytes of each frame following the flag contain the information required for multiplexing across the link. The last two bytes of the frame are always

X.25, the Granddaddy of Packet Networks

In "ancient times" (up until the mid 1970s), digital devices such as mainframe computers and terminals had a difficult time communicating with each other over long distances, since the fledgling Internet connected only a few large universities at the time and the largest network then in place was the analog PSTN, which was engineered for voice.

Thus, each remote user at a computer terminal needed to use an expensive analog modem to dial up and establish a continuous circuit-switched connection to another modem at the mainframe (although a continuous leased line could be purchased). The modem in general use at that time was the Bell 103 that AT&T had introduced back in 1962, which had a capacity of a mere 300 baud. Businesses were billed for the entire continuous connection even though such asynchronous communications are "bursty" in nature — keystrokes and file transfers tend to occur in bursts, with much idle time during which nothing is transmitted.

Packet-switched networks solve these problems. They support low-speed, asynchronous and bursty communications between computer systems. Unlike circuit-switched connections, packet-switched network usage billing is based on the number of packets transmitted, and offers error detection and correction at each of the packet switches, or nodes.

The first of these packet-switched digital networks was X.25. In 1976 the CCITT (ancestor of the ITU) adopted the X.25 standard under the designation: "Interface Between Data Terminal Equipment (DTE) and Data Circuit Terminating Equipment (DCE) for Terminals Operating in the Packet Mode on Public Data Networks."

X.25 is a peer-to-peer network that covers Layers 1 to 3 of the ISO Communications Model. Equipment must be connected to the network via PADs (Packet Assembler/Disassemblers), a kind of packet modem which converts back and forth between the protocol used by the device and the X.25 packet protocol.

generated by a *Cyclic Redundancy Check (CRC)* of the rest of the bytes between the flags and is used for error correction. The rest of the frame contains the user data.

Frame Relay provides connectivity at speeds from 56 Kbps to 45 Mbps. This service employs permanent virtual connections (PVCs) between predetermined end points. Ideally suited for medium-speed, bursty, LAN interconnection applications. Prior to the rise in popularity of IP, many VPNs were (and still are) based upon Frame Relay service. Frame Relay, along with it's relative, IBM's SDLC, has been used extensively for serving IBM's SNA (System Network Architecture) environment that han-

---

Each X.25 packet typically contains 128 or 256 bytes (or "octets") of payload (user data); packet sizes of up to 4096 bytes are possible in some networks, though the typical upper limit is 512 bytes or 1024 bytes. Each packet's header includes a packet address field of 8 bits (4 bits for the calling DTE and 4 bits for the called DTE), which is the Logical Channel Identifier (LCI) or Logical Channel Number (LCN) that aids the the network's packet switching nodes in routing each packet over an appropriate path to the intended device. Control data (8 to 16 bits) in the header contains information enabling the target node and terminal equipment to identify errored, corrupted or lost packets, and to resequence the packets should they arrive out of order. The header control data also includes the number of the virtual circuit (4 bits) and virtual channel (8 bits) over which the data will travel, if a network path has already been dedicated. Error control data is contained in 16 bits in the packet trailer, which is subject to a highly reliable CRC check.

X.25 uses a buffered, "store-and-forward" technology, so two DTEs communicating with each other don't have to use the same line speed. For example, a host connected at 56 Kbps can communicate with various remote sites connected with cheaper 19.2 Kbps lines (X.25 packets can even travel over a 9.6 Kbps portion of the little-known 16 Kbps "D" or "Delta" channel of Basic Rate Interface ISDN). However, this same store-and-forward mechanism leads to a turn-around delay of about 0.6 seconds. This has no effect on large block transfers, but in conversational, back-and-forth "flip-flop" types of transmissions, the delay becomes noticeable.

X.25 and the well-known TCP/IP packet protocol used in LANs and the Internet are vaguely similar but differ in several ways: TCP/IP has only end-to end error checking and flow control, while X.25 is error checked from node to node. TCP/IP has a more sophisticated flow control and window mechanism than X.25, to compensate for the completely passive nature of the TCP/IP network. Also, electrical and link levels are rigorously specified in the X.25 specifications, while TCP/IP is designed to travel over many different kinds of media, with many different types of link service, such as Ethernet, Frame relay, X.25, ATM, FDDI, etc.

dles high-speed data transfers between IBM mainframes.

Frame Relay is also very cost effective, partly because its network buffering requirements are scrupulously optimized. Compared to X.25, with its store and forward mechanism and full error correction, network buffering in Frame Relay is minimal. Frame Relay switches packets end to end much faster than, say, X.25: the frames are switched to their destination with only a few "byte times" delay (the time it takes to transmit a byte at wireline speed), as opposed to several hundred milliseconds delay on X.25.

But the biggest difference between Frame Relay and say, X.25, is that X.25 guarantees data integrity and network managed flow control at the cost of some network delay, while Frame Relay offers little or no guarantee of data integrity.

The FR network delivers frames, whether the CRC check matches or not. The network does not even necessarily deliver all frames, since Frame Relay has no flow control — frames are simply discarded whenever there is network congestion. Thus, to be certain that all of the information transmitted is getting through, one must use an upper layer protocol that runs above Frame Relay and has built-in error correcting abilities, such as HDLC, IPX or TCP/IP.

This is not quite as big a problem as it sounds, since modern digital networks (unlike the noisy analog communication lines originally used for X.25) have very low error rates. Very few frames are discarded by the network, especially networks that have excess capacity.

Frame Relay provides indications that the network is becoming congested by means of the Forward Explicit Congestion Notification (FECN) and Backward Explicit Congestion Notification (BECN) bits in data frames. These are used to tell the application to slow down, hopefully before packets start to be discarded.

Frame Relay is also well-suited to handle multi-protocol communications. Since multiple virtual connections can be established over a single physical access line, Frame Relay can reduce the number of customer premises (e.g., router) ports required for wide area network communications (in comparison to dedicated private lines).

Switched Multi-Megabit Data Service (SMDS): Suited for dynamic environments, SMDS is a connectionless data service that operates at speeds of from 56 Kbps to 45 Mbps. Like ATM (see next), SMDS uses cell relay transport: both services use fixed-sized 53-byte cells for transport and can accommodate packet lengths of 9188 "octets" (an octet is another name for an eight bit byte). However, the maximum packet length for SMDS is 9188 octets and the maximum length for ATM is 65,535 octets.

Because SMDS is a public service using a universal addressing plan, any SMDS customer can exchange data with any other SMDS customer. This facilitates high-performance, inter-enterprise networking that is more flexible, manageable, and cost-effective than previously available.

Asynchronous Transfer Mode (ATM): I've saved ATM for last on this list of fast packet services because, at the moment, it's the packet-switched network of choice for nearly 90% of the VoDSL service providers — the other two being Frame Relay and the increasingly popular IP. Indeed, about 80% of the world's carriers use ATM in the core of their networks, since it is incredibly flexible. ATM serves as a bridge between legacy equipment and "next-gen" systems, and supports a huge array of other transport technologies, including DSL, IP Ethernet, Frame Relay, SONET/SDH and wireless platforms.

A typical communications network configuration may have a mix of TDM, Frame Relay, ATM and/or IP. Within a network, carriers often extend the characteristic strengths of ATM by blending it with other technologies, such as ATM over SONET/SDH or DSL over ATM. Since ATM freely interfaces with all of these technologies, carriers can inexpensively extend the management features of ATM to other platforms, thus maximizing their infrastructure investment. In fact, because of its *Quality of Service (QoS)* capabilities, much of "voice over DSL" technology is really "voice over ATM."

ATM's Architecture

Let's dig further into ATM's architecture. . .

The ATM network is a connection-oriented fast packet services network. The highest bandwidths yet achieved over the ATM network run a close second to SONET (ATM scales from T-1 to OC-48 at speeds that average 2.5 Gbps in operation, 10 Gbps in limited use and up to 40 Gbps in trials). ATM offers a plethora of internationally recognized standardized QoS definitions, which allows the network to easily support many real-time services, including voice-over-DSL (VoDSL). Further, it allows for a simplified switch design since all ATM traffic is segmented into fixed-length 53-byte long packets (called "cells") rather than variable-sized packets or frames. But, like X.25's packets or Frame Relay's frames, each ATM cell contains header information that is used to direct the cell through the network. Each cell contains 48 payload bytes and five header bytes. Information to be transmitted is buffered and placed in a cell. When each cell is "full" it is sent through the network to the destination specified within the cell's header.

The advantage of this fixed-length cell design is that any type of communication can be multiplexed into an ATM transmission, regardless of intrinsic data rate or packet size.

ATM resulted as a combination of the concepts of cell relay switching technology (supporting both variable and constant bit rates) and the Synchronous Digital

Note: ATM networks can be used in both LAN or WAN environments, and some rackmount computers even have ATM backplanes over which peripheral resource cards communicate with each other. ATM as a network transfer protocol technology originated with the Consultative Committee on International Telephone & Telegraph (CCITT) International standards organization, an ancestor of the International Telecommunications Union (ITU). The ITU adopted ATM as a standard in 1988. Today, ATM and SONET form the basis for a host of network services known collectively as Broadband ISDN (B-ISDN) or ATM Broadband Networking, available commercially in major metropolitan areas.

Hierarchy (SDH) over optical fiber transmission technology (supporting large amounts of data to be transmitted over a network efficiently). Thus, ATM's cell-switching technology combines the best advantages of both circuit-switching (for constant bit rate services such as voice and image) and packet-switching (for variable bit rate services such as data and full motion video) technologies. The result is the bandwidth guarantee of circuit-switching combined with the high efficiency of packet-switching.

ATM's originators thought that ATM was flexible enough so that in the future there would be just one network for all traffic types: voice, data and video. At one point it was thought that a 25 Mbps LAN version of ATM would bring ATM to the desktop and with it, multimedia. But, as fate would have it, 100 Mbps and 1000 Mbps "Gigabit" Ethernet ended up dominating the LAN, while ATM rules outside the office in the WAN and network core. Still, ATM does offer *LAN Emulation (LANE)*, which is a set of services, functional groups and protocols (operating at the Link Layer or Layer 2 of the OSI Reference Model) that provide for the emulation of Ethernet and Token Ring LANs over an ATM backbone and is used to internetwork LANs over ATM. LANE takes over the MAC (Medium Access Control) layer function found on Ethernet and Token Ring NICs (Network Interface Cards), and supports connectionless service in either a broadcast or multicast mode. Even more sophisticated than LANE is the newer *MPOA (MultiProtocol Over ATM)* which operates at the Network Layer (Layer 3 of the OSI Reference Model) and overcomes some of LANE's limitations by supporting multiple network protocols such as IP, IPX, and AppleTalk.

As alluded to earlier, ATM differs from most other packet-switched networks in that it does provide QoS by ensuring "cell sequence integrity" whereby cells arrive at their destination in the same order as they left the source. This is because nearly all of the possible ATM topologies are connection-oriented, establishing a temporary, virtual, "logical" circuit through a network of switches from end to end when a cell needs to be sent so that cells sharing the same source and destination are guaranteed to travel over the same route. This is unlike a network such as the Internet, where each succeeding packet can take a different route to a particular shared destination.

But unlike telephone switches that dedicate circuits end to end, the ATM connection is not dedicated to one conversation. When the ATM channel does not use the reserved bandwidth of the connection path, information from other channels can use this spare capacity. If the ATM network is idle, the logical circuits' unused bandwidth can be used to transport unassigned cells. A lull during a telephone conversation or a videoconference can be exploited by the system by sending data cells or some other type of cells — the system ensuring that there is no "space" or "gaps" between cells.

Thus, ATM is "asynchronous" in the sense that although cells are relayed synchronously, data need not be sent at regular intervals, i.e. ATM frames are synchronous, but the circuits are not allocated specific time slots within the ATM frame and the slots may vary in bandwidth. Cells from multiple sources and multiple destinations are asynchronously multiplexed between multiple packet switches. Thus, ATM uses a random or "statistical" multiplexing scheme, rather than the fixed-time or "deterministic" multiplexing scheme that characterizes the older *Synchronous Transfer Mode (STM)*, which is a newer term encompassing TDM, used in almost all current non-ATM switches. This sharing or statistical multiplexing can handle high bandwidth bursty traffic such as compressed video, which can have a peak bit-rate ten or more times its average bit-rate.

ATM is highly scalable. ATM switch manufacturers have made scalability a key tenet of their design efforts, allowing service providers to easily augment their ATM switches as needed. ATM can support transmission speeds as slow as 9.6 Kbps between ships at sea or increments of 1.5 Mbps, 25 Mbps, 100 Mbps, 155 Mbps, and 622 Mbps (OC-12) full duplex. ATM can also provide 1.224 Gbps (OC-24) and 2.488 Gbps (OC-48) services, and even higher OC-n (n x 51.84 Mbps). Yet, ATM is not tied to any particular physical medium, so it can be applied to any of the existing physical networks such as optical fiber, coax, or twisted pair.

Since the cell is defined independently of speeds, framing or physical media, all networks — LANs, WANs and public networks — can use the same cell format. A 45 Mbps DS-3 line in a CO can receive cells generated by a multimedia application on a 155 Mbps LAN (either shielded twisted pair or fiber) and switch them onto a WAN-based OC-12 (or faster) SONET system. ATM devices such as network cards are simply routing cells. This means that ATM technology blurs the distinction between the local and public networks since routers and bridges aren't needed, just switches residing on a fast, optical backbone.

By the early 1990s commercial ATM equipment was being developed for large backbone telco applications and private network vendors also became interested in the high-bandwidth possibilities of ATM. These events resulted in the founding of the ATM Forum in 1991 to replace the ITU as the specifications setting body for ATM.

ATM Specifications

The ATM specifications consist of the length and format of the ATM cell, adaptation layer functions, and signaling.

The large or variable length cells such as the kind used by X.25 and Frame Relay networks gives a better payload-to-overhead ratio than ATM, but at the expense of longer, more variable delays. In order to maintain QoS for time-sensitive voice and video communications having a *variable bit rate (VBR)*, packetization delay variance (the time it takes to fill a cell with data) had to be reduced as much as possible, which can be accomplished by specifying small cells. A small ATM cell ensures that voice and video can be inserted into the stream at a high rate of periodicity, enough to support real-time transmissions. Smaller cells reduce the amount of information carried per cell, but larger sized packets are required by data networks to increase bandwidth efficiency. Hence, ATM's 53-byte cell format was something of a compromise.

Still, the fact that ATM cells are fixed-length allows for the construction of very fast switches, since any processing associated with variable-length packets is eliminated — the system doesn't need to be looking for the end of a frame, for example.

The ATM Reference Model

ATM has its own protocol reference model independent of the standard seven-layer OSI model (see Appendix A), consisting of a control plane, user plane and management plane. Yet, the ATM model does have some components resembling equivalent ones in the OSI layered model. The User plane (for information transfer) and Control plane (for call control) are structured in three main layers: Physical Layer, ATM Layer and ATM Adaptation Layer (AAL), with the Physical and AAL Layers further divided into sublayers.

The Physical Layer (or "PHY," corresponding to the OSI Physical Layer) is usually taken to be SONET/SDH (which itself consists of four layers) but can involve other underlying transport media.

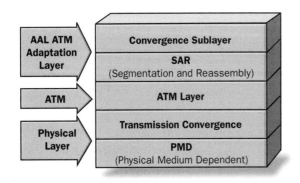

Fig 2.3. ATM Layers.

Above the ATM Physical Layer sits the AAL and the ATM Layer. The AAL formats the 48 byte cell payload so as to adapt ATM's cell switching abilities to the attributes of many different higher layer protocols, each having varying traffic characteristics. This service-dependent layer roughly corresponds to the OSI Data Link Layer (data error control above the Physical Layer), though some argue that it's really similar to OSI transport, as it involves end-to-end connections.

Of the original five types of AAL, two (AAL3 and AAL4) were combined. The ATM adaptation sublayers are now defined as follows:

AAL1: Standardized in both the ITU-T and ANSI since 1993, AAL1 is still used by large, traditional telephone networks for placing Voice-over-ATM, for uncompressed video and for other isochronous traffic. AAL1 is used for connection-oriented services that demand Constant Bit Rate (CBR) traffic as well as specific delay and timing requirements, such as DS-1 and DS-3. AAL1 has been for many years the only standardized way to provide timing recovery, optional forward error correction, and minimization of the effects of cell loss in the network.

Chracteristics	Class A	Class B	Class C	Class D
	Constant bit rate	Variable bit rate	Connection-oriented data	Connectionless data
Synchronization between Source and Destination	Required	Required	Not Required	Not Required
Bit Rate	Constant	Variable	Variable	Variable
Connection Type	Connection Oriented	Connection Oriented	Connection Oriented	Connectionless
Adaptation Layer	AAL1	AAL2	AAL5	AAL3/4

Fig 2.4. ATM Adaptation Layer. Adapts different kinds of information streams to ATM and makes possible various service types.

AAL1 is a connection-oriented service that can supply a constant bit rate. Synchronization must exist between the source and the destination. These factors make it a "Class A" ITU-T Service Classification.

AAL1 figures into two of the three standard methods used to transport voice traffic over ATM networks: AAL1 Unstructured Circuit Emulation, AAL1 Structured Circuit Emulation, and AAL2 Variable Bit Rate Voice-over-ATM (VoATM).

By extending AAL1 to allow replacement of 64 Kbps traditional digital voice circuits, ATM can be used to convey voice on ATM backbones instead of TDM infrastructures.

AAL1 Unstructured Circuit Emulation allows the user to establish an AAL1 ATM connection allowing ATM to replace TDM circuits to carry plain DS-1 at fixed rates (such as a full T-1 or E-1) end-to-end over the ATM backbone, with bits in equaling bits out (no robbing of bits), and with clocking information carried within the cells.

AAL1 Structured Circuit Emulation is for "loop timed" N x 64 Kbps services, such as a fractional T-1 or E-1 where the ATM network switches the N x 64 circuits over the ATM backbone.

In either case, emulating a circuit inside an ATM network tends to consume more bandwidth than necessary. Because of the overhead in the ATM cell and in the AAL1 layer, the ATM connection needs an overhead of 12% more bandwidth than the circuit it is carrying. A 1.536 Mbps DS-1 circuit carried across an ATM backbone requires 1.73 Mbps of ATM bandwidth. AAL1 thus tends to take a "brute force" approach to sending anything over ATM because of the permanently allocated bandwidth that is poorly utilized and inefficient, making AAL1 a wasteful solution for Voice over ATM. Still, in the early days of ATM, AAL1 became a *de facto* standard in the absence of a real, optimized specification that could handle Voice-over-ATM.

AAL2: Previously known as Composite ATM or AAL-CU, AAL2 is a relatively new connection-oriented ATM Adaptation Layer. AAL2 is specified in ITU-T Recommendations I.363.2 (1997), I.366.1 (1998), and I.336.2 (1999); it carries the specific mandate to provide highly efficient means of mapping voice and other bursty transmission real-time into ATM cells.

Originally AAL2 was targeted for video, but AAL5 seems now to occupy that niche. It was found that it was not necessary to link a cell's bit format (specified by the adaptation layer) to its priority requirements. Instead, AAL2 uses a scheme called *Real-Time Variable Bit Rate (rt-VBR)* wherein AAL2 sends voice-oriented minicells of variable size (up to 64 bytes in length) packed into the normal fixed size cells, allowing for multiplexing within the cell and cutting delays. Its structure also provides for the packing of short length packets into one (or more) ATM cells. The variable size packets have an end-to-end timing relationship as can be found in private enterprise or internal public trunking. This leads to a major improvement in bandwidth efficiency over either the structured or unstructured circuit emulation of AAL1, since AAL2's variable bit rate makes use of the more statistically multiplexible variable bit rate ATM traffic classes, allowing users to utilize more available bandwidth, which can be further increased using voice compression, silence suppression, and idle channel removal. Access connections using AAL2's new techniques can transport voice circuits over the same facilities as data circuits.

AAL2 also enables multiple user channels on a single ATM virtual circuit and varying traffic conditions for each individual user or channel.

AAL3/4: This adaptation layer is the result of a merger of what was originally two distinct adaptation layers — AAL3 (connection oriented VBR) and AAL4 (connectionless VBR) — both of which are subject to considerable overhead. AAL3/4 places four bytes of overhead in each cell, plus an additional minimum of eight bytes of overall overhead for each stream or "datagram." It is a connectionless data and variable bit rate service. Like AAL5 (and unlike AAL1 and AAL2), there doesn't have to be any synchronization between the source and the destination.

AAL5: Perhaps the most commonly used adaptation layer, AAL5 is data-oriented and supports connection-oriented variable bit rate data services (but not connectionless ones, which would have allowed channel sharing).

AAL5's low and good error detection has made it popular for the efficient transport of TCP/IP over LANs, classical IP-over-ATM and frame relay traffic.

Unlike AAL3/4, AAL5 puts no overhead in each cell aside from changing a bit in the cell header. AAL5 uses variable length *Protocol Data Units (PDUs)* consisting of user data that can be from one to 65,535 bytes long. (A PDU is essentially an OSI term for packet.) A *Segmentation and Reassembly (SAR)* sublayer of AAL5 divides the PDU into 48-byte chunks for cell transport over the network. During this process a PDU may be padded so to always be a multiple of 48 bytes. Finally, AAL5 adds an eight byte trailer to the end of the PDU. The PDU is closed with a 32-bit *Cyclic Redundancy C=Check (CRC)* performed over all of the PDU's contents.

Furthermore, AAL5 is more efficient to process than AAL3/4 (one need only look at a bit in the header in each cell, rather than examine the data contents) and has better error detection properties.

So, AAL5 is "leaner" and needs less overhead than AAL3/4, but this comes at the expense of error recovery and built-in retransmission rules demanding that, unless the packet is fully received, the application cannot receive any part of the packet and must ignore the segment of the packet already correctly received.

AAL5 is suitable for sending a constant data stream from one location to another, but it tends to have problems unless some form of synchronization pattern for interoperability can be embedded into or otherwise sent along with the data. If a service lacking timing clock signals tries to make end-to-end connections between synchronous narrowband equipment such as PBXs or devices with H.261 codecs, frame slips can occur since the equipment is designed to be used on networks delivering very little packet jitter. One or two frame slips is interpreted as a fault and will cause the equipment to go careening out of service.

One might wonder, then, why the ATM Forum long ago endorsed using the real-time AAL5 variable bit rate class of service for real-time video, even though most MPEG streams use fixed packets running at a constant bit rate. The answer is that real-time VBR was chosen over CBR since MPEG already has its own time base, the Program Clock Reference, in its transport stream and didn't need the time stamp found in AAL1's CBR class of service. By using real-time VBR, each video connection's jitter, latency, cell-loss and cell-error rates can be specified to satisfactory levels in each user contract. While video is sensitive to jitter and latency, the decoding process can compensate for any picture frames lost as a result of cell loss and errors.

Moving on from the ATM adaptation layer, perhaps the most important layer is the ATM Layer, which is responsible for creating cells, formatting the cell header (five bytes) and actually transporting information across an ATM network. Some claim that this too corresponds to the OSI Physical Layer, since it deals with bit transport, while other maintain it actually corresponds to the OSI Data Link layer, since it involves such things as addressing, formatting and flow control.

An ATM cell header can be one of two formats: *UNI (User Network Interface)* or *NNI (Network Node Interface* or *Network-to-Network Interface)*. The UNI header is used for communication between ATM endpoints and ATM switches in private ATM networks. The actual user network interface can be an interface point between ATM end users and a private ATM switch, or between a private ATM switch and the public carrier ATM network. The NNI header is used for communication between ATM switches, generally between two public network pieces of equipment.

An ATM cell's five byte header consists of six fields containing information used in the correct routing and QoS of the cell's 48 byte data payload.

1. Generic Flow Control (GFC): This four-bit field has only local significance, since its value is not preserved from end-to-end and it is overwritten in ATM switches. The GFC is used locally to identify multiple stations that share a single ATM interface and to give the UNI a way to negotiate with the shared access networks about

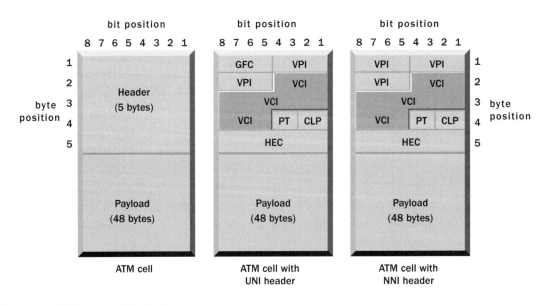

Fig. 2.5. ATM Layer: ATM Cell Formats.

how to multiplex the shared network among the cells of the various ATM connections, thus controlling the traffic flow on ATM connections from a terminal to a network. The GFC is also used to reduce cell jitters in CBR services. Otherwise, this field is typically not used and is set to its default value of 0 (binary 0000).

2. Virtual Path Identifier (VPI): The VPI field identifies multiple circuits destined for the same endpoint, greatly reducing the number of entries in the translation table of each intermediate switch and minimizing the call-setup delay. The length of the VPI field is eight bits for UNI and 12 bits for a NNI. If you consider a telephone number analogous to the VPI/VCI combination, then the VPI is the "area code."

3. Virtual Channel Identifier (VCI): This is a unique 16-bit field. The VCI is analogous to a local seven-digit telephone number. The VPI and VCI address fields store cell routing information concerning the network path. In conjunction with the VPI, the VCI identifies the next destination of a cell as it passes through a series of ATM switches on the way to its destination.

The main difference between a phone number and the VPI/VCI combo is that the VPI/VCI has only local significance between two ATM interfaces. This means that the VPI/VCI for each individual ATM cell passing through an ATM network can (and probably does) have a different VPI/VCI identifier between each switch between the source and destination. The combination of the VPI and VCI fields uniquely identifies each of the possible 2^{12+16}=268,435,456 channels which may be asynchronously transmitted across a shared link.

ATM is connection-oriented only in that, prior to sending data between two endpoints, a virtual / logical connection has to be secured between the two ports. The virtual connections ATM uses for information transport are the *Virtual Path Connection (VPC)* (also called the *Virtual Path or VP*) and the *Virtual Circuit Connection (VCC)* (also called the *Virtual Channel or VC*). Both the VPC and the VCC make cell multiplexing possible.

The VCC is the actual connection between the source and destination endpoints in an ATM network. It consists of a series of Virtual Circuit (VC) links extending between VC switches. These links of the VCC comprise the path along which cells for a particular call are transmitted between the two endpoints. The VCC can be identified in the cell by the value of the VCI.

Communicating ports can have multiple VCCs, but only one VPC. The VCI holds the address of the VCC, while the VPI stores the address of the VPC.

As cells reach each switching point at the end of each link during the course of their journey in the network, their VCI and VPI fields are examined, their values being used to determine where the cell should best be forwarded next towards the destination.

VCIs are not centrally managed, so it is usually impossible for the VCI to rely upon the same numeric value to guide the cell through every link in its journey. Instead, a unique number is used for each leg of the journey and the VCI value changes at each link along the way. At the end of each link a cell enters a new VC ATM switch where it encounters routing translation tables that alter the cell's VCI (and perhaps VPI) to the correct value for the next leg of the cell's journey. The ATM switch then builds the new cell header having the new VCI value and sends it along the next link.

The Virtual Path is essentially a bunch of virtual circuit links that follow the same route and have the same endpoints. As with VCs, virtual path links can be strung together to form the Virtual Path Connection. VPs provide logical direct routes between switching nodes via intermediate cross-connect nodes, and can establish logical links between switches not directly physically connected. Such flexibility can be used to reconfigure the logical network structure when the network's data traffic characteristics change. Indeed, VPs can be managed together and some central switches need only be capable of switching whole virtual paths, not individual virtual channels.

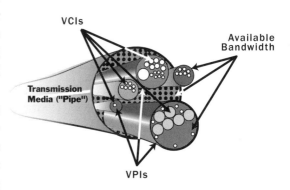

Fig. 2.6. ATM Bandwidth Utilization.

As with VCs, VPs are identified in the cell header with an identifier, the VPI, which is usually either eight bits long for UNI or 12 bits for NNI. The VPI field identifies multiple circuits destined for the same endpoint, drastically reducing the number of translation table entries of each intermediate switch. The combination of the VPI and VCI fields uniquely identifies each of the possible 268,435,456 channels that can theoretically be asynchronously transmitted across a shared link.

Every circuit on each link of an ATM network is thus identified by the unique integer fields of the VPI and the VCI. Both of these are used by the switch to ensure that circuits and paths are routed correctly and enable an ATM switch to distinguish the type of connection. As it is, ATM switches have quite a bit to do, since they are responsible for switching cells between ports, buffering cells, translating VPIs and VCIs, guaranteeing QoS, and performing both connection set-up and tear-down.

4. Payload Type (PT): This three-bit descriptor field indicates what kind of data is contained in the cell, such as user data, OAM data (network internal, signaling data). It also gives information about any ATM cell affected by traffic congestion. The first bit of the field indicates whether the cell contains user data or control data. If the cell contains user data, the bit is set to 0. If it contains control data, it is set to 1. The second bit indicates congestion (0 = no congestion, 1 = congestion), and the third bit indicates whether the cell is the last in a series of cells that represent a single AAL5 frame (1 = last cell for the frame).

5. Cell Loss Priority (CLP): This one-bit field is used to show whether a byte of information should be discarded during network congestion. If the CLP bit of an ATM cell is 0, the cell has high priority compared with a CLP bit set to 1. As a result, when congestion occurs, cells with the CLP bit set to 1 are dropped before a cell that has a CLP bit set to 0.

6. Header Error Control (HEC): The HEC eight-bit field does a CRC check of the first four bytes of the cell header contents to correct single bit errors occurring in the cell header and to detect multi-bit errors. This allows the cell to be preserved rather than be discarded.

IP and the Internet

The Internet Protocol has probably had more of an effect on everyday life than any other protocol in history, as it makes possible the Internet, which is both a transport network that moves data as well as a network (literally a network of networks) of computers and appliances.

The TCP/IP Suite

IP is the fundamental part of the TCP/IP family of protocols describing software that tracks the Internet address of nodes, routes outgoing messages, and recognizes incoming messages. *TCP (Transmission Control Protocol)* and *UDP (User Datagram Protocol)* are both protocols that use IP. Whereas IP allows just two computers to communicate with each other over the Internet, TCP and UDP allow applications (called "services") on those computers to talk to each other. They are also used in gateways to connect networks at OSI Network Level 3 and above.

The TCP/IP suite of network protocols was adopted worldwide for Internet use in 1983. It was developed by Vinton G. Cerf and Robert E. Kahn. The original Cerf/Kahn paper described just one protocol, called TCP, which provided all the transport and forwarding services in the Internet.

Kahn had intended that the TCP protocol support a range of transport services, from the totally reliable sequenced delivery of data under the virtual circuit model to a datagram service in which the application made direct use of the underlying network serv-

Fig. 2.7. Vinton G. Cerf (left) and Robert E. Kahn (right) developed what ultimately became the TCP/IP suite of protocols.

ice, which might imply occasional lost, corrupted or reordered packets.

The implementation of TCP, however, was something that allowed only virtual circuits, which was well-suited for file transfer and remote login applications, but some of the early work on advanced network applications, in particular packet voice in the 1970s, made clear that in some cases packet losses should not be corrected by TCP, but should be left to the application to deal with. This led to a reformulation of TCP into two protocols, the simple IP which provided only for addressing and forwarding of individual packets, and the separate TCP, which handled service functions such as flow control and recovery from lost packets. For those applications not needing TCP services, UDP was introduced as an alternative protocol that could provide direct access to the basic service of IP.

Today, the TCP/IP suite includes the following protocols: IP, or Internet Protocol (operating at OSI Layer 3); TCP, or Transmission Control Protocol (OSI Layer 4); UDP, or User Datagram Protocol (OSI Layer 7); FTP, or File Transfer Protocol (OSI Layer 7); TELNET, or TELecommunications NETwork (OSI Layer 7); SMTP, or Simple Mail Transfer Protocol (OSI Layer 7); and SNMP, or Simple Network Management Protocol (OSI Layer 7). TCP/IP protocols are not used only on the Internet but on private networks and LANs, called intranets.

Using the terminology of the OSI networking model, IP can be described as providing a "Connectionless Unacknowledged Network Service," which is a "no promises" or "send-and-forget" paradigm for sending packets. IP allows for the actual packets of data (called "datagrams") to arrive out of order, damaged, in multiple copies, or perhaps not arrive at all. With IP, there are few guarantees. This is why applications rely on other related transport protocols such as TCP or UDP to provide assurances of data integrity and, in TCP's case, complete and fully ordered packet data delivery.

IP does try to protect each packet's IP header by calculating a checksum on the header fields (containing the 32-bit binary value representing the destination IP address) and including that number in the transmitted packet. The receiver verifies the IP header checksum before processing the packet. If the checksums don't match, the packet is damaged and must be discarded. The same checksum algorithm is used by TCP and UDP, but they are more comprehensive in that they also include the data portion ("datagram" or "payload") of the packet in their calculations.

Since IP makes so few guarantees, it is a very simple, "lightweight" protocol that can run on computers and appliances having minimal processing power and memory. It also makes few if any demands from the underlying physical network over which the IP packets travel, and so IP can be deployed on various networking technologies. On a LAN, for example, IP is carried in the data portion of the LAN frame and the frame header contains additional information that identifies the frame as an IP frame. Different LANs have different conventions for carrying that additional information. On an Ethernet network the Ethertype field is used; a value of 0x0800 identifies a frame that contains IP data. FDDI and Token Ring use frames that conform to the IEEE 802 Logical Link Control, and on those LANs IP is carried in Unnumbered Information frames with Source and Destination Link Service Access Points (LSAPs) of 0xAA and a Subnetwork Access Protocol (SNAP) header of 00-00-00-08-00 (SNAP is an Internet protocol that operates between a network entity in the subnet and a network entity in the end system and specifies a standard method of encapsulating IP datagrams on IEEE networks).

One of the few things that must be specified per network for IP is the maximum size of a "frame" or packet that can be carried on the particular physical medium being used. Very large bursts of data must be broken down or "fragmented" into smaller packets, called "fragments." At the destination a "fragment reassembly" process occurs, and the original IP packet is reconstructed.

Both IP and TCP fragment data, but in different ways. TCP is far more efficient at the fragmentation game than IP. Therefore, TCP "stacks" (software) are always trying to figure out the largest acceptable packet size for the current network, a process called "Path MTU (Maximum Transmission Unit) discovery."

Let's consider two computers communicating with each other over an Ethernet LAN. Ethernet has a maximum frame size of 1500 bytes, but in this case imagine that a router on some intervening link limits the maximum IP packet size to 500 bytes. This means that all 1500-byte IP packets sent will be fragmented by the routers on that link into three 500-byte fragments. Since it's much more efficient to fragment packets at the TCP layer, the TCP software stack will attempt to discover the MTU. It does this by setting what's called the "DF" (Don't Fragment) bit to a "1" in all the packets it transmits. As soon as one of these big packets encounters a router that can't forward a packet that large, the router will send back an error message using yet another protocol in the TCP/IP suite called ICMP, or the Internet Control Message Protocol (whereas TCP and UDP carry data, ICMP deals purely in control messages). When the TCP stack receives the ICMP error message, it will know that it must fragment packets when they reach 500 bytes in size.

It probably comes as a surprise to many people that TCP/IP didn't become the official protocol of the Internet until January 1, 1983. Back in the days of the Internet's "predecessor," the U.S. Government's Advanced Research Project Agency Network (ARPANET), the first functionally complete Host-to-Host network protocol appeared in December 1970, when the Network Working Group (NWG) working under UCLA's Steve Crocker finished what was called the Network Control Protocol (NCP). ARPANET sites implemented NCP during the period 1971 to 1972, which made it possible for network users to begin to develop network-based applications. The first major example of this was e-mail; the basic version of this was developed in March 1972 by Ray Tomlinson at Bold Baraneck and Newman (BBN). In July 1972, Lawrence G. Roberts of MIT augmented its usefulness by writing the first e-mail utility program to list, selectively read, file, forward, and respond to messages. From then on (until the development of the browser) e-mail became the largest and most frequently-used network application, replacing phone calls and bringing forth the annoying advertising medium known as "spam."

In the case of TCP/IP, users access applications (e-mail, newsgroups, web pages with ActiveX or Java applets) over the network, connecting to Internet servers which are named after the applications they house, such as e-mail servers, FTP servers, News servers and World Wide Web servers. Internet servers run on Unix / Linux and various advanced versions of Microsoft Windows.

Unlike the older NCP protocol, each application running over TCP or UDP uniquely identifies itself from other applications running on a particular computer by reserving and using a 16-bit "port" number. Destination and source port numbers are encoded in the UDP and TCP headers by the packet's originator before handing it over to IP, and the destination port number allows the packet to be delivered to the intended recipient at the destination device. In this way a computer can run multiple applications (e.g. e-mail, file services, web services) using the same IP address and the applications' data traffic won't "collide" with each other (e-mail data won't interfere with web data).

Port number assignment is in theory arbitrary, but tradition combined with semi-official allocations by the Internet Assigned Numbers Authority (IANA) have led to accepted port numbers such as 80 for web (HTTP) communications, 23 for FTP, and 25 for e-mail (SMTP).

From IPv4 to IPv6

The current version of the IP protocol is Version 4, or IPv4. The newest version of IP, which is expected ultimately to be used worldwide, is Version 6, or IPv6. (There is no IPv5 — the version number "5" in the IP header was assigned to identify packets carrying an experimental non-IP real-time stream protocol called ST. ST was never widely used, but the version number 5 had already been allocated, so the new version of IP was given the next available version number, 6.)

IPv6 uses larger addresses (128 bits instead of 32 bits in IPv4) so we'll never run out of available IP addresses. IPv6 includes features like authentication and multicasting that had been retrofitted to IPv4 in a series of "extensions" done over the years.

The IP / ATM Debate

Whereas IP has few Quality of Service features when it comes to real-time transmissions such as Voice over IP (VoIP) and videoconferencing, Frame Relay fares a little better in this regard, and ATM fares much better still. This has led some enterprises and Voice over Packet (VoP) service providers to encapsulate IP packets in ATM cells, thus in effect running one network layer protocol over another network layer protocol. This has the additional advantage in that ATM is found throughout the continental network backbone. But although ATM has a good QoS, when voice is transmitted as *Constant Bit Rate (CBR)* AAL1 traffic over ATM, about 20% of your bandwidth must be reserved for voice even if you're not actually sending it. If you send the voice as IP packets over ATM, then that 20% of the packet availability is lost to any other application. For ATM-based VoDSL, voice packets are generally sent over ATM AAL2 while IP data packets are sent over ATM AAL5. We will examine this in greater detail in Chapter 5.

Various exotic schemes for efficiently encapsulating IP packets into ATM cells have been formulated, represented by equally exotic acronyms such as LLC/SNAP, VC-multiplexing, TULIP, TUNIC, etc. At the moment, it appears that LLC/SNAP encapsulation is the most popular encapsulation scheme as it is the least inhibitive and allows the maximum multiplexing onto a single virtual circuit. Any set of protocols that may be uniquely identified by an LLC/SNAP header can be multiplexed onto a single VC. This is why the ATM authorities decided to make LLC/SNAP the default encapsulation for IP over ATM (so it must be supported, no matter what). Still, it is also possible to bind virtual circuits to higher level entities in the TCP/IP protocol stack, examples of this being single VC per protocol binding, TULIP, and TUNIC. Moreover, some "encapsulation negotiation" is possible between equipment. For example, the calling ATM "endsystem" (endpoint) could prefer to use VC multiplexing, but is willing to agree to use LLC/SNAP encapsulation instead, if the called ATM endsystem only supports LLC/SNAP.

Other more general models and protocols include the LANE protocol for LAN emulation, the Integrated P-NNI model for an integrated services network and the MPOA protocol for establishing Layer 3 virtual LANs.

One increasingly popular protocol contender is *Multiprotocol Label Switching (MPLS)* which creates business-class IP services on ATM networks by combining Layer 3 connectivity (routing) and Layer 2 switching performance. While most carriers use ATM backbones, most enterprise networks prefer IP. MPLS integrates the two.

MPLS' QoS is achieved by assigning a label to each packet designating a priority used during transport, something like, "data, voice, video" or "silver, gold, platinum." MPLS works best in large enterprise office-to-office communications over ATM and frame-relay backbones for Virtual Private Networks (VPNs).

One possibility, of course, is to eliminate ATM entirely and simply send IP over high-speed optical backbones with some kind of additional QoS protocol or process (e.g., marking bits in a packet header to designate priority level of packet delivery). This is part of the IP / ATM controversy that we will discuss later in the book.

Router and Voice-over-IP proponents claim that good QoS can already be delivered over IP-routed networks, obviating the need for more expensive ATM technology. At the moment, however, unlike IP, ATM actually can guarantee quality transmissions of voice and video. The probability of an ATM switch losing a cell (called the cell loss probability) is kept below one in 100,000,000 and delay values within the range of 10^{-8} to 10^{-10}.

Still, while it's true that TCP/IP doesn't yet provide good quality of service, the slow but steady transition to the IP6 protocol as well as the adoption of other protocols (such as MPLS and RSVP) will allow the current version to support much larger address spaces, encryption, and quality of service. In a few years the distinction in quality between Voice-over-IP and Voice-over-ATM networks is expected to be negligible.

The IP versus ATM controversy and its impact on VoDSL is examined in greater detail in Chapter 5.

Summary

Conventional circuit-switched networks such as the PSTN were unsuitable for transporting the increasing amounts of data generated by and traveling among the ever-growing population of computers. The development of various kinds of packet-switched networks have revolutionized business and enable the development and deployment of exciting new bandwidth hungry applications.

Of all the packet networks, ATM is perhaps the most versatile and complex form of network communication for data and voice, but it is also the most reliable and stable. ATM's competitor, IP is immensely popular because it makes possible the Internet. Although seen as competitors, ATM and IP are often used in conjunction with each other. As we shall see, both have a place in the world of Voice over DSL.

CHAPTER THREE

DSL Explained

For more than a hundred years the local loop was occupied solely by analog POTS. Now, however, Digital Subscriber Line (DSL) technology is breathing new life into the PSTN. DSL technology makes it possible for service providers to furnish high speed communication connectivity over the world's universal network medium — the ordinary pair of twisted copper wires that stretches across the "last mile" from the customer's premise to the nearest telco central office.

Mostly as a result of its guaranteed high bandwidth carrying capacity, DSL technology supports not only content-rich data, but also, as we shall soon discover, voice. It can do this at a cost that's significantly less than the price of the former king of heavy-duty digital business communications, the T-1/E-1 line, which offers a guaranteed bandwidth of 1.5 Mbps (T-1) to 2.048 Mbps (E-1).

DSL is thus a more cost-effective option when compared to other PSTN-based high-speed access technologies, such as the *Integrated Services Data Network (ISDN)*, which not only costs more than DSL, but delivers less. For example, DSL transfers many times the data in the same time interval than does Basic Rate ISDN (from 384 Kbps to 8 Mbps *versus* 128 Kbps). And unlike ISDN, DSL is easier to configure with a dedicated connection, doesn't demand complicated configurations, and there's generally no per minute usage charge. You pay the same flat rate every month.

As another example, unlike what's provided by the more volatile cable TV systems, DSL offers a guaranteed amount of bandwidth.

As we'll soon see, the most popular DSL, called Asymmetric DSL or ADSL, offers speeds that range up to 8 Mbps for downstream transmissions and up to 1.5 Mbps for upstream traffic. Yet, most individuals and small businesses find a less costly "lite" variant of ADSL, which offers speeds of around 384 Kbps, adequate for their daily needs. The type of DSL provisioned can impact your VoDSL service options. For instance, to achieve the maximum number of bi-directional channels for Voice-over-DSL (VoDSL) one should use a symmetrical flavor of DSL.

DSL's Story

With the Internet's overwhelming rise in popularity during the latter half of the 1990s, websites seeking ways to build a customer base often resorted to bandwidth-intensive eye candy to entice the Internet populace to visit their site. At the same time, the corporate community found web-based, bandwidth-hungry applications such as multimedia presentations, audio and video conferencing and e-learning to be cost-effective methods for disseminating information. All of this led to a clamor throughout the Internet community for access to the Internet at increasingly higher bandwidths. Fortunately, DSL technology happened on the scene to "fill the bill."

Remarkably, twisted pair copper wiring can transmit frequencies up to a little over 1 MHz; thus these unused higher frequencies are an "unclaimed territory" that can be used to support data traffic. DSL is a communications technology that utilizes the additional, normally unused bandwidth that's available within analog voice-grade twisted pair copper wiring. Some DSLs divide the copper bandwidth into frequency bands, with the lowest band (0 to 10 kHz) ignored by ADSL (for example) since it has been used since the 1870s for regular analog voice traffic.

Since DSL technology sends digital packets of data over conventional copper telephone lines via higher frequencies than those used for voice, with the proper technology in play (frequency filters, splitters, etc.), it's possible to send and access data while speaking on the telephone or sending a fax.

Since 1999, service providers have made enormous efforts to provide multiple forms of DSL service in most major markets. But it takes more than just new DSL technology to provision DSL to the customer premises. Although twisted copper wires have a reasonably large bandwidth capacity, the legacy POTS infrastructure with its primitive analog switching equipment was designed only for voice transmission at frequencies below 3400 cycles per second (3.4 kHz). This placed severe restrictions on the amount of digital traffic that could be transmitted over the PSTN. Today, through the use of hardware, software and ingenious application of these technologies, DSL service providers have managed to overcome most of these roadblocks.

Hertz vs. Bits

In analog communications, bandwidth is typically measured in Hertz or cycles per second. In digital communications, bandwidth is generally measured in bits per second (bps). A voice conversation in analog format typically ranges up to 3300 or 3400 Hertz, carried in a 4000 Hertz analog channel.

In digital communications, this same analog waveform would be "digitized" or encoded into an roughly equivalent digital bit stream using a technique such as Pulse Code Modulation (PCM) which results in a digital channel carrying 64,000 bits per second.

Here's how it's done: First, the amplitude (height of the waveform) of the voice conversation is sampled at regular time intervals (a process called PAM, or Pulse Amplitude Modulation) and the sample is then coded (quantized) into a binary (digital) number consisting of zeroes and ones. A scientist named Harry Nyquist (1889-1976) discovered that if the sampling is done at a rate twice as fast as the highest frequency on the channel, then when the digital stream is turned back into analog sound waves, the quality of the sound will be acceptable to human listeners. Since the voice channel ranges up to 4000 Hertz, sampling is done at 8000 times a second. Each sample is then turned into an eight bit number (represented by a "byte"), so 8000 x 8 equals 64,000 bits per second.

So, bandwidth is roughly related to the data speed in bits per second — the greater the data speed, the larger the bandwidth. But don't assume that data speed is the same thing as bandwidth. This is because digital transmissions through copper are not purely digital; the zeroes and ones must be transmitted by waves, which possess a fundamental analog nature. For example, an analog modem delivering data at a speed of 28,800 bps is encoding waves at frequencies ranging up to just a few thousand kHz. As it turns out, the number of bits per second actually transmitted depends on how the waves of the signal are modulated to convey information, not just on the actual raw number of individual electromagnetic waves per unit of time.

Also, don't confuse "bandwidth" with band. Let's say a "digital" modem is communicating over copper in the 1 MHz band. What's its bandwidth? That's the space above 1 MHz that it's using to transmit. Let's say it's occupying from 1 MHz to 1.1 MHz. This means that it's occupying the frequency range from 1,000,000 Hz to 1,100,000 Hz, so its bandwidth is one hundred thousand cycles per second or 100 kilohertz (kHz). If the arrival of one wave cycle equals one bit, then 100,000 bits per second can be transmitted. If the transmitting and receiving equipment can assign additional bits to each cycle depending on a change in the amplitude (waveform height) or phase (arrival time) of each wave cycle, then more than one bit per cycle can be transmitted. In fact, when DSL technology takes advantage of these unused frequencies on the standard telephone line, it makes use of specially modulated channels to maximize its information-carrying capacity.

In many ways the DSL story is not just a story about bandwidth. DSL is the PC revolution of data communications — it's modular, inexpensive, and built using products and technology developed by the marketplace. However, to tell the full story of DSL, you need to learn how DSL technology can provide all of this incredible speed over the existing copper-based PSTN.

The Birth of DSL

DSL technology arose out of continual research into how regular analog telephone service could be either improved or supplemented with some other service, such as a data transmission service.

Although voice band modems were developed as early as 1962 to transport digital data over the analog telephone network, their capabilities are limited. Such modems modulate data at the transmitter and transmits the signal bi-directionally over one or two copper wire pairs through the PSTN, where another modem at the destination site demodulates it, reconstructing the data (that's why the word "modem" is an abbreviation for *Mod*ulator and *demod*ulator). Wildly different modulation techniques have been used over the years, beginning with Frequency Shift Keying (FSK) in the ITU-T V.21 standard, Differential Phase Shift Keying (DPSK) in the V.26, V.27 and V.29 standards, to Quadrature Amplitude Modulation (QAM) in the V.22bis, V.33 and V.34 standards.

Duplexing methodologies have also varied: Full duplex transmission over a single pair with separated Frequency Division Duplexing (FDD) or overlapping Echo Canceling (EC) frequency bands for both directions of transmission, dual simplex transmission over two copper wire pairs, and half duplex transmission over one pair.

The available bandwidth afforded by these technologies increased at a snail's pace from 1962 to the early 1990s, starting with a bit rate of 300 bps in the V.21 standard and progressing to 33.6 Kbps in the V.34 recommendation that was ratified by the ITU-T on September 20, 1994. In 1997, bit rates of 56 Kbps were achieved with the V.pcm standard, which became the V.90 standard of 1998. Although as pointed out in a previous chapter, to help stamp out crosstalk in the U.S. copper network, the Federal Communications Commission placed a limit on the amount of power that phone companies can use to send signals over the network. This limitation results in the maximum data throughput of 56 Kbps modems being reduced to 53 Kbps.

Moreover, even the 53 Kbps bit rate can only be achieved in the downstream direction (upload speeds are still limited to 33.6 Kbps) and then only when an Internet service provider (ISP) has installed a digital connection to the central office and where just one analog-to-digital (A/D) converter exists in the connection chain.

Thus, the limit for transmitting data over a plain analog phone line seemed to be that of the V.90 standard modem, with download speeds of up to 53,000 bps and upload speeds of up to 33,600 bps. Also in the works is an all-digital standard called V.91 which will allow users with digital phone lines, such as those at corporate installations with PBXs, to use "Switched 56" or 56 Kbps dialup links in both directions.

Yet, engineers knew that the local loop's twisted pair copper wire was capable of transmitting much higher frequencies and, therefore, has a far greater data carrying capacity. What had to be overcome was the PSTN's legacy switching systems, which internally encodes voice connections at 64 Kbps. This meant that different kinds of modem devices would have to be installed at both the customer's premise and the telco's local loop termination point (usually a central office), along with the formulation of a new method to divert the resulting high data rate transmissions away from the PSTN and toward a different high-speed data network.

In the mid 1980s, Joseph Lechleider (now retired), while working at Bell Communications Research (Bellcore, which is now Telcordia), used mathematical analysis to demonstrate the feasibility of sending broadband (higher frequency) signals over POTS copper wires. Lechleider and his colleagues originally worked on a way to carry the combined 144 Kbps signal for ISDN Basic Rate service over the existing telco copper phone lines. But, Lechleider pondered the prospect of something even more impressive — sending 1.544 Mbps T-1 service over the same copper wires. He succeeded, even though the original "digital" T-1 service needed "repeaters" to reamplify the signal every 6000 feet (1.83 km).

Note: Basic Rate Interface ISDN, capable of transmitting two 64 Kbps channels and a 16 Kbps control or "delta" channel, never really caught on in the U.S. although it did become popular in Europe, particularly Germany.

Lechleider and his colleagues persevered in their search for an inexpensive high-speed digital service and in late 1986 found success with a modified DS-1 / T-1 implementation ultimately called High bit-rate Digital Subscriber Line (HDSL) which could run over not one but two twisted pairs of copper wire. Two twisted pairs were necessary to match the capacity of a T-1, since by spreading the data among more wires a lower frequency could be used, thus enabling the signals to travel a greater distance before they had to be amplified (three wire pairs could carry the even higher bandwidth of an E-1, though this requirement was later eliminated when better signal modulation techniques were used). Another trick was to encode the transmission using a more bandwidth-efficient coding system called 2B1Q rather than AMI. The 2B1Q format was originally used for ISDN Basic Rate Interface (BRI), and it yields a transmission frequency that is one quarter of the transmission rate. Thus, a 784 Kbps transmission encoded in 2B1Q requires frequencies ranging up to 196 kHz.

HDSL could achieve full-duplex (bidirectional) transmission at a rate of 1.544 Mbps over distances of up to 12,000 feet (or 3.6 km, a "reach" considered equivalent to a telco's so-called CSA or Carrier Serving Area) on local loops having a wire thickness of 0.5mm (24 gauge). A reach of 26,000 feet (5.0 miles or 7.93 km) can be achieved either with heavier wire gauges (22 gauge) or HDSL repeaters (also called "doublers" since they double the loop length; equipment manufacturers ultimately introduced range extenders that allowed the carrier to triple the transmission range to 36,900 feet). Prototype HDSL systems first appeared in 1989 and became commercially available three years later.

HDSL was the first successful DSL technology to be deployed by the RBOCs. Indeed, after 1992 when a business ordered "T-1" service, most often they were really getting HDSL.

While all of this work did serve to provide digital communication capability to the consumer, there were many limitations and costs associated with both the IDSN and T-1 digital services. Consequently, the Lechleider and Bellcore group, along with others, continued their work on perfecting DSL technology.

A number of DSL "flavors" (variants) evolved from Lechleider and Bellcore's original invention. On the heels of the original HDSL came a one-pair copper wire version known as HDSL2 or single-line symmetric DSL (SDSL), which became a leader in the ensuing stampede of DSL flavors that began to proliferate the telecommunications community.

ADSL

In 1989, while HDSL was still in the prototype stage, speculation began on what other DSL variants could be developed. It came to light that in order to reduce interference in this new type of digital transmission the data should be transmitted asymmetrically, with an upstream rate in the Mbps range and a downstream rate in the 100+ Kbps range. Thus was born the concept of Asymmetric Digital Subscriber Line, or ADSL.

Lechleider found that multi-megabit transmission rates on ADSL could be achieved by a *Discrete Multi-Tone (DMT)* modulation technique that was being developed by John Cioffi working at Stanford University. DMT is a form of Frequency Division Multipexing (FDM). Cioffi learned that by using a technique known as Discrete Fast-Fourier Transform, a copper wire's total available 1.1 MHz bandwidth could be separated into 256 subchannels or subcarriers (called "tones" by DSL engineers) of 4.3125 kHz each — think of 256 virtual modems operating simultaneously over the same line. Each subchannel is separated from its neighbors by an empty 4.3125 band. Thus, each subchannel has the same bandwidth but a different center frequency, with the subcarrier frequencies being multiples of one basic frequency, spread over the available bandwidth spectrum that ranges from about 20 kHz to 1.104 MHz (anything below 20 kHz is reserved for analog voice service, which in fact occupies only the bottom 4 kHz). Each of the many virtual modem subchannels are modulated with a form of *Quadrature Amplitude Modulation (QAM)* which, ironically, is the same modulation technique used by conventional analog modems.

Surprisingly, this concept of sending data over many channels each with a very narrow bandwidth actually has several advantages: All subchannels become independent regardless of line characteristics so subchannels can be individually decoded, the optimum decoder for each subchannel requires no memory capacity and is thus easy to implement, and the theoretical maximum data carrying capacity of the line (as calculated by Claude Shannon's information theory) can be approached by using this technique.

Lechleider found that one can also minimize the problem of line noise by using this technique, since the system can also test which subchannels have the least interference and then concentrate its transmit power on those portions of the spectrum to send data.

To support bi-directional channels, two bandwidth allocation methodologies are included in the ADSL standard to separate the up- and downstream transmissions. The first method uses *Frequency Division Duplexing (FDD)* in which case non-overlapping frequency bands are assigned for the up- and downstream data. The upstream band ranges from about 25 kHz to 138 kHz (subcarriers 6 to 32) while the downstream band extends up to 1.104 MHz (subcarrier 256). The second method allows overlapping spectra for up- and downstream transmission but compensates for this by employing *Echo Cancellation*. Amazingly, this seemingly defective idea actually results in improved downstream performance because by allowing overlapping bands, the lower portion of the channel spectrum with highly desirable low attenuation characteristics now becomes available for downstream transmissions.

In 1987, Lechleider at Bellcore encouraged Cioffi and his Stanford students to bring DMT to fruition. Bellcore and Bell Atlantic (now Verizon) funded Cioffi's DSL research at Stanford in the early 1990s, allowing the development of ADSL from concept to prototype. Field tests began in 1995. Alcatel became the first company to license the technology.

Cioffi founded Amati, a company that in 1993 designed DSL equipment using DMT that had dramatically better results than all competing technologies when they were subjected to testing by Bellcore. DMT subsequently became the most common modulation technology standard for DSL service. Amati was acquired by Texas Instruments in late 1997.

In spite of ADSL's promise of affordable high-speed communications for the home and small business marketplace, DSL technology was not immediately adopted. Indeed, by the mid-1990s few real products had appeared. This was mainly because even though the RBOCs were installing high bandwidth digital technology to connect central offices, they moved at glacial speed when it came to bringing such technology to the local loop.

Attempts to "jump-start" ADSL acceptance included the formation of the Universal ADSL Working Group (UAWG), an alliance between telecommunications and computer companies led by Compaq, Intel and Microsoft. UAWG members united around a standard for a slower, but relatively inexpensive and easy-to-install DSL-enabling technology called "ADSL Lite" or "G.Lite," which only required a "plug and play" *DSL modem* — plug one end of it into a phone jack and the other into a PC (via a USB or Ethernet connection) and you can start using it immediately.

The DSL Modem

The DSL modem is situated where data from the customer's computer or network enters and leaves the DSL line. The DSL modem device can connect to the customer's equipment in various ways, though most residential installations use USB, 10Base-T Ethernet or wireless connections. Unfortunately, a cable to a serial port, the time-honored way of connecting a PC to a modem, can't be used with DSL modems since they can't handle DSL's high bandwidth.

Actually, the term "DSL modem" is something of a misnomer, the device is actually a DSL transceiver that can even be used as a type of bridge that connects two segments of the same network, forwarding packets from one segment to the other. Nearly all DSL modems exist as hardware and look very much like some of the older 56 Kbps modems; although in April 2000, Motorola demonstrated a *software DSL*

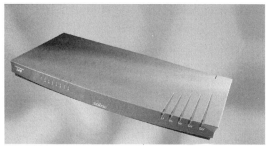

Fig 3.1. Alcatel / Thomson Multimedia ADSL modems. In February 2001, the market research firm Dell'Oro Group identified Alcatel as the world's No. 1 DSL modem maker with a 34.9% market share, selling about 1.7 million units in 2000. At top is the stylish Speed Touch Home, which uses a USB connection to the PC. In the middle is the Speed Touch Wireless, which uses a "Wi-Fi" or 802.11b connection to the PC. The bottom photo is the high-end Speed Touch Integrated Access Device (IAD).

modem or *"SoftDSL modem"* at the WinHEC show in New Orleans. The SoftDSL modem uses the processing power of a PC or Apple Mac instead of a dedicated modem chipset to process DSL signals. This eliminates the need for costly fixed-function hardware as well as the need for service providers to distribute and support customer premise equipment. At the same time, it gives the customers a flexible product that's easily adaptable to new features, functions and standards with simple Internet software upgrades.

DSL modems can interconnect networks that use different networking protocols (TCP/IP, Appletalk, IPX, IBM, XNS, etc.), without requiring specific network information and configuration on the equipment. Many ADSL transceivers sold by ISPs, and telcos to the home and SOHO market are simply transceivers, though more sophisticated versions of these devices used by businesses may combine network routers, network switches or other networking equipment in the same device. The service provider technicians call such customer premise devices an ATU-R, which means "ADSL Transmission Unit Remote." At the CO end it's called the ATU-C, or "ADSL Transmission Unit Central Office." Thus, the ATU-R is the customer premise equipment (CPE) and the ATU-C is the equipment on the phone company's end. The ATU-C is sometimes called "the transmitter" and the ATU-R "the receiver" though both devices obviously can send and receive signals.

The copper line distorts the signal amplitude and phase, a distortion that changes from carrier to carrier and varies over time. DSL modems continually analyze the communications link with the CO and adapt to line conditions, always compensating for ongoing changes (such as those owing to temperature changes and interfering noise that may be present). The modems contain advanced *Digital Signal Processing (DSP)* algorithms that produce mathematical models of the distortions caused by the line and produce automatic corrections, so that data is passed (in both directions) on the line as fast as the line will allow.

Of course, a DSL modem *per se* is not what you would use for a serious multichannel VoDSL installation having a high degree of quality of service. At best, only a hobbyist would use this to do some scratchy "PC as IP phone" type of phone calls over the Internet.

The IAD

Most DSL installations with a VoDSL overlay use an *IAD,* or *Integrated Access Device.* Prior to DSL, an IAD was considered a device that multiplexes various forms of communications (analog voice, ISDN, T-1, etc.) in the customer's premises onto a single telephone line for transmission to the carrier. The IAD also demultiplexes the incoming streams into their respective channels. Today, the term IAD most often refers to a device that deals with DSL, mixing both voice and data over the DSL line to the carrier's *DSL Access Multiplexer (DSLAM)* and then to an access switch where voice is forwarded to a voice gateway and then to the PSTN while the data traffic is diverted to a data network (ATM, IP, etc.). IADs can also have routers built into them. The list of IAD vendors includes many of the major telecom manufacturers, such as Cisco Systems (San Jose, CA — 408-526-4000, www.cisco.com) and RAD Data Communications (Tel-Aviv, Israel — +972-3-6458181, www.rad.com).

Regulatory Boosts

Aside from these technological developments, the DSL industry, in general, received a helping hand from the passage of the U.S. Telecommunications Reform Act of 1996. This Act obligated U.S.-based Incumbent Local Exchange Carriers or ILECs (your local telephone service company) to allow their network infrastructure and services to be accessed and resold by competitors. Some of these resellers are known as Competitive Local Exchange Carriers (CLECs). So, in theory, ILECs, CLECs, long-distance carriers, cable companies, radio/television broadcasters, Internet/online service providers, and telecommunications equipment manufacturers in the U.S. are all in competition with each other. In actuality, the ILECs were very uncooperative in allowing CLECs access to their central offices. Attempts were made to build separate "cyberhotels" to counter this, but the economic downturn at the dawn of the 21st century led to many CLECs filing for bankruptcy or suffering from severe financial difficulties, such as Northpoint, Covad and Rhythms Netconnections.

In the U.S., the government once again stepped into the arena and changed the rules with its famous 1999 Federal Communications Commission (FCC) ruling. Previously, CLECs had to lease separate lines to offer their services, raising their cost for service. That FCC decision meant they could use existing lines, which became instrumental for quickly deploying DSL services because it leveled the playing field. If the CLEC wanted to start a DSL service that the ILEC refused to supply, the CLEC could now get access to the raw copper wiring of the local loop and set up the infrastructure necessary to establish the new DSL service. This seemed like a godsend for the CLECs, and trouble for the ILECs/RBOCs.

In Europe, governmental regulators have also prodded the incumbent telecommunications operators to relinquish their monopoly access to the telco customer. Since it was far too expensive for would-be competitors to build their own local loop networks to link individual premises to the backbone, incumbents had no incentive to upgrade their vast existing network and were dragging their feet. To force competition, European regulators were forced to insist that the incumbent telcos give competitors access to their networks. Regulators call this phenomenon "local loop unbundling." This bold regulatory move has helped with DSL roll-out — telcos are rushing to upgrade their networks, especially in the more lucrative locales, to prepare for the competition that lies ahead.

Service providers began to see a real benefit from DSL technology — the opportunity to leverage customer demand for faster data access into a healthy profit margin. The race to provide high-speed bandwidth to the masses is now underway. The DSL industry has the perfect recipe for explosive growth, and the executive boards of many corporations now comprehend that DSL has not only the potential to deliver high-speed data access, but much more in the way of profit-making applications.

The Different Flavors of DSL

There are several competing forms of DSL, which are typically referred to as "flavors or variants"; each of which is adapted to specific needs in the marketplace. The most prominent DSL flavor today is ADSL. The so-called "full-rate ADSL" supports theoretical speeds up to 8 Mbps (some experts say 9 Mbps) continuously in one direction over short distances. But the continuous rate of megabits per second ADSL can deliver decreases as it reaches a typical maximum range of 18,000 feet. ADSL provides a much larger downstream signal than upstream, thus increasing efficiency and reducing interference. How this is accomplished will be explained later.

There have been numerous varieties of DSL developed and deployed over the years — many aim specifically toward the business community. But as things are currently shaking out, the symmetrical flavors of DSL have won a top position in the business market because they offer symmetrical data rates high enough to support bandwidth-intensive digital services such as video conferencing, VoDSL, VPNs and web hosting.

At the same time, the Asymmetric DSL (ADSL) family of technologies has found favor among the predominantly web-surfing (residential and SOHO) user group due to its lower cost, and ease of implementation. ADSL variants offer a high downstream rate, and a relatively low upstream rate, which matches this group's data stream needs — they typically download more information than they send (e.g. application graphics, video, audio, MP3 files, downloads with few or low bandwidth instances of upstream transmissions such as requesting a web page or sending data files to another user).

ADSL: Its Advantages and Disadvantages

ADSL (Asymmetrical Digital Subscriber Line) is the most common term used to denote "full-rate ADSL" (ITU reference code G.992.1). It expands the usable bandwidth of existing copper telephone lines by supporting up to 8 Mbps bandwidth downstream (some experts say it can do 9 Mbps under the right conditions and one chpset developed in 2002 claims to support 12 Mbps) and up to 1.5 Mbps upstream.

The word "asymmetrical" holds the key to ADSL — the bandwidth from the provider

Note: The reader may wonder how ADSL, which can send 1.1 million cycles per second over a phone line, can transfer not 1.1 Mbps, but 8 Mbps. The answer is that, through clever modulation, ADSL (and all DSLs, for that matter) send more than one bit for the duration of a single cycle of the wave form (referred to as the "symbol time"). "Spectral efficiency" is a measure of the number of digital bits that any given modulation technique can encode into the symbol time. The most primitive way of doing this is simply by increasing the voltage, and hence the height of the wave. Different heights would mean different combinations of bits are associated with each cycle (00, 01, 10, 11). At least two waves would be sent so that the receiver can determine with certainty the height of the waveform. In the case of full-rate ADSL, more complicated forms of modulation allow up to 15 bits of information to be sent per symbol time.

to the user (downstream) will be at a higher rate of speed than bandwidth from the user to the provider (upstream). The asymmetrical properties of ADSL serve to limit crosstalk (compared to a symmetrical configuration), while at the same time efficiently bring full-motion video, effective teleworking or remote local area network (LAN) access, and high-speed data transmission to the home or business over the same in-place twisted pair copper wire that provisions voice-based POTS.

The Distance Equation

Not everyone is so fortunate as to live close to a central office (CO) and thus not everyone can enjoy optimum ADSL downstream speeds of around 8 Mbps (available at up to a distance of about 6000 feet or 1820 meters) and upstream speeds of up to 1.5 Mbps. As the length of the copper local loop between the user and the CO increases, the signal quality decreases, which in turn leads to a decrease in available bandwidth. At this time, the maximum range for ADSL service is about 18,000 feet (5460 meters) from the local CO, at which point customers encounter bandwidths far below the maximum possible. This is why many service providers deliberately specify a lower limit on the available distance so as to guarantee delivery of quality service, or at least service good enough so as not to rankle their customers.

The speed that can be attained via symmetrical DSL technology is also impacted by the distance of the customer's premise from the local loop termination point — typically the CO.

xDSL

xDSL is a term used interchangeably with the acronym DSL to represent all variations of the technology. As such, it is the catchall term that signifies the various similar, yet competing, versions or flavors of DSL technology: ADSL, HDSL, HDSL2, IDSL, SDSL, G.SHDSL, VDSL, etc.

Another way to explain the xDSL moniker is that xDSL represents the technology in total, without regard to any specific format or protocol. xDSL is simply the acronym for that general type of technology that pushes a large number of bits through twisted pair copper wires used typically for "last mile" telephone connections — i.e., small gauge copper wire of lengths less than 18,000 feet.

Splitters and Filters

The impedance of the wiring within the copper plant varies significantly. It mainly depends on the length and gauge of the wire, but also the telephone itself is a factor, especially when the line is short. To keep high ADSL frequencies from damaging or otherwise interfering with conventional phone equipment at both the CPE and the CO, filtering and divergence mechanisms must be installed both at the central office and at the customer premises. This is a non-trivial operation, particularly when a service provider tries to fit just one design to all cases. Yet, the echo return loss, which affects the voice band quality of service, can not be compromised.

"Full rate" (8 Mbps) ADSL implementations normally require someone to install a "splitter" or "POTS splitter" at the customer's premise, which separates the low frequency voice from the higher frequency data transmissions that might interfere with or damage the electronics of analog phones and fax machines.

Passive Splitters: Most POTS splitter designs are passive because they maintain lifeline telephone access even if the modem fails (because of a power outage, for example) since the telephone is not powered by external electricity. Passive filters are also much better protected from the lightning that may be coupled through the line.

Active Splitters: In some countries, performance requirements require active filters in the POTS splitters.

Active POTS splitters consume electrical power, thus if the power fails or the modem fails with an active POTS splitter, then the analog telephone service also fails.

The splitter is a three-node device:

One node is a low pass filter. Since the analog POTS signal is located in the low frequency band and all the ADSL signals reside at frequencies higher than about 25 kHz, the splitter must provide a low pass filter between the copper line and the telephone.

Another node provides a high pass filter between the copper line and the ADSL modem. Thus, all signals are attenuated (eliminated) between the ADSL modem and the telephone.

Finally, there's a node that blocks impulsive electrical noise coming from the home telephone or narrow band switch at the central office into the delicate electronics of the ADSL modem. Severe impulses can be generated by the ring trip signal. The filter also blocks the ADSL modem signals from going into the telephone set, which can reduce line quality and impair the functions of certain phones.

A typical splitter installation is done by first deploying the splitter, then a new twisted pair copper wire line is installed inside the customer premise to connect the splitter with an ADSL modem, DSL router or Integrated Access Device (IAD). The main disadvantage of using a splitter is that only a single wall jack (the one connected to the splitter) is available as an outlet for ADSL service.

So much for the customer premises. Over at the central office, there are many ways that a DSL (in particular ADSL) circuit can be terminated:
- POTS splitter located with the ATU-C (ADSL transmission unit at the network end).
- POTS splitter located near the ATU-C.
- POTS splitter located on the Main Distribution Frame (MDF).
- POTS spitter located near the MDF.
- ATU-C remote.
- POTS splitter located near the PSTN switch.
- An integrated POTS and ADSL line card.

Each of these configurations has its pros and cons.

ADSL Lite

ADSL Lite is a lower bandwidth alternative to "full-rate" ADSL. It was first proposed as an extension to the ANSI (American National Standards Institute) standard T1.413 by the Universal ADSL Working Group (UAWG). The term "ADSL Lite" is generally used interchangeably with Splitterless DSL, G.Lite and UADSL (or Universal ADSL). The ITU established a Universal ADSL working group (ITU Study Group 15) to study ADSL Lite technologies. The group studied various ADSL Lite schemes and eventually proposed a standard it designated as G.Lite (to differentiate the ITU approved technology from other "lite" delivery methods). The cost-effective G.Lite standard has brought about increased interest in and installations of DSL from mid-1999 onward and remains the most popular DSL technology used by residences and small businesses.

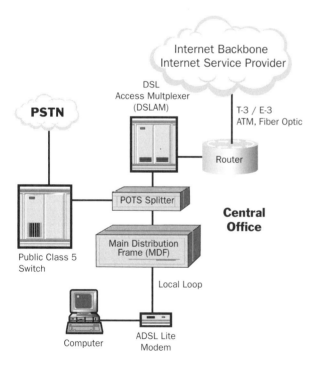

Fig. 3.2. ADSL Lite.

ADSL Lite provides bandwidth that peaks at 1.5 Mbps downstream and 640 Kbps upstream. ADSL Lite cuts the maximum number of DMT subcarriers from 256 to 128, and whereas full-rate ADSL can send 15 bits per "symbol time," the Lite version reduces this to eight bits. As with full rate ADSL, ADSL Lite provides service up to a maximum range of 18,000 feet (about 3.4 miles, or 5.5 km) from a telco's CO. By using even lower bandwidth, say 0.7 Mbps downstream, the service could be extended out as far as 22,000 feet from the CO.

Why Asymmetric?

Many flavors of DSL are asymmetric because the technology designers determined that an asymmetric technology could aid in the reduction of a certain kind of signal interference. Between different wires in the same cable there exists capacitive and inductive coupling. This coupling increases as the wires are brought closer together, and it leads to undesirable crosstalk between the wire pairs (this crosstalk is usually worse between two pairs in the same binder than for wires in adjacent binders). It can be reduced somewhat by the optimization of the twist of the individual pairs, which has led the European telcos to twist the cable binders as well as the wires themselves: Wire pairs are often combined in "quads" consisting of two pairs twisted around each other. These quads are then grouped into binders of some tens of pairs. Finally, several binders are grouped into a single cable. However, this technique is not used in the U.S., and when twisted pair copper cables from users in a Carrier Serving Area all converge at the CO they're simply grouped into large bundles where electromagnetic leakage can jump from one cable end to another, causing crosstalk to appear on the lines. This phenomenon is called near-end crosstalk or NEXT. By reducing the upstream bandwidth going to the CO, NEXT is reduced. Since there's usually just one or two twisted copper pairs at the far end (the end user's premises), NEXT isn't a problem there, so the downstream bandwidth can be pushed to the maximum possible limit. Fortunately, the most popular DSL signal modulation technique, DMT, is also biased in favor of greater bandwidth downstream than upstream — a few low frequency channels go upstream, while many high frequency channels come downstream.

VoDSL and ADSL

ADSL suffers from latency problems (voice packets arriving late, resulting in gaps in audible speech) that degrades the quality of VoDSL and videoconferencing transmissions. If you turn off ADSL's error correction capabilities, the latency problem improves, but then you have to deal with packets that are damaged or don't show up at all! Also, since ADSL is asymmetric, there's more bandwidth coming downstream than can be sent upstream to the CO.

For a robust VoDSL facility, you would probably want the capability of having the same number of voice calls coming in as leaving the premises, which suggests adoption of one of the symmetrical DSL technologies. As we shall see, G.SHDSL, also known as G.991.2 and SHDSL, is a much better choice for VoDSL.

Yet, vendors, manufacturers and the engineering community are working on methods to enable ADSL to delivery quality VoDSL services. For example, Integrated Device Technology, better known as IDT, (Santa Clara, CA — 408-492-8314, www.idt.com) and VoicePump (Santa Clara, CA — 408-986-4320, www.voicepump.com) have announced a partnership aimed at the development and delivery of VoDSL via an ADSL architecture.

VDSL

Whereas IDSL (as described later) is the "slowest" DSL, VDSL (Variant Digital Subscriber Line or Very high-bit-rate DSL) offers the fastest bit rate (highest bandwidth) DSL technology to date. It's an ADSL flavor that may, in the future, provide the optimum asymmetrical technology for deployment of VoDSL. VDSL is an asymmetrical DSL technology that offers data transfers from a lightning 13 Mbps to a screaming 52 Mbps of downstream data and from 1.5 to 2.3 Mbps upstream. Like all DSL technologies, it provides an inverse relationship between bandwidth and distance: the shorter the distance, the higher the achievable data transmission rate. VDSL is limited to a maximum range of 1000 to 4500 feet (about .2 to .9 miles, or 300 to 1371 meters) from the CO, depending upon the speed, though some designs can supposedly reach about double these distances.

Although VDSL can provide up to 52 Mbps, at the time this book went to press, it requires shorter connection lengths than are generally considered practical (within 1000 to 4500 feet of the CO or other access point). Realistically, VDSL is still in the experimental phase.

Symmetrical DSL

HDSL (High Bit-Rate Digital Subscriber Line) is the pioneering high-speed symmetrical DSL format that is standardized through ETSI and ITU and is the most established of the DSL technologies. Since its creation in the late 1980s, it has delivered symmetrical DSL service at speeds up to 1.544 Mbps over two copper pairs and up to 2.3 Mbps over three pairs at a maximum range of 20,000 feet (about 3.8 miles, or 6.1 km) from the CO. However, HDSL, unlike ADSL, doesn't permit line sharing with analog phones.

Yet, HDSL emerged in the early 1990s as a viable, relatively low cost alternative to

T-1/E-1 service and fiber optics links. It provides the bandwidth to accommodate LAN internetworking, video conferencing, and PBX interconnects. HDSL's primary benefit is that (unlike T-1/E-1 service) it requires no signal repeaters for distances under 15,000 feet. Although HDSL has a comfortable distance limit of from 12,000 to 15,000 feet (3657 to 4572 meters), its range can be extended with the use of signal repeaters, high quality lines and a third set of copper wires. Under normal conditions, HDSL can transmit upstream and downstream data at speeds between 768 Kbps and 2.3 Mbps, depending on the wire gauge and number of wire pairs used.

Most T-1/E-1 lines installed today utilize HDSL technology which has a 1.5 Mbps rate that's identical to the old T-1/E-1 data rate standard, since HDSL is easier to deploy than T-1/E-1 and runs for greater distances. HDSL can be used for both data and voice applications, although because of the electrical effects of its high-speed bi-directional traffic, there are limitations on where and how HDSL can be deployed.

IDSL

IDSL stands for ISDN DSL, and in many ways IDSL is similar to ISDN technology. A hybrid of DSL and ISDN technologies, IDSL bypasses the congested PSTN and travels along the data network using a *Digital Loop Carrier* (*DLC*) — a remote device often placed in newer neighborhoods to simplify the distribution of cable and wiring from the phone company.

The best way to describe IDSL is "ISDN without the telephone switch" — the two 64 Kbps "B" channels of an ISDN BRI (Basic Rate Interface) are multiplexed to offer a dedicated 128 Kbps of bandwidth for data only (144 Kbps in total when including the D channel). (Some might argue that IDSL is not really a form of DSL in that it does not allow for analog voice.)

Although IDSL is the slowest of the DSL family members and can be used for only limited VoDSL installations, it does have its advantages: It's often available when its brethren aren't; it offers an always-on, unswitched version of ISDN; and it provides symmetric download and upload speeds from of approximately 144 Kbps on a single pair of copper wires.

IDSL service is often billed on a monthly basis compared to ISDN, which is per packet or minute. This allows for a more predictable monthly cost.

SDSL

SDSL (Symmetric or Single-line Digital Subscriber Line) delivers high-speed bandwidth over a single-pair of copper phone lines, at the same speed in both the upstream and downstream directions. "SDSL" is actually an umbrella term for a number of supplier-specific implementations that provide varying rates of symmetric service over a single copper pair. The actual speeds can range from a high of 2 Mbps down to 160 Kbps at a maximum range of 24,000 feet (about 4.5 miles, or 7.2 km). The problem with this family of symmetrical DSL is that the higher rates of data exchange are limited to relatively short distances because it suffers NEXT interference limitations since the same frequencies are used for transmitting and receiving.

Therefore, although SDSL's symmetric transmission of data can match the speed of HDSL using only a single pair of wires, the customer's premise must be no more than about 10,000 feet from the telephone company's local loop termination point or CO.

This symmetrical DSL variant can also provide support for an analog voice channel (like its asymmetrical brethen and unlike the pioneering HDSL), although if POTS service is provided simultaneously with DSL data transmission, the top speed achievable is limited to 784 Kbps or half the bandwidth of a T-1 line.

When SDSL first became available, it was considered an attractive alternative to symmetrical high-speed links already in use, such as leased lines or frame relay service. It can handle branch-office applications needing symmetrical send/receive channels, such as transaction processing and LAN-to-LAN connections.

SDSL uses the same kind of line-modulation technique employed in ISDN (2B1Q), but SDSL transmission over a single wire pair is technically challenging because SDSL generates more NEXT than ADSL, so maximum speeds can't easily top 2 Mbps. However, the telcom industry is expected to begin moving towards the higher performing and standardized G.SHDSL technology developed by the ITU with support from the T-1/E-1 standards committee, T1E1.4 (USA) and ETSI (see a subsequent entry for more on this new technology).

HDSL2

Prior to the availability of G.SHDSL, HDSL2 (Second-Generation High Bit-Rate Digital Subscriber Line) was the symmetrical DSL flavor of choice for business installations. Unlike its forerunner, SDSL, HDSL2 is designed to handle the ultimate worst-case installation scenario — a long copper loop afflicted with a couple of annoying bridged taps, a binder group containing cables running other interfering DSL services, and a huge (99 percentile) degree of electromagnetic coupling between the possible "crosstalker" cables and the HDSL2 circuit itself.

HDSL2 achieves this by using an efficient, modulation technique which is a modified version of *Pulse Amplitude Modulation (PAM)*, called *Trellis Coded Pulse Amplitude Modulation (TC-PAM)*. The PAM line coding represents digital information on pulses of different amplitude, hence representing more bits per pulse. To combat the increase in the error rate that would normally occur, an error-checking bit is added for every three data bits. The error checking bit uses a trellis coding technique that was first employed with voice band modems in the late 1970s. PAM offers low transmission latency when this so-called "trellis coding" is used. Indeed, approximately half the latency (latency is another name for delays in the time that data packets are received) found in an equivalent, older QAM system can be eliminated. This is a distinct advantage for HDSL2 and the main reason that HDSL2 can operate over a single wire pair. More importantly, the "folded spectrum" that occurs during PAM demodulation can be exploited with the clever use of spectral shaping and excess bandwidth. Since HDSL2 must provide quality service in cable binders that may contain ADSL, T-1, and/or HDSL, it was designed to be compatible with other services.

There seems to be a rapidly growing market for HDSL/HDSL2-enabled business connections which, at $90 to $200 per month for basic service, are significantly less expensive than the typical $900 to $2000-per month full T-1/-1 connection.

HDSL2 offers the same options as HDSL, delivering a full-duplex (symmetric) T-1 or E-1 payload (1.544 Mbps or 2.048 Mbps) up to about 4.5 km (2.8 miles), and up to 2.3 Mbps for 4 km (2.49 miles). HDSL2 is optimized for transmission at practically any data rate from 160 kbps to 2.3 Mbps and it can be programmed to transmit in regular fixed rate mode or "rate adaptive mode" (where the bandwidth changes according to prevailing line conditions)

Unlike standard HDSL, however, HDSL2 uses only a single-wire pair (it also gives you the option of using a passive POTS splitter for simultaneous data access and analog telephone service). This single-wire pair capability is a distinct advantage in certain geographic areas where unused copper pairs are becoming a rare commodity.

Originally HDSL2 didn't officially support range extenders, so the maximum span reach was 12,300 feet. But, ways of using signal regenerators have been demonstrated by Orckit Communications (Folsom, CA — 916-351-5600, www.orckit.com) and HDSL2 can provide quality service to customers well beyond the normal 12,300 foot range. In many cases, however, instead of repeaters, if the service provider connects HDSL2 with two twisted copper pairs (just like the older HDSL), then the data rate can be doubled (i.e. 4 Mbps or more) for the same distance of 4.5 km. More importantly, if the data rate of two-pair HDSL2 is lowered back down to about 2 Mbps, then one can transmit data at nearly double the distance as that of ordinary HDSL. So, HDSL2 technology allows the service provider to either double the symmetric high-speed capability of a given cable binder, or else keep the same capacity and instead nearly double the distance over which the HDSL2 service can be offered.

HDSL4

Furthermore, the industry and standards groups have developed yet another form of HDSL. Called HDSL4 (also known as HDSL2 Issue 2), it takes the TC-PAM coding developed for one-pair HDSL2, and uses it in a two-pair topology that's optimized for DS-1 transport (equivalent of a T-1), allows for a 15 to 20% increase in the basic range and will allow for signal range extenders to be used, so customer circuits can be extended beyond the standard *Carrier Serving Area (CSA)* range. The other major development in the new design is spectrum compatibility, which means better control of crosstalk so that HDSL4 can coexist with ADSL and other services that might wind up in the same cable (to be specific, HDSL4 maintains full compliance with the T1.417 Spectrum Management Standard, which specifies an allowable level of crosstalk that new technologies may generate when deployed with ADSL and other DSLs).

At first glance, HDSL4 resembles a minor reinvention of the modified forms of HDSL2 we looked at previously. Indeed, when developing early versions of SDSL standards, many semiconductor and equipment companies thought that there was a need to rapidly develop a simple successor to HDSL that satisfied only the U.S. T-1 bidirectional data bit rate (1.5 Mbps), using two pairs of copper wires instead of the four specified in the original HDSL specification. HDSL2 research surged forward, but Trellis-coded PAM turned out to be extremely difficult to develop, and by the time that chipsets were ready to be deployed in real equipment, many carriers and equipment vendors had lost interest and wanted instead to leapfrog directly to a more sophisticated multi-rate two-wire HDSL solution. European carriers and manufacturers, in particular, expressed concern that there were too many competing technologies, urging that what was needed was a single, unified, two-wire standard that would serve a variety of bidirectional rates.

Around this time, however, even the strongest proponents of HDSL2 and HDSL4 became enamored with a new, adaptable multi-rate technology called G.SHDSL that was rapidly moving through the ITU acceptance process...

G.SHDSL aka SHDSL — The "New" Standard

Single pair High-bit-rate Digital Subscriber Line (G.SHDSL) — also known as Symmetric High-bit-rate DSL or SHDSL — is the latest standard for high-speed access and is considered one of the best ways to implement VoDSL. G.SHDSL was approved by the ITU in February 2001 (ITU reference code G.991.2). G.SHDSL supports bit rates from 192 Kbps to 2.312 Mbps (some equipment tops off at 2.304 Mbps) on one wire pair and 384 Kbps to 4.624 Mbps on two pairs. Greater reach is possible with the four-wire configuration. Like HDSL2, G.SHDSL can use bandwidth-efficient pulse amplitude modulation with additional trellis coding to improve error performance (Trellis-coded PAM-16), which results in a standard operating range from 15 to 20 percent greater than HDSL when uploading and downloading data over existing copper telephone lines. G.SHDSL technology allows the use of repeaters, enabling service providers to reach subscribers up to around 24,000 feet (7800 meters) from the service node. G.SHDSL modems and range extenders can be powered from the central office over the link. Such "span powering" is available in both the two-wire and four-wire implementations.

G.SHDSL can also be intermixed with ADSL lines in the same cable binder without crosstalk affecting performance.

In Europe, it appears that G.SHDSL is aimed at providing E-1 replacement, while in the U.S. the G.SHDSL standard is aimed at serving the symmetrical portion of the Internet access and VoDSL market segments. Indeed, G.SHDSL transceivers don't support the use of analog splitting technology for coexistence with "voice under DSL" POTS, so G.SHDSL is not only optimized for VoDSL, it pretty much compels you to use it if you want any kind of voice service over the same twisted pair.

Like HDSL, G.SHDSL supports an 8 Kbps embedded operations channel, or *Embedded Overhead Channel (EOC)*. This allows for control and monitoring capability similar to Extended Superframe Format (ESF) used in DS-1 framing. But unlike ESF formatting, which is embedded in the DS-1's 1.544 Mbps transmission, the EOC is added by the G.SHDSL modem. As a result, it will provide the same type of diagnostic capability but will not require any change to the user's PBX or router equipment. Hence, in the U.S., the line rate of a 1.544 Mbps G.SHDSL link should actually be 1.552 Mbps.

G.SHDSL is just the impetus needed to spur deployment of VoDSL since G.SHDSL enables service providers to provide symmetric, high-speed data transmission over existing copper pairs. The ITU G.991.2 standard defines G.SHDSL as a multi-rate DSL technology that provides symmetric data transmission rates at various speeds, from 192 Kbps to 2.304 Mbps, thereby maximizing the requisite data transmission rate for each customer. G.SHDSL can transport T-1/E1, ISDN, ATM and IP signals.

G.SHDSL achieves 20% better loop-reach than older versions of symmetric DSL, and at the same time causes much less crosstalk into other transmission services running in the same cable. And at any given range, G.SHDSL promises 35% to 50% higher data rates than existing symmetric DSLs. Note that for operation on longer loops, two pairs of copper wire may be combined to maximize the transmission rate. For example, with two pairs of wire, 1.2 Mbps can be sent over 20,000 feet of twisted pair copper wire consisting of 26 AWG (26 gauge wire).

Improved spectral compatibility was one of the key drivers for the new standard. The older 2B1Q modulation used in proprietary SDSL systems begins to severely interfere with ADSL when deployed at data rates above 784 Kbps. The newer HDSL2 technology, on the other hand, uses advanced modulation and coding techniques to limit the transmit spectrum and assure spectral compatibility with ADSL. ANSI's criteria for developing HDSL2 included the proviso that signal degradation should be no worse than the pre-existing and widely-deployed HDSL service. However, HDSL2 is not entirely appropriate for access networks. First, HDSL2 was primarily developed for North American feeder networks. It uses a shaped *Power Spectral Density (PSD)* mask that has been optimized for noise conditions in North America.

Additionally, loop conditions and noise environments in Europe and the rest of the world are significantly different, and therefore require different shaping characteristics.

Recognizing the need for a simplified symmetric service prompted the initiation of an SDSL project in the ITU. In the fall of 1998, the G.SHDSL project was launched. After evaluating the alternatives, the ITU determined that G.SHDSL would use the same advanced signal modulation and data encoding techniques as HDSL2. *Carrierless Amplitude/Phase (CAP)* and *Discrete Multi-Tone (DMT)* modulation schemes were also investigated, yet the basic technique settled upon, *Pulse Amplitude Modulation 16 (PAM-16)* used in HDSL2 proved to be the best alternative in terms of complexity, performance, and latency, especially when trellis coding was added for error correction.

Most HDSL2 silicon solutions have flexible DSP cores. One of the ITU's objectives was to define the standard in such a way that HDSL2 silicon could be re-used for multi-rate applications. However, to minimize power consumption, the signal shaping needed to be simplified.

Across the Atlantic, Committee TM6 of the European Technical Standards Institute (ETSI) had also begun the work of defining a multi-rate SDSL for business and residential use. ETSI wanted to take advantage of the foundation established by HDSL2, and based their system on trellis coding and PAM-16 modulation. While Europe and North America share many service requirements, the noise and loop conditions of each region are different. Consequently, the two regions require slightly different standards. Ultimately, the ITU adopted the ETSI work for the European environment and then produced a North American version.

The G.SHDSL specification has been organized into a base document that describes fundamental modem start-up and operation, and within that document are two Annex documents. Annex A describes transmission and performance requirements for North America and utilizes the basic PSD defined by ETSI, while specifying a different set of loop conditions and performance requirements. Annex B describes performance and transmission requirements for Europe and also adopts the ETSI's draft SDSL standard.

The G.SHDSL standard's transmit power is much lower than HDSL2 (which has a maximum transmit power of 16.5 dBm) with only 13.5 dBm at up to 2.048 Mbps. At and above the 2.048 Mbps data rate, the power will in all likelihood need to be increased to 14.5 dBm. For instance, at the 2.048 Mbps data rate and under worst-case conditions, G.SHDSL can reach 2.4 km (1.5 miles) and still provide 6 dB of margin and a 10e-7 *BER (Bit Error Rate)*.

The power spectral density (PSD) template scales with the data rate, and the -3-dB lowpass cut-off is placed at half the baud rate. A sixth-order Butterworth filter roll-off attenuates the out-of-band noise and provides good spectral compatibility with ADSL.

While the primary objective of G.SHDSL is to produce a multi-rate symmetric solution for access networks, with Annex A and B of the specification there are also optional PSDs for DS-1 transport in feeder networks, although the optional PSD in Annex A of the specification is essentially the HDSL2 solution. ETSI defined two optional PSDs: one for E-1 transport and the other for carrying synchronous digital hierarchy (SDH) tributary payloads. In the G.SHDSL's optional PSDs there is the ability to use higher transmit power to provide added performance in severe noise conditions. Also, to compensate for the higher transmit levels, the optional PSDs use excess bandwidth as well as asymmetric upstream and downstream PSD characteristics, which can improve spectral compatibility to ADSL.

But, all-in-all, while the G.SHDSL standard uses a more simplified PSD and lower transmit power than HDSL2, many of the functional "blocks" are still the same. Thus, most HDSL2 implementations designed with flexible DSP cores may be re-used for G.SHDSL deployments. For instance, the analog functions; the core of the transceiver (the scrambler, 16-level symbol mapper, transmit precoder, and transmit filter, which is programmable and can realize the different PSD shapes), all carry over directly from HDSL2 to G.SHDSL. Note, however, that the lower transmit level of the G.SHDSL standard does allow the line driver to be optimized for lower power consumption.

Yet, even with the ability to re-use some HDSL2 technology, G.SHDSL does require some new functions to enhance its multi-rate capability. For example, the CPE will need to be able to negotiate line rate and configuration details; and in access networks, the loop lengths and noise conditions will differ significantly from pair to pair.

The standard also provides for 13.5-dBm transmit power, if required in a worst-case noise condition, although many lines will not require such power levels. Then the G.SHDSL standard also includes a power back-off algorithm that enables CPE to adjust transmit power according to conditions on the line.

To negotiate all of these features, G.SHDSL adopted the ITU G.994.1 handshake procedure, allowing equipment at either side of the line to trade configuration and line-rate details prior to activation. Although in most applications, the operator will specify the line rate. Note that G.SHDSL also provisions for automatic rate negotiation via ITU's G.994.1, which uses a spectrally benign technique of narrowband frequency tones and *Differentially encoded Phase-Shift-Keying (DPSK)* modulation for negotiation.

Thus, G.SHDSL is superior in many ways to older SDSL technology because it supports higher data rates at greater distances from the telephone company's CO. G.SHDSL also uses a single, twisted pair of copper wiring. These factors let service providers offer less expensive DSL services to a greater proportion of business customers.

Furthermore, G.SHDSL's multi-vendor interoperability should facilitate the adoption of the ITU standard version of G.SHDSL. This is significant because it allows vendors to make SDSL equipment that can work on any network.

Integrated circuits or "chipsets" supporting the G.SHDSL standard have been on the market since the third quarter of 2000. Moreover, silicon chips that can handle G.SHDSL, HDSL2 and HDSL4 are also now in production. For example, the PeakSHDSL octal SHDSL chipset from Tioga Technologies (San Jose, CA — 408-434-5300, www.tiogatech.com) meets all SHDSL standards, including ITU-T G.991.2 (G.SHDSL), ITU G.994.1 (G.HS), ETSI RE/TM-06011 (SDSL) HDSL2, HDSL4, and HDSL. Aimed at enabling DSL central office applications such as DSLAMs, *Digital Loop Carriers (DLCs)*, and *Multiple Tenant Units (MTUs)*, the chipset includes the TS1080, an octal ADSL transceiver and the TS2040, a quad *Analog Front End (AFE)*. The total power consumption of the TS1080 is 140mW while total port power consumption, including line drivers, is below 600mW per port for symmetrical masks at all rates up to 2.3 Mbps, and below 800mW for all asymmetrical masks including HDSL2.

In 2002 Nokia Broadband Systems (Santa Rosa, CA — 707-535-7000, www.nokia.com) and other vendors, such as Adtran (Huntsville, AL — 256-963-8000, www.adtran.com), Copper Mountain (Palo Alto, CA — 858-410-7305, www.coppermountain.com), Efficient Networks (Dallas, TX — 972-852-1000, www.efficientnetworks.com) and Netopia (Alameda, CA — 510-814-5100, www.netopia.com) have tested and deployed G.SHDSL equipment.

Clarifying How DSL Works

Unlike private leased lines or frame relay, DSL is not an end-to-end WAN service or protocol. It's an access line technology used only on a Local Exchange Carrier's (LEC's) copper wire local loop connecting a remote site to the nearest central office or remote terminal location that can serve as a CO. From there, the DSL data channel is converted to some other transmission method (most likely ATM), which the LEC connects to an ISP (Internet Service Provider) or to a central site serving a private DSL network.

In a nutshell, since DSL is a "local loop technology," if one end of the DSL link is at the customer's premise then the other end must be at the end of the local loop; this usually (but not always) means the local telephone company's central office (CO). For many flavors of DSL, the POTS twisted pair copper wire goes into a splitter (installed at either the CO or the user's premise) that splits the data frequencies from the voice frequencies. The voice frequencies are wired into a traditional POTS switch and enter the normal telephone switching network. The data frequencies are wired into a corresponding DSL modem-like device (the ATU-C discussed earlier) located at the telco's local loop termination point and the resulting high-speed digital data stream coming from (or going to) the end user is then handled as normal data rather than analog voice. Consequently, it may be hooked into any number of networking technologies for further connection to the data's destination. The data never enters the standard telephone switching system.

Typically the data will be routed over a LAN/WAN connection (10 Base-T Ethernet, T-1, T3, ATM, Frame Relay, IP, whatever) to a business office. The business may be an ISP (the ISP can but might not be the local telephone company), which may then route the data onto the Internet, thus providing the end user with Internet connectivity. Or the business may be the company the end user works for and the connection provides high-speed access from the end user's home direct to his or her company's network. Note that if the connection is made to an ISP, the end user is not connecting to the ISP over its standard modem bank, instead the end user's signals are coming in over some sort of LAN/WAN data connection that the ISP has arranged with the local telco. This is the only way an ISP can provide DSL-connected ISP service for its customers.

A telephone or other telecom device, such as a fax machine or analog modem takes the acoustic signal (a natural analog signal that varies in lock-step with the originating source of sound), and converts it into an electrical current that consists of a signal amplitude (the electrical equivalent of volume, which is the height of an analog waveform) and frequency of wave change (the electrical equivalent of pitch, which is the spacing between waves). Since the telco's local loop signaling is set up for analog, it's easy to use the analog signal to exchange data within a POTS system, and that is why a digital device, such as a computer, needs a modem to convert analog data into a digital form it can understand.

However, the maximum amount of data that an analog modem can transmit is around 44 Kbps because of the telco's equipment limitations. These limitations are related to the filtering of data when it arrives in digital form, translating that data to analog format, then changing the data back to digital format so a computing device can read it. This unwieldy process creates a bottleneck that greatly slows down data transmission.

DSL is a technology that cuts out the analog step and thus leverages direct digital communication, a process that allows greater bandwidth speeds for data transmission. Also, with most flavors of DSL, the analog line signal can be isolated so that some of the bandwidth can be used to transmit an analog voice phone call. This is simply the same analog phone service that we're all used to, but now it has some companions on the line — the higher frequency transmissions used by DSL services. You won't hear these frequencies as you make your normal phone conversations, thanks to devices, such as "splitters" and "microfilters" that prevent these higher frequencies from reaching your telephone handset or fax machine. This data / voice separation thus allows the user to transmit voice and data over the same line, using different frequencies.

Encoding and Upper-layer Protocol Stacks

Today's broadband products employ different encoding schemes and "upper-layer protocol stacks" — they're proprietary.

DSL manages to transmit digital signals over the length and gauge of the typical local loop, i.e., 12,000 to 18,000 feet (or 3.7 to 5.5 km), without requiring repeater equipment by using advanced modulation techniques (also called "line-coding schemes"). Modulation techniques define methods for encoding (or modulating) digital (or analog) signals onto a waveform (the carrier signal) and then using an inverse process, called decoding (or demodulating), to recover the original signal.

DSL's predecessor, ISDN, encountered the same challenge. Indeed, the original definition of "digital subscriber line" doesn't refer to a "line" at all — it's the modem specification for ISDN. On a single twisted pair copper wire, ISDN modems use a modulation technique called two binary one quaternary (2B1Q). A different technique, called *Time Division Multiplexing (TDM)* allows three channels to be integrated (or "multiplexed") so they can share the same wiring. Although ISDN transmission can be viewed as a form of DSL, there is one important difference — ISDN *integrates* voice and data communications on the phone network, while DSL *relegates* voice and data to *separate networks.*

As discussed previously, *Discrete Multi-Tone (DMT)*, the modulation technique developed by John Cioffi, divides upstream and downstream data into 256 separate subchannels (sometimes called "bins" or "tones"), each is 4.3125 kHz wide and modulated by an encoding techique not too dissimilar from standard QAM. About 247 of these channels are used for actual data. Low-frequency subchannels typically carry more bits per hertz (cycle) than high-frequency subchannels since low-frequency subchannels are less affected by attenuation. By using many narrowband carriers, all transmitting is done at once in parallel with each carrying a small fraction of the total data.

Channel 1 (bin 1) starts at zero frequency (0 to 4.3 kHz), which is unused and is allotted to analog POTS service. Upstream DSL data usually travels over low frequency channels 6 to 33 (spanning about 25 kHz to 163 kHz). Downstream data usually occupies channels 33 to 255 (from 142 kHz to 1.1 MHz). Bins 16 (69 kHz) and 64 (276 kHz) are generally used for pilot tones.

DMT measures line integrity, it can avoid or compensate for crosstalk or interference by monitoring each subchannel and, if the quality is impaired, the signal is shifted to another subchannel. DMT constantly shifts signals between different subchannels, searching for the best subchannels for transmission and reception. This allows DMT to allocate data so that the throughput of every single subchannel is maximized. DMT can also use some of the lower channels (starting at about 8 kHz) as bidirectional channels, for upstream and downstream information. The monitoring and sorting out of information on the bidirectional subchannels, and keeping up with the quality of all 247 channels, makes DMT more complex to implement than CAP, but gives it more flexibility when used on copper lines of differing quality. If one subchannel can not carry any data, it can be turned off, optimizing the use of available bandwidth.

DMT is the signal modulation technique in favor at the time this book was written. Both the American National Standards Institute (ANSI) and the International Telecommunications Union (ITU) selected DMT modulation as the standard for ADSL. DMT is closely related to what's called OFDM orthogonal frequency division multiplexing, which was selected for Europe's Digital Audio Broadcast (DAB) to broadcast CD quality sound and multimedia over the airwaves for in-car and mobile applications.

To sum it up, DMT uses more "smarts" to tailor its signals to the copper wire. In the past this was not feasible due to the sophistication of the electronics necessary and the cost. Moreover, in some applications, such signal modulation technique isn't useful — but in many flavors of ADSL it is a powerful, efficient and cost-effective technique to achieve optimum performance of digital signal transmission over POTS.

The DSLAM

The DSL access multiplexer (DSLAM) keeps the digital data from entering the POTS' voice network when it reaches the CO. The DSLAM diverts the voice channels (typically with the aid of a POTS splitter) so those signals can be sent over the regular PSTN, and the remaining data channels are then terminated on the DSLAM — essentially a large bank of DSL modems.

More specifically, one side of a DSLAM connects to the local loop that leads to the customer premises and the customer's communications equipment having DSL network interface devices (NIDs). The other side of the DSLAM interfaces with high bandwidth, long distance networks that lead to an ISP or elsewhere.

After removing the analog voice signals, the DSLAM collects all of the many DSL modem signals from the end users and consolidates them into a single high bandwidth signal, via multiplexing. This aggregate signal is then channeled by powerful backbone switching equipment as it travels at many billions of bits per second through an *Access Network (AN)* (also known as a *Network Service Provider (NSP)*) which provides the Internet access services that make DSL links so popular. The signal travels across the Internet or other network, whereupon it reappears at a destination CO where another DSLAM awaits. This other DSLAM then fragments the signal into its component parts and transmits them via its individual DSL modems to the designated residential or commercial DSL-enabled devices.

DSLAMs are flexible and able to support multiple variants of DSL in a single CO and different varieties of protocol and modulation, i.e., both CAP and DMT, and may even provide routing or dynamic IP address assignment for the end users.

If there's no room in the CO (or if the distance between the CO and your business or home is too great), a remote DSLAM can be placed somewhere between the customer's endpoint and the CO.

The DSLAM is responsible for the primary difference in end user quality of service when comparing DSL technologies against cable modem technologies. Cable modem end users share a network loop that runs through a neighborhood; as more end users sign on and begin using the bandwidth, performance can degrade. DSL provides a dedicated connection from each end user direct to the DSLAM. This means that DSL end users don't see a degradation in performance as new DSL users appear in their neighborhood, at least until the total number of users begin to saturate the CO's single, high-speed connection to the Internet. At that point, an upgrade by the service provider can provide additional performance for everyone connected to the DSLAM.

When a CO runs out of space or the customer's premise is too far away, local loops are terminated at intermediate points called Remote Terminals (RTs) using *Digital Loop Carriers (DLCs)*. The DSLAM doesn't send traffic directly to the DSL modem of an end user when the signal must transverse a DLC-enabled local loop. Nor does the DSLAM send DSL traffic through the DLC, because sending data at 8 Mbps, or even 1.54 Mbps, per end user, is too much data for a DLC to handle. Instead, through various ingenious methods (discussed more fully later in this chapter), the DSL signal is intercepted prior to transmission to the DLC. Still, there are new technologies and

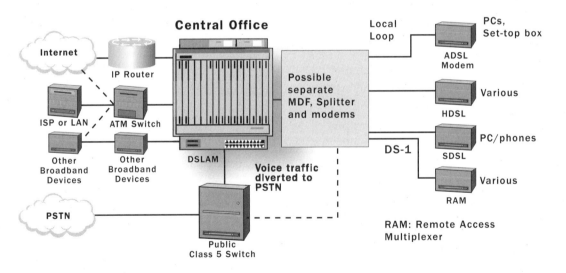

Fig. 3.3. Full Service Network Architecture Using DSLAM and ATM.

new generation DLCs on both the drawing board and the production line that can provide greater access to bandwidth.

DSL communications only operate between the CPE, the ATU-R, and the DSLAM's ATU-Cs. All other communications, once they pass the DSLAM (whether toward the

ISP or across the core network generally) use high-bandwidth ATM, though, depending on the vendor, the DSLAM can access lines to Ethernets, T-1 lines, serial lines, or Frame Relay.

The DSL "Modem" or ATU-R

Here is how a full-rate ADSL ATU-R works. It divides the available 1.1 MHz bandwidth into three segments, each of which is divided into 4.1 kHz channels:
- 0 to 4 kHz range (The bottom six channels are left for your POTS)
- 26 kHz to 138 kHz (channels 7 to 32 used to transmit data upstream)
- 138 kHz to 1.1 MHz range (channels 33 to 250 used to transmit data downstream). Channels 16 and 64 are reserved for pilot tones, which are used to recover timing. The total number of usable channels is generally around 247.

The ANSI and ITU specifications call for operation rates of up to 8 Mbps downstream and up to 640 Kbps upstream when operating over telephone lines at a distance of up to 18,000 feet, though the author speculates that perhaps a maximum reach of 24,000 feet can be attained if heavy gauge wires, the best DSP microcode, and absolutely optimum conditions can be brought to bear.

In addition to higher-speed data capabilities, modems for full-rate ADSL can easily distinguish between voice calls and data calls because each occupies different frequency bands. This is essential since most of the current telephone company equipment can't distinguish between voice and data transmissions. A data transmission usually lasts much longer than a voice call. Because telephone company switches were designed to handle a volume of calls based upon the average length of a voice call (just a few minutes), they become inefficient and "congested" if they must deal with a large number of data calls, many of which can last for hours. One of the great advantages of full-rate ADSL is that it allows telcos to take data traffic off the voice switch network, thus alleviating a potential bottleneck.

Conditioning the Local Loop for Optimal DSL Service

As the reader should now understand, DSL is simply a modified form of the local loop. In the telco's central office, the DSL line first passes through splitters to split off the low frequencies that make up the POTS voice channel. The voice analog signal will continue to follow the same kind of path it has followed for more than a century, to a large CO switch and the rest of the PSTN. The data part of the DSL signal will be sent over one of the newer data networks.

As you are reading this book there is a massive struggle taking place within the telco industry to replace an old infrastructure with a new one — a difficult and expensive proposition. Take, for example, the snarl that ensues when the telecommunications industry begins to install new wires and/or fiber optic cables — the integration of the new with the old. City streets are plagued with trucks parked by open manholes, as workmen pop in and out with new cables. During the transition phase the new services often become unpredictable, at least until all of the bugs are worked out.

A legacy telecommunications infrastructure is a roadblock where many of the problems that plague DSL service providers can be found. The offending culprits include old copper circuits, wire that's too thin, bad connectors, interfering electrical "hum" leakage from power lines, to name just a few. These legacy technologies are major obstacles that must be overcome before there can be widespread adoption of DSL. For example, to provide quality voice services, telcos deal with the long loop problem by bringing to bear a variety of technologies:

- As discussed in Chapter 2, inductive load coils are used on longer POTS lines to extend distance and limit bandwidth for standard telephone service (i.e., voice).
- Bridge taps are formed when copper wire pairs are reassigned over the years and the existing copper wiring is simply tapped with a branch instead of rerouted to a new location. The wires that hang off of a main wire cause electrical disturbances and degrade the performance of DSL.
- Digital Loop Carriers (DLCs) that terminate local loops at intermediate points closer to the end users' premises so that multiple end user lines can be consolidated onto a few transports lines that run back to the CO.

Finally, there's the build-out of a new telco infrastructure. State-of-the-art fiber optic cables also cause problems at the present stage of DSL's technical development because DSL signals can't pass through the conversion process from analog to digital and back to analog if a portion of your telephone circuit is diverted over fiber optic cable.

Load Coils

In Chapter 2, we examined how load coils (sometimes referred to as "impedance matching transformers") are the reason that sheer distance doesn't have the same limitation for POTS as it does for DSL. This small piece of technology is used to stabilize voice signals. Load coils are placed at 6000 foot intervals along POTS lines.

With POTS, frequencies above the highest attained by human speech (above 4 kHz) are considered interference. Load coils create what are called "in-line inductances" that improve the voice frequency transmission characteristics of an analog telephone circuit. Like an electronic Robin Hood, a load coil robs energy from the high frequencies and gives it to the lower voice frequencies. Thus, a load coil both acts like an amplifier, boosting the range that voice can travel along the line, and like a filter, since it attenuates (reduces) the interfering higher analog frequencies.

The problem is that in today's era of DSL, the "interfering" high frequencies are no longer random interference, but are actually used for sending data traffic. Load coils are a godsend for voice, but they limit the frequency spectrum available to end devices working at the higher DSL frequencies. Indeed, placing a load coil in a local loop reduces the effective bandwidth by cutting out the top 25% of the available frequencies. This can be shown by applying a tone generator to the line, which will reveal that the highest quality signal can be found at approximately 1000 to 2000 Hz, but when the tone generator reaches 2900 Hz transmission efficiency significantly declines.

This means that load coils will completely block DSL service by passing only the low voice frequencies and filtering out all of the high frequency channels used by DSL's DMT encoding technique. Therefore, it is critical that the service providers identify these devices and remove them before implementing most DSL services. It takes time.

This places the service provider in a Catch-22 position. While it is obvious that load coils can cripple a high-speed signal such as DSL, simply removing the load coils can make the signal worse instead of better! Removing the load coils will just cause exponentially more distortion across the copper wire both to the original voice circuit (there was a reason why those coils were there in the first place) and to the DSL signals too. This is because load coils also have the functions of maintaining both the proper "capacitance" and "impedance termination" — vital for absorbing the maximum possible power on the line.

In a long copper loop, load coils are placed at the 3000 foot mark and then at 6000 foot intervals at strategic points to reduce the negative effects that capacitance (the ability of a capacitor to store an electrical charge) will have on a signal's characteristics. The theory is that three discrete segments of 6000 feet of copper wire has less total negative capacitive (or electrical charge) effects than a single copper segment that is, for example, 18,000 feet long. Capacitance is also considered an enemy of high baud rate applications, i.e., DSL. A loss in amplitude (or signal intensity) can be overcome with amplifiers but capacitance is much harder to control, at least in a cost-effective way.

Load coils are not necessarily placed at exact intervals in the circuit; their locations can vary greatly depending on the wire gauge and other variables related to the cable bundling.

Some of these variables relate to "impedance" or the total opposition to electrical flow in an alternating circuit. If the impedance of the load coils are not properly matched to the signal, they will only absorb part of the signal, causing an improper signal transfer to the switches (i.e., most commonly, the CO or DLC) receiver circuitry. And, aside from the electrical resistance in the copper wire that makes up the local loop, there are also additional "terminal impedance loads" on the loop that are added by the terminating devices at either end of the line; namely, the line card at the telco end and the telephone or analog modem at the customer end. All of this equipment is engineered so that the termination loads are either 600 or 900 ohms (units of electrical resistance) to match the amplitude and frequency characteristics expected of the analog signal. By following these standards the twisted pair copper wire of an analog line can present what's called a "balanced pair" interface to the devices at both ends, which means that both wires in the transmission line are electrically identical and symmetrical with respect to a common reference point, usually the Earth.

Maintaining proper impedance termination is vital to absorb the maximum possible power on the line, since all the energy that's not absorbed by the termination load reflects back on to the copper pair wiring, causing interference within the original signal because such a reflected signal is usually out of phase from the original signal. This results in "common mode rejection" or cancellation or loss of amplitude which, in turn, results in the original signal becoming degenerated by its own reflection.

Also the reader must keep in mind that impedance matching, in theory, only works when it is possible to predict the signal characteristics of the waveform being impeded. Therefore, if the characteristics of the waveform are severely altered, the circuits termination will be less effective in absorbing the maximum signal energy, resulting in reflection of the signal.

What this all means is that if a service provider removes load coils it must completely re-engineer the way signals are presented to the twisted pair copper wire.

Bridged Taps

Bridged taps, which consist of bits and pieces of leftover wiring attached to the local loop, are perhaps the most annoying and offensive of all the anomalies found in a telco copper plant. Most loops contain at least one bridged tap, and the effect of multiple taps is cumulative. Bridged taps are sections of copper wire that extend off the main loop. In old neighborhoods, the phone wiring will have been used by a succession of customers through the years. If a customer moves away or a phone is removed but the wire remains, then this unconnected "spur" of wiring is a bridged tap on the currently connected circuit.

Bridged taps are the main reason end users using analog service have trouble getting modems to connect and stay connected at a relatively high speed. DSL and bridged taps have similar difficulties. Unfortunately, bridged taps are pervasive throughout most copper plants, meaning they are in most residential neighborhoods and corporate parks alike.

Short bridged taps have the greatest impact on high-speed service (DSL), while long bridged taps have a greater impact on narrowband services (POTS).

Here is where bridged taps cause problems: when an electrical signal hits the end of a copper wire it must go somewhere and if there is no impedance load to absorb the signal, then the signal (in its entirety) gets reflected back over the entire copper segment. The signal that comes from your modem headed for the CO arrives at a specific time interval and the reflected signal coming back off the unterminated copper extension enters just behind yours causing your signal to appear phase distorted. When you get two out-of-phase signals at a certain impedance load, "common mode rejection" begins and so at certain "standing wave" points the original signal undergoes interference by its own reflection — worse, if the reflected signal so happens to arrive 180 degrees out of phase, your signal can be canceled out completely. The more the second reflected signal is closer to 180 degrees the more the signal will be attenuated and phase distorted.

Now, at lower frequencies this doesn't cause many problems since the reflections are only fractions of a waveform out of phase. But when the waveforms are smaller, as is the case with the higher frequencies used by DSL, the problem becomes exponentially more apparent. In other words, phase is much more an issue with smaller, shorter or higher frequency waveforms. This isn't so much because of the high frequency itself as much as it is that the bridged tap wire is long in relation to the signal's wavelength — bridged taps will actually act as band pass filters if they're about 1/4 wavelength of the signal, so certain subchannels of a DMT-based DSL modem might bounce back, rather than be passed along with the others to the receiving modem, thus reducing the bandwidth for that particular line.

Digital Loop Carriers

Remote Terminals (RTs) using *Digital Loop Carriers (DLCs)* is a solution many telcos implement in answer to the distance quandary. DLCs are placed along local loops at intermediate locales that are closer to the end users' premises than the CO, thus allowing signals to be terminated at a convenient intermediate point.

Around the early 1980s telcos began to install DLCs (especially in fast growing areas) to save money, since copper loops are relatively expensive. Although the older

DLCs support ISDN, they don't support DSL — this is a problem. Almost one-fourth of the telephone service in the U.S. is served from DLCs and around 50% of those are original installations using technology incompatible with DSL. The rest, installed with technology capable of supporting DSL, are just now being *enabled* for DSL. Some good news — telcos have found that they can retro-fit the older DLCs by plugging in new cards that provide DSL compatibility.

Rather than just amplifying the signals and passing them along the same wires to the CO, various technologies exist at the DLC to combine multiple voice channels together onto just a few transport lines that run back to the CO (typically multiplexing 24 subscribers onto a single T-1 line; although some DLCs can handle up to 1000 lines). In this way, the telco can save both time and money, since fewer and smaller cables are required to expand services. DLCs also make the task of serving growing business and residential areas quicker and easier by eliminating the need to design and build an entirely new telecommunications infrastructure for each new housing addition, suburb, and business complex.

These consolidating technologies involve ways of putting more than one voice transmission on a single pair of wires. An analog technique would be to shift voice frequencies up or down and combine them with those of other voice transmissions. A digital technique takes the voice signals, digitize them, and divides the telephone circuit among these voice signals by either TDM or special encoding technique. The general term for this consolidation process is "pair gain," as discussed in Chapter 1.

Other problems that telcos must overcome with legacy DLCs residing within the basic structure itself are backplanes, common units, etc. that don't allow the entire DLC to support bandwidth beyond a single T-1 (1.54 Mbps, or 64 Kbps/circuit x 24 circuits), far below the 6 to 8 Mbps needed for DSL.

Not all telcos use DLCs on their local loops. Some utilize a serving wire center (SWC) or "wire center," which may not have high-end switching devices but has transmission equipment that connects to a CO. The "backhaul" to the CO could be via T-1/E-1 circuits using copper- or fiber-based technologies.

Remote DSLAMs

For DSL-based service to connect to a RT, the DSL portion must terminate at the RT, where the DSL transmission is converted to a format compatible with the DLC. This is precisely where a DSLAM might come into play.

Although DSLAMs are an essential DSL ingredient, since they're expensive, they represent an obstacle to DSL implementation. The remote DSLAM is equipment that's

Chapter Three

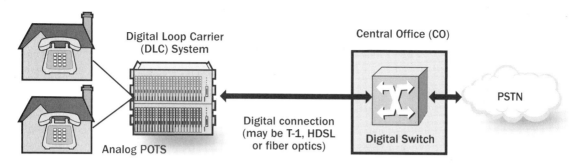

Fig. 3.4. Typical DLC Deployment.

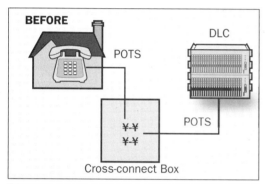

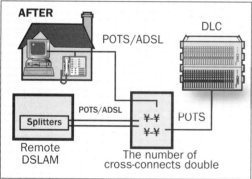

Fig. 3.5. Cross-Connects Before and After DSL Deployment.

ruggedized or made as impervious to environmental damage as possible and placed inside an environmentally hardened cabinet so it can then be installed in the field, typically close to a DLC. From there, the remote DSLAM negotiates the transmission of the DSL-based traffic between the DSL modems and a WAN (on its way to the Internet), while the POTS voice traffic is sent back to the DLC which in turn routes it to the CO.

Remote DSLAMs can offer the greatest flexibility for the largest number of subscribers, but their deployment demands a considerable commitment of cash and resources. It's an investment that can easily be amortized over a large subscriber-base, but in small line-size DLC environments where the potential number of subscribers is limited, a remote DSLAM may never be cost justified.

Even the actual installation of the remote DSLAM poses problems, such as obtaining right of way, pouring the concrete needed for a pad, installation of the cabinet, power for the electronics, and deploying wiring to and from the DLC. Additionally, significant problems can arise concerning the size and configuration of the one or more "cross-connect" boxes that sit close to the DLC cabinet where all the subscriber wire pairs are cross-connected to the wire pairs going to the RT cabinet.

Because DSL service is transmitted over the same copper wires as POTS service, rerouting at least some of the pairs is necessary. As I've said, the copper wire pairs carrying DSL/POTS traffic are routed to the remote DSLAM where the POTS and DSL signals are split. The POTS traffic must then be routed back to the cross-connect for connection to the DLC cabinet.

There are advantages and disadvantages to using remote DSLAMs. Their capacity can scale up easily — a typical remote DSLAM can serve 60 to 100 DSL subscribers. Remote DSLAMs require no additional management systems or personnel training because element management is similar to that of a larger CO-based DSLAM. Also, remote DSLAMs can be used with any DLC system with no impact on the quality of POTS service because they're independent from the DLC system. The remote DSLAM simply splits POTS traffic off and sends it back to the DLC while it's still in its analog form.

The DSL service provider might come up against cross-connect boxes that have been designed to support the number of pairs the DLC supports, with limited or nonexistent spares. If this is the case, then to expand the system the service provider must add cross-connects or resize the existing ones. The situation is further complicated in cases where RTs have incorporated the use of multiple cross-connect boxes since the service provider cannot predict how many of its current customers will request DSL service.

Line Cards

An alternative to remote DSLAMs is the *channel bank line card* that plugs into an open slot in the DLC. Although line cards avoid many of the cost issues that plague remote DSLAM solutions, they introduce other difficulties. Line cards must interface to and juggle signals from various incompatible systems. In addition, they can add significant constraints to the DLC's capacity from a POTS perspective.

That being said, the service provider dealing with legacy DLCs will usually opt for the considerable mechanical stability of channel bank line cards. And, even if not burdened with legacy DLCs, the service provider might go with a line card that comes as a pre-integrated piece of the DLC system. With this option, the DSL and voice transmissions share the same backplane (similar to a "motherboard" in a computer in that it has sockets or "slots" that special cards, i.e., boards, can plug into). In this way, both analog and digital traffic are consolidated by the system, and can share the same transmission facilities to reach the CO. Note that integrated line cards are the norm in newer-generation DLC systems.

The use of line cards has pros and cons. The good news is that line cards can take advantage of unused card slots inside the DLC, mitigating the costs associated with remote DSLAM solutions. Also, an integrated line card solution can obviate the need to cable or rewire components within the cabinet. But the bad news is that line cards can cause administration problems if the network access provider is using DLCs from different vendors. This means that the signaling characteristics of proprietary DSL modems must match up with those of the proprietary line cards developed by various vendors.

Today, the selection of the DSL-related semiconductor "chip set" used on the line card determines the type of DSL modem that can be used at the end user's premises. As a result, if an end user moves or perhaps even changes service providers, the end user will most likely be required to purchase a new DSL modem since the new provider may use a different DLC.

Yet another line card issue arises when the DLC's card slots become full (i.e., no room at the inn). The initial decisions regarding the size and type of DLC cabinet was probably based on a growth plan that didn't include DSL service. Trying to retrofit this service affects the network access provider's ability to provide additional POTS services in the future.

Finally, there may be some electrical inconsistencies that force a service provider to rewire the DLC. There are also restrictions in place that govern the placement and quantity of line cards that can be installed in a DLC, which further complicates the engineering activities associated with the line card solution.

Remote Access Multiplexer (RAM)

RAMs provide a low-cost solution for extending DSL services to remote end users fed by DLC systems. They are cost-effective and easy to deploy while providing a universal solution that works with all DLC systems. At the same time, RAMs provide much of the same functionality as a remote DSLAM, although they are designed primarily for deployment inside DLC cabinets. RAMs differ from DSLAMs in that a RAM integrates into existing DLCs without a costly infrastructure upgrade. The RAM is installed inside the DLC cabinet, so the service provider can avoid the kind of cost and rewiring issues that pop up when installing remote DSLAMs, and there are no problems with rights of way, concrete pads, power, or cross-connect boxes. The current RAMs combine the advantages of both remote DSLAM and line-cards, while avoiding some of the disadvantages of both.

Since RAMs are independent of DLC systems, they have the flexibility to work with any DLC system without impacting POTS capacity, thus avoiding the interoperability problems associated with line cards. If the DLC cabinet is full, RAM technology can

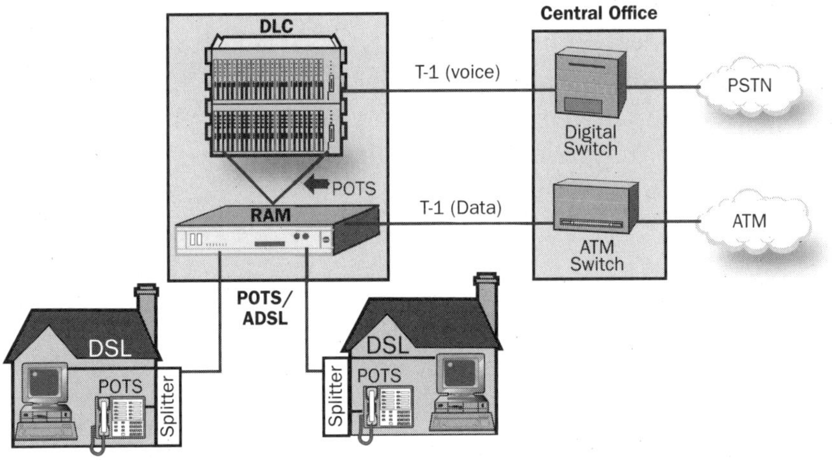

Fig. 3.6. Typical RAM deployment.

take the form of a small cabinet that attaches to the side of an existing DLC cabinet.

RAM technology can utilize transport back to the CO in one of two ways — a proprietary method that requires termination on a DSLAM, or a standard transport method, such as a digital service level 1 (DS-1) user-network interface. If the service provider opts for termination on the DSLAM there must be available ports on the DSLAM, but if standard transport is used, termination can occur directly on a long distance ATM switch or, in many cases, it will be termination on other vendors' DSLAMs.

But RAMs have their own issues. For example, RAMs are not scalable — at least not at this point in time. They are best suited for small line sizes, which means that as more lines are required, more RAMs must be installed. Also, interoperability between DSL chip sets isn't available yet, thus RAM compatibility with DSL modems is a necessary consideration. The author notes, however, that there exists on the drawing board newer, denser RAMs that take advantage of *Digital Signal Processor (DSP)* technology to share DSL transceivers. These will probably on the market in the near future.

Attenuation

As discussed in a previous chapter, attenuation arises partly as a result of the legacy infrastructure that DSL service providers inherited. When digital signals are transmitted at high frequencies (such as DSL) the signal loses power quickly, which can result in not all of the transmitted signal arriving at the ATU-R. So, the use of higher frequencies to support higher-speed services (such as DSL) necessitates shorter loop reach.

To minimize attenuation or the drawing out of energy, the service provider can use lower-resistance wire — thick gauge wires have less resistance than thin wires (i.e. less signal attenuation) — allowing the signal to travel a longer distance. Telephone companies (more so in the U.S. than in other countries) designed their cable plant using the thinnest gauge wire (26 AWG) they could get away with and still support the required voice-based POTS services. Fortunately, 26 AWG is just thick enough to support DSL service.

Another way to override attenuation is to install repeaters along the copper wire lines to repeat or replenish the signal. While this method works, it's not often used because it's expensive.

Crosstalk

To successfully deploy DSL service, your friendly DSL service provider must also address the environment inherent in the bundle of cables leading from the CO to the end users' neighborhoods. In the telco's network, multiple insulated twisted copper pair wires (the local loop of each user) are bundled together into what's called a cable binder. But, the electrical energy transmitted as a modulated signal across each twisted pair copper wire also radiates energy onto adjacent copper wire loops in the same cable bundle. This cross coupling of electromagnetic energy is called crosstalk.

Adjacent systems within a cable binder that transmit or receive information in the same range of frequencies create crosstalk interference, because crosstalk-induced signals combine with the signals that were originally intended for transmission over the copper wire loop. What results from this crosstalk is a waveform shaped differently than the one originally transmitted.

Crosstalk is generally categorized as follows:

• Near end crosstalk or NEXT that occurs at the central office is the most crucial since the high-energy signal from an adjacent cable or system can induce relatively significant crosstalk into a primary signal.

• Far end crosstalk or FEXT that occurs at the end user's location is typically less crucial because the far end interfering signal is attenuated as it traverses the loop and is thus quite weak.

Crosstalk is a dominant factor when trying to obtain optimal performance from many types of telco-related systems, including DSL. DSL system performance can be calculated relative to the presence of other systems running at other frequencies within the same cable binder, known as a "disturber" (a high-speed data service e.g., ISDN, T-1, DSL) can cause crosstalk which can seriously hamper DSL system performance. Fortunately, it is unlikely that a service provider would be so reckless as to actually deploy DSL service in a 50-pair cable that happens to have many of these disturbers concurrently running in the same cable binder.

If the effects of the attenuation and crosstalk are not too significant, the DSL system has the capability to accurately reconstruct the signal back into a digital format. (However, when the interference becomes too great, the signals are misinterpreted at

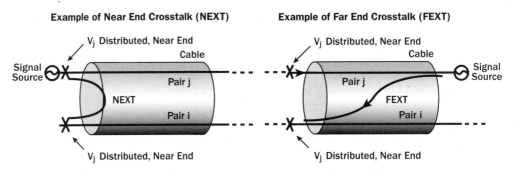

Fig.3.7. Types of Crosstalk.

the far end and bit errors can occur.) To create multiple channels without interference, both CAP and DMT ADSL modems use one of two ways to partition the available bandwidth of a telephone line — echo cancellation or Frequency Division Multiplexing (FDM). With either technique, of course, ADSL allows for a 4 kHz region for ordinary analog POTS at the lowest end of the frequency band.

With echo cancellation, the bands for both upstream and downstream data overlap, but the two are separated by an echo cancellation technique that has been around since the days of V.32 and V.34 analog modems.

On the other hand, FDM assigns different frequency spectra for the separate transmit and receive, upstream/downstream signals, but the downstream path is divided by *Time Division Multiplexing (TDM)* into one or more high-speed channels and one or more low speed channels. The upstream path is multiplexed into corresponding low speed channels.

The advantage of a FDM-based system over an echo-canceled system is that NEXT can be eliminated since this type of DSL system is not receiving in the same range of frequencies in which the adjacent system is transmitting. Even so, FEXT is still present — through it's not a big problem — a FEXT signal is at the far end of the loop, which means it's attenuated and weak. Therefore, FDM-based systems often provide better performance than echo-canceled systems, at least in terms of crosstalk from similar adjacent systems.

Echo-canceled systems are also subject to an interesting form of interference called "self NEXT," which appears when too many similar echo-cancelled systems are next to each other in the same cable binder. Self-NEXT degrades the performance of all like-type systems within the cable binder. An example: a single CAP-based system may achieve a targeted 6000 foot loop reach, but if additional CAP-based systems are added to the cable binder, the loop reach of the first system and the subsequent systems may be reduced by a substantial number of feet per each added service. This phenomenon is true of basically all echo-canceled systems, such as 2B1Q, CAP HDSL and SDSL, and echo-canceled DMT ADSL systems.

Therefore, in the face of self-NEXT, service providers must keep an eye on system performance as they add more and more DSL lines to their cable binder. Using FDM technology instead of echo cancellation forces an engineering compromise in that the separated upstream and downstream signals of FDM occupy a greater range of frequencies than the overlapped transmit and receive signals used in systems based on echo cancellation, resulting in less reach.

Thus, where little crosstalk is expected and NEXT is moderate to low, an echo-canceled system may be the better answer for that DSL service provider. In other cases where crosstalk is expected to be the norm and NEXT is likely to be more assertive, an FDM-type system may perform better.

POTS Equipment

While one would think simple attenuation would be the most significant factor in DSL performance, crosstalk becomes the critical factor as more and more systems are deployed.

Another consideration is interference with and errors caused by POTS equipment. After all, ADSL and POTS equipment must share the same twisted pair. POTS equipment produces all kinds of electrical perturbations on the line as it goes about the business of signaling, alerting, ringing, answering, and going off hook and on hook, all of which might generate some high frequencies that reach up into the ADSL bands and interfere with ADSL performance.

In Summary

DSL technology is designed and implemented to meet increased user demand for high-speed, multimedia communications based upon an ability to support voice, data, and video over the installed base of twisted-pair copper telephone wires.

Although telcos must make changes in equipment and infrastructure to accommodate DSL technology, DSL doesn't demand that telcos perform a massive and expensive rebuild of their infrastructure. Indeed, DSL has been referred to as "copper optics," since it can provide multi-megabit bandwidth via the twisted-pair copper plant, allowing telcos to defer the hefty cost of building a more advanced (and expensive) fiber optic infrastructure.

While data services were the first to be deployed over DSL, lucrative voice services are now vying for top billing. DSL technology offers a "win-win" scenario for both the service providers who provide high-speed, value-added services, such as VoDSL, web serving, VPNs and CENTREX and the customers who benefit from these value-added services and who at the same time realize dramatic cost and performance benefits.

CHAPTER FOUR

VoDSL - The Basics

Voice-over-DSL or VoDSL refers to the techniques used to transport voice and data in an integrated way over a single DSL circuit. Basically, VoDSL technology can support multiple voice calls, either inbound or outbound, over a single DSL circuit, while simultaneously supporting high speed Internet access, with all transmissions being in digital format.

VoDSL is a non-standard technique, though it's based on a number of standards.

As CT Labs' founder Chris Bajorek once wrote: "I cannot escape the pretzel logic of voice and data traffic multiplexed onto DSL, which is itself encoded onto twisted-pair analog (voice) circuits. Let's see, that's voice- and data-over-data-over-voice — VADoDoV. Yes, it does make sense, but ... whew!"

To date, DSL technology has been used almost exclusively to transport data over the Internet at high speeds through the PSTN's existing copper plant, where typically the only voice carried is analog voice (that's not the "voice" referred to in "voice-over-DSL"). But, it became apparent, after the initial success of DSL, that DSL technology could be used to offer a greater range of services (e.g. additional voice services, VPNs, video conferencing, video on demand) targeted at specific markets. The high intrinsic bandwidth that DSL provides, backed up by new innovations in voice compression, echo cancellation, digital signal processing, and silicon technologies, in general, makes all of this possible.

Data accounts for most traffic over today's DSL networks, but it's voice that brings in the most revenue for the typical service provider; though the "voice" we talking about here is just standard analog POTS lurking on the same line with DSL (you might call it "voice-under-DSL"). By many reports, voice via POTS accounts for up to

80% of the average service provider's revenue. It's no wonder that out of the growing list of new services being bundled with DSL service, the most attention has been given to provisioning digitized, packetized voice, thus positioning VoDSL to become the next "must have" application.

Service providers looking for compelling business opportunities are moving quickly to upgrade their DSL access networks so that both voice and data service can be integrated onto a single copper line (reminiscent of how ISDN integrates both voice and data services into digital channels). Two main market segments are expected to have a high rate of interest in VoDSL.

The first is the SME (small/medium-sized enterprise), a significant percentage of which need to be able to send and receive data at around 500 Kbps and which have voice needs that are typically met by eight to twenty-four phone lines. Companies with small or informal call centers would fall into this category.

The second group consists of SOHOs and residential customers with need for less than six outgoing phone lines. The high end residential market, in particular, appreciates the extra two to four voice lines that VoDSL technology offers, since there's been an increasing trend for "upscale" residences to install two, three or even four phone lines. For them, VoDSL connectivity should look attractive, too, if it results in significant savings over conventional phone line and Internet connectivity costs.

Other advantages for both target markets include IDSN voice quality, automated provisioning (which greatly reduces the time taken to add or remove services), one-stop-shopping via bundled data and voice services, a single monthly bill, and a common help desk for all their telephony needs.

The VoDSL Upgrade

VoDSL (more accurately described as voice-over-IP-over-DSL or voice-over-ATM-over-DSL) solutions use technology which overlays onto the DSL network, enabling the provisioning of voice services on what started out as a data network.

If the DSL networks have the proper technology at work in the background, voice-over-DSL solutions can be provisioned. Yet, just as the local loop must be upgraded to support DSL, the DSL access network must be further upgraded to deliver VoDSL service. Specialized equipment at the customer's premises and in the service provider's network (both of which we will examine in detail in later chapters) converts voice into digital data packets and then transmits these packets as multiple calls (lines) over a single copper line. The equipment chops up the available DSL bandwidth into phone-circuit sized streams — from a few Kbps up to 64 Kbps each, depending on the type of compression / decompression algorithm protocol or "codec" used — to transport as many as 24 calls at a time. Thus, VoDSL technology takes the extra bandwidth on the copper wire made possible by DSL and dynamically allocates it for additional voice channels, but at the same time leaving some bandwidth for a data channel. As call volume increases, data capacity decreases; and when there isn't any voice traffic, the entire bandwidth is available to carry data. This enables the service provider to sell multiple services over a single access line.

All of this totally transforms the economics for delivering telephony services to small businesses, satellite offices, SOHOs and residences. The average small business spends about 10% of their communication budget on data and 90% on voice. The convergence of voice and data services over DSL not only saves money on telephone bills, but also leaves enough margin to make it a profitable converged business model for sellers of voice and data. Thus, the service provider now has a means to get the most Return on Investment (ROI) from its DSL upgrade.

Getting to the VoDSL paradise isn't easy, even with the present rush of VoDSL equipment coming onto the market; for even as it takes hold, VoDSL technology continues to evolve. This continuing evolution makes picking the right equipment one of the most challenging decisions for carriers deploying VoDSL service, something akin to hitting a moving target. On one important level, the decision boils down to a choice between whether the service provider will go with a centralized, circuit-switched or a distributed, packet-based platform.

Yet, most of these high-level system architecture challenges are overlooked at first because, down at the component level, it's exciting and relatively easy for carriers to translate an analog voice signal into a digital format and then send it across a data network like any other piece of information.

With VoDSL, once the service provider has a presence in an incumbent carrier's central office and has deployed its network gear — a DSLAM, ATM or IP network, and a voice gateway — it can order an unbundled copper line from the incumbent carrier, and *voila!* — it's in business.

QoS

Challenges remain about whether the DSL access networks can effectively deliver voice traffic to the standards that customers, particularly business customers, demand. First, the service provider must determine if the VoDSL equipment can deliver the same call quality and feature sets found with PSTN equipment. Second, a determination needs to be made as to which vendor products do the best job for the service provider's ecosystem. Then, an architecture that can consistently deliver high call quality needs to be set in place.

Underlying all this is the need for the service provider to assure that sufficient bandwidth is available to transmit voice packets in a timely manner. This can be difficult; especially on Internet protocol (IP) networks, where packets are scrambled as they travel haphazardly across a network.

Many carriers are addressing the problem by layering their advanced DSL services on top of *Asynchronous Transfer Mode (ATM)* networks, which already have in place Quality of Service (QoS) features that ensure a clear path from point to point. Indeed, most of the DSL lines currently deployed use ATM as the underlying data-transport protocol beneath TCP/IP.

Limitations in the DSL Technology

As is the case with any technology, DSL has limitations. Customers must be located within a few thousand feet of the DSLAM and the copper-wired circuit between the central office and the customer must be free of the impedances that phone companies have always allowed to exist on analog voice lines: attenuation, crosstalk, bridged taps, changes in wire diameter, loading coils, amplifiers and other devices.

Deploying the IAD — Complexity vs. Ease of Use

Another challenge to deployment is the need for carriers to install *Integrated Access Devices (IADs)* at the customer's premise. IADs consolidate multiple information feeds onto a single line. Such devices were originally designed to carry data traffic only, but new multiservice IADs are making their way to the market.

As far as VoDSL goes, the primary job of the IAD is to take a data connection and let you connect standard phone sets into it, ideally with multiple lines. It will then packetize the voice traffic and multiplex it with data traffic and send it over a single line having a high-speed DSL connection into the public network.

With the demise of the CLECs, RBOCs have become interested in the VoDSL IAD market. It's a simple process to physically swap out a small or medium-sized enterprise's DSL modem for an IAD, then connect an existing PBX or key system as well as the company's LAN to the IAD, and thus onto the DSL line. This permits the service provider to continue delivering high-speed data transfer and offer multiple derived phone circuits all over one set of copper wires.

It's even more compelling to convert businesses without PBXs or key systems to

VoDSL IADs. This market allows the telco provider to offer its business customer value-added services such as those normally supplied by a PBX. Gradually many SMEs will move away from their PBXs and key systems and will be using these services within a broadband environment.

Yet, the cost of the IAD can be a barrier. Currently, each device costs the service provider between $800 and $1500. Who pays for this equipment — the customer or the service provider? Fortunately, this hurdle should become less of a problem in the future — IAD prices are decreasing.

Of course, larger businesses have different needs than smaller ones, so vendors have taken basically three directions in IAD development: scaleable, modular devices not exclusively DSL in nature find their way into higher-end environments (50 to 200 users) that are fortunate enough to own multiple WAN lines as well as IT staff.

A less complicated fixed-configuration device is most often used with lone DSL connections to serve the lower-end market, which consists mostly of SOHOs and residential users.

Then there's the flexible configuration IADs that try to service all markets through the use of compelling software innovations.

Scalable IADs

One example of a scaleable high-end IAD is the InstantOffice 5500 system from Vertical Networks, Inc. (Sunnyvale, CA — 408-523-9700, www.verticalnetworks.com) which can serve small offices having up to 84 employees (84 station telephone ports, 84 data ports, 78 trunk ports and five WAN access ports). The InstantOffice 5500 integrates the functions of a PBX, voice mail system, automated attendant, LAN hub, voice-over-IP gateway, and multiprotocol router. It supports a variety of WAN access types — DSL, T-1/E-1, Frame Relay, ISDN Primary Rate Interface (PRI), 56/64K DDS, and analog lines. It's so sophisticated that they don't' even call it an IAD, but an Integrated Communications Platform (ICP).

Fixed-configuration IADs

Carriers that deploy broadband access services to SOHOs and residential customers typically only require a relatively simple, fixed-configuration device that can juggle multiple telecommunications functions and to integrate multiple lines for voice, data and the Internet onto converged connections. Some of the simpler of these devices are quite similar to the DSL modem that might already be in place.

Flexible-configuration IADs

SMEs and others with one or a few DSL connections and a diminutive or nonexistent IT staff will use either the fixed-configuration IAD or a flexible-configuration IAD, which can match a fixed-configuration unit's ability to integrate voice and data and handle a number of telecommunication functions.

In the past, providers that targeted SMEs have often struggled to come up with a platform that allowed enough complexity to be expandable and configurable without sacrificing reliability and ease of use. Of course, this means that simplicity is often sacrificed in the name of flexibility since carriers need an easy and affordable means of implementing the VPNs, high-speed data and other enhanced services that SMEs demand.

Nevertheless, most in the SME community need easily deployable products that quietly and economically aggregate voice and data traffic at the customer premise onto the access loop and aren't difficult to use or maintain. These kinds of IADs are produced by companies such as Advanced Fiber Communications (Petaluma, CA — 707-794-7700, www.afc.com), with its PreMax 410; AG Communications Systems (Phoenix, AZ — 623-582-7000, www.agcs.com), with its SuperLine IAD; Cisco Systems (San Jose, CA — 408-526-4000, www.cisco.com), with its IAD2400 Series Integrated Access Devices supporting G.SHDSL; and Vpacket Communications (Milpitas, CA — 866-872-2538, www.vpacket.com), with its Vpacket 6100 that supports converged voice/data services over IP (over DSL or T-1), and which has a Network Management System (NMS) that provides comprehensive voice services quality measurement for precise records of the actual customer voice QoS. The Vpacket 6100 is even large enough to fully support both the voice and data requirements for an office or floor of a building.

The Current Trend

In the past, the acceptance of such manageable, user-friendly IADs has been hampered by the lack of standard interfaces, since interoperability between these devices and equipment at the customer site and central office is often limited.

Fortunately, the current trend in the IAD market has been the introduction of devices with autoconfiguration and autosensing management capabilities. These functions speed up and simplify IAD installation at the customer premises, since a nontechnical end user can install the device, then the service provider can configure and provision the box remotely, often without customer involvement. This eliminates the expensive "truck-roll" process, which requires a skilled technician to visit the customer premises. Vendors developing products in this area include Cisco,

Advanced Fiber Communications or AFC (Petaluma, CA — 707-794-7700, www.afc.com), and VINA (Newark, CA — 510-492-0800, www.vina-tech.com).

The Softswitch

Vendors have been testing their IAD products to make sure that different devices can exchange information, but on a time-consuming, case-by-case basis that has inhibited the deployment of some VoDSL equipment. Conditions will improve as new and more flexible equipment at the CO is installed over time. For example, a conventional CO switch is "hard-wired" to exclusively support voice-grade telephone service on a circuit-switched basis. But, so-called *softswitches* (essentially computer servers that can handle call processing functions) are gaining popularity. Softwitches allow carriers and service providers to quickly create new services and have the advantage of supporting open software application programming interfaces (APIs). They support multiple protocols, including IP and ATM, and multiple QoS and *Grade of Service (GoS)* levels, and can serve as gateways between the circuit-switched PSTN and the packet-switched network of the service provider, by resolving any protocol issues internally.

So, when connecting to an *International Softswitch Consortium (ISC)* architecture, which consists of a call agent (aka media gateway controller, softswitch); media and signaling gateways; feature, application and media servers; and management, provisioning and billing interfaces or a *MSF or Multiservice Switch Forum* (a technical organization formed in late 1999 to develop and promote an open architecture for multiservice broadband systems to enable end-to-end interoperability) architecture at the central office, the softswitch can interoperate with the PSTN by signaling to PSTN switches and *Intelligent Network (IN)* platforms. The softswitch also may access network elements in the PSTN for routing, billing, emergency services and other value-added services as needed.

Why this discussion on softswitches? Because IADs and other customer premise equipment such as small gateways are controlled by an access softswitch or "call agent" instead of a conventional PSTN switch. This allows the CO equipment and software to take care of the interoperability issues that can arise with a PSTN gateway.

Still, the CPE vendors' worries are generally concentrated at the voice bearer level, bringing up questions such as "which codec is to be used for voice or video compression / QoS?" The *Real Time Transport Protocol (RTP)* developed by the IETF (Internet Engineering Task Force) adds a layer to the Internet protocol and addresses the problems caused when real-time interactive exchanges such as voice and video are transported over what were formerly data networks. RTP is fairly stable and is pretty much compatible between different vendors. Current industry uncertainty comes from the call signaling and control protocols (such as SIP, H.323, MGCP, H.248/Megaco, which we will examine in more detail in Chapter 5). This is where the softswitch interoperability process occurs; but the IAD itself, typically, doesn't have to participate in this process, although some service providers are attempting to push the intelligence in the network all the way past the network edge to intelligent IADs. Normally, if the call agent or softswitch at the CO that the IAD is communicating with handles interoperability, then the IAD can talk to multiple gateways from multiple service providers. Also, if a customer premise is equipped with an IAD which doesn't have any QoS features, then any call agent (softswitch) should be able to provide telephony services to that customer premise.

The unique, proprietary feature that a connection provider can provide their customers is a particular type of QoS, at least if they are technically capable of doing so.

The Biggest Headache

The most vexing problem faced by many of today's local VoDSL service providers is worrying about and dealing with how the incumbent carrier moves the customer's voice service from its network onto its competitor's (the VoDSL service provider's) network. Very often, the VoDSL service provider doesn't discover problems until the customer's service is supposedly switched over, and everyone suddenly discovers that the VoDSL service doesn't work. Therefore, the local VoDSL service provider must have in place a process that allows it to control the customer's conversion step by step.

This process involves getting all the necessary services up, running and fully tested before turning anything on at the customer's premise. What "fully tested" means, is that the service provider should install the IAD at the customer's premise, tune up the data circuit and make sure the voice lines are working. What it doesn't mean is everybody standing around idly, waiting for the incumbent carrier to do the work. Only after the VoDSL service provider knows everything is up and running does it switch the customer over to VoDSL service.

VoDSL's Ancestors can Retire Now

In the latter part of the 20th Century, voice networks relied on analog lines and trunks as well as IDSN and digital T-1 circuits to serve all markets — enterprise, SME, SOHO and residential market. Although still in wide use today, these technologies offer the service providers and their customers a Hobson's choice between low efficiency and high cost — analog lines are inefficient, T-1 circuits are expensive, ISDN is a little of both.

This hybrid analog/digital transport network's problems (inefficiency and high costs) are further compounded by requirements for separate connections to deliver multiple voice lines and end-user data services (with ISDN being a limited exception). The advent of DSL has substantially improved providers' options. DSL lets providers use a single physical connection to carry both voice and data traffic. VoDSL lets providers further improve efficiency and reduce costs for voice service by replacing edge access and core transport components of the legacy voice network with elements that offer lower capital expense and lower recurring costs.

ISDN

Let's take a look at the *Integrated Services Data Network (ISDN)* first since not only does it have a presence in the SME market, especially in Europe, but it also could be considered a close "cousin" of VoDSL and was used in many of the situations in which VoDSL now finds itself applicable. For instance, Nortel Networks was one of the first adopters of full-service teleworking for its employees, which began with the use of ISDN in 1994. Today, only around 25% of Nortel's teleworkers use ISDN (down from 90% at its peak in late 1998); they are migrating to DSL as quickly as it becomes available in their area.

Symantec (the company that produces Norton anti-virus software) also initially considered ISDN as a way to extend PBX functionality to its employees' home offices, but DSL is now the company's technology of choice for this task.

T-1/E-1

Mixing voice and data over a digital access line isn't exactly new — telcos have been doing it over channelized T-1/E-1 circuits for years. Yet, in many ways VoDSL and T-1/E-1 voice services are like oil and water. T-1/E-1 technology is based on *TDM (Time Division Multiplexing)* and DSL technologies are packet-based, and make use of *STDM (Statistical Time Division Multiplexing)* and the bandwidth allocation is dynamic, which means that it can shift between voice and data (but it always prioritizes voice or whatever is considered to be real-time traffic).

For the large- to enterprise-sized businesses, a pair of T-1/E-1 lines — one for voice and one for data — were (and probably will continue to be) used to service their private networks. This is a very expensive solution and out of reach for most of the SME market — which just happens to be one of the target markets for VoDSL. This group (and others) now can run two or three DSL lines instead of the costly T-1/E-1, and use the Internet backbone for both their VPN, WAN and PBX trunking, thus completely changing the economics of communications.

Analog

SOHO and residential customers have almost always opted for multiple analog lines to meet their voice and data access requirements. But even with this user group, there should be a demand for DSL service. This market, while happy with the quality of their POTS service, stop being happy when they can't use their telephone because they're stuck online due to uploading or downloading of megabyte files over a slow analog modem. To have a product that alleviates the slow Internet service provided by modems, while allowing simultaneous use of the same line for voice and data, will be of enormous interest in the home.

How VoDSL Works

As we've seen, DSL lets providers use a single physical connection to carry both voice and data traffic to and from their customers. VoDSL uses the extra DSL bandwidth dynamically, which means that voice calls only consume bandwidth when a call is active on a line. Thus, if there happens to be no calls on the line at a particular time, all of the bandwidth is made available for other services, such as Internet access.

You've been given the who, what, when and where of VoDSL, so let's now delve deeper into how VoDSL generally works. . .

The DSL Technology

First there's the DSL line itself that's used to transport the data and packetized voice (utilizing the existing copper plant) to the nearest carrier facility. Of course, the copper wire first must be conditioned so it can support the distance and quality requirements for the DSL variant being offered. Once that's accomplished, there are several DSL options over which VoDSL runs to choose from.

SDSL

As the reader now knows, SDSL (Symmetric DSL) is a perfectly symmetrical version of DSL. A derivative of HDSL, with equal amounts of bandwidth both downstream and upstream. SDSL runs at signaling rates up to 784 Kbps (half a T-1) on one wire pair or as high as 1.544 Mbps on two pairs, which is equivalent to T-1 speed, over distances up to 2 miles or so. SDSL also can be adjusted downward to run at rates of 384 Kbps and 128 Kbps, for example.

VoDSL implementations typically support as many as 16 voice conversations over a full-rate SDSL link running at 1.544 Mbps, although some manufacturers support as many as 24 conversations. The balance of the bandwidth is reserved for data communications purposes. Although SDSL was not originally intended to operate simultaneously with POTS, various vendors have managed to achieve POTS sharing so that residential or small office users can use the same telephone line for data transmissions, voice or fax.

G.SHDSL

The first DSL technology to be developed from the ground up as an international standard is G.SHDSL, ratified by the ITU in February 2001. Also known as G.991.2 and SHDSL, G.SHDSL is an international standard for symmetric DSL, which provides for sending and receiving high-speed symmetrical data streams over a single pair of copper wires at rates up to 2.3 Mbps. It was specifically developed to incorporate the features of other DSL technologies, e.g. ADSL and SDSL, and can transport T-1, E-1, ISDN, ATM and IP signals. G.SHDSL is what brings VoDSL to the forefront of business communications.

G.SHDSL allows users to have multiple telephone connections, along with a fax and broadband data channel. With G.SHDSL, multiple voice and video channels can be embedded in the data payload thanks to its low latency, which is less than 1.2 ms. ADSL uses *Reed-Solomon Forward Error Correction (FEC)* coding (the ADSL FEC can be turned off to bring transmission latency to acceptable levels, but it results in ultimately reducing the performance margin and bit error rate of the line), resulting in a relatively large transmission latency of 20 ms. This latency makes it difficult to transport voice and real-time video within the ADSL payload since they are very time-sensitive.

G.SHDSL, however, already has sufficient QoS and G.SHDSL equipment has already begun appearing. By reducing the latency to less than 1.2 ms G.SHDSL makes DSL technology suitable for digital voice transport and real-time video conferencing. Since the various services are handled in the digital domain, bandwidth can be dynamically allocated between voice, video, and data. By supporting a variety of line rates and payload configurations, G.SHDSL allows different service applications to be tailored for a wide variety of users — satellite offices, the SME market, SOHOs and residential users.

The G.SHDSL standard brings a common definition for the many variants of symmetric DSL. Now vendors worldwide can build interoperable equipment. Interoperability helps minimize cost and deployment time; issues that are critical to the success of new services.

Other Options

There are many DSL flavors or variants over which VoDSL can run. For instance, recognizing that there are a lot of ADSL installations out there, Integrated Device Technology (Santa Clara, CA — 408-727-6116-www.idt.com) and VoicePump (Santa Clara, CA — 408-986-4320, www.voicepump.com) have partnered in the development and delivery of a voice-enabled ADSL reference platform, called the 79RP355V, for customer premises equipment applications that offers support for both data-only gateways and IADS.

Basically, all that's needed is enough bandwidth running in each direction to support quality voice service and data service running simultaneously — many DSL variants fill the bill.

Power Considerations

Power is also a concern when providing voice service. Many VoDSL service providers are required by the regulatory community to provide a minimum of one voice circuit in the event of a power outage at the customer's premises. This implies that the CPE, whether DSL modem, IAD or a small gateway, must be powered over the twisted-pair cable. However, for safety reasons, regulators have traditionally limited the voltage and current that can be delivered on the line.

If the customer's premise is located on a long loop, it's probable that more than half of the delivered power is dissipated in the resistance of the twisted-pair cable. Considering that power at the remote end of the line must be budgeted between the IAD, voice codec, and telephone ringer, it becomes clear why "span powering" places tight constraints on CPE power consumption.

Because of these power budget restrictions, the defining VoDSL enabling standard, G.SHDSL, has restricted the modulation technique and degree of DSP processing that can be employed.

The IAD Connection

Now that we've got the line in ship shape, introduced the primary CPE (the IAD), the power and the type of DSL is taken care of, it's time to dig deeper into the first (and sometimes the only) piece of equipment the customer encounters — the IAD. It's here that analog voice is first digitized at 64 Kbps using the *Pulse Code Modulation (PCM)* format, or perhaps compressed into a reduced format (typically to a 32, 12 or even an 8 Kbps bandwidth), and placed into the appropriate packet type (ATM, IP or Frame Relay). Digital voice from a PBX is already in PCM format, so the PCM conversion step is unnecessary.

VoDSL implementations can typically support from 16 to 24 voice conversations over a full-rate SDSL or G.SHDSL link running at 1.544 Mbps. The balance of the bandwidth is reserved for data communications purposes, ensuring that voice demand won't stall data traffic. However, should the entire 1.544 Mbps be in use for data communications (an unlikely scenario in most DSL environments), then the data traffic will be throttled back in favor of voice, but even under the worst circumstances the available data bandwidth won't fall below around 128 Kbps.

Encoding and Compression Techniques

The exact amount of bandwidth required for a voice conversation is not specified, so the specific encoding techniques and compression algorithms employed are selected by the manufacturer of the IAD and the CO equipment. Assuming that the approach uses ATM's AAL1 (or, more likely, AAL2) to maintain a high quality of service, and digitizes the voice using PCM at 64 Kbps per voice conversation (the so-called G.711 standard), there would be sufficient bandwidth available for 16 voice conversations and 512 Kbps left over for data. This approach is typical, if any implementation of such a new technology can be characterized as being typical. Standard voice compression codecs (or "vocoders") of the kind often used in Internet telephony can reduce the bandwidth needed for each voice stream to a value far less than 64 Kbps. Such vocoders make up the "G Series" of ITU voice compression standards and include G.722, G.723.1, G.726, G.729, G.729A, G.729B, G.729AB.

IP or ATM is typically used as the Layer 2 protocol in VoDSL implementations. At the customer premises, both voice and data equipment connect to the DSL circuit through a modem, bridge, router, or an IAD. At that point, outbound analog voice is converted to digital PCM format through a codec (coder/decoder).

If the transport mode is ATM, the PCM samples are formed into ATM cells using AAL1 or AAL2, depending on the vendor's specific implementation. In either case, the ATM voice cells take precedence over the IP-based packet data transmissions, which are generally segmented AAL5 cells (sometimes AAL3/4 cells).

A top-notch IAD can serve multiple functions, including those of a router and an

ATU-R (otherwise known as the "DSL modem"). The IAD serves as the interface between the DSL network service and the customer's voice and data equipment. Voice and data traffic is converted to IP packets or ATM cells and crosses a DSL link into the carrier's network. The IAD prioritizes the voice packets over data calls to ensure quality voice delivery and then sends the packets over the DSL line.

The IAD, to a large extent, determines which DSL features the customer can use, so it's important to ask your service provider in advance what IAD it offers. See chapter 6 for a more detailed discussion of IADs and other Customer Premise Equipment.

The PSTN's Role

Then the DSLAM at the central office comes into play to terminate multiple DSL lines and aggregate traffic from them. At the central office or other *POP (Point Of Presence)*, the service provider's DSLAM typically demultiplexes the voice cells (i.e. separates the data from the voice packets). The data is sent to a data network (most commonly the Internet), while the voice packets are then converted into a form suitable for circuit-switched calls by a voice gateway, which can convert voice cells back into a pure PCM byte-interleaved format (i.e. depacketized and converted to a standards-based format — GR-303, TR-08, or V5.X), presented to a Class 5 circuit switch as a standard 64 Kbps (DS-0) digital stream, and the switch then sends the voice over the PSTN. The Class 5 telephony switch not only provides the dial tone, call routing, and other services, but it also generates records used for billing.

Alternatively, if one wants the voice traffic to bypass the PSTN entirely, the public voice gateway could be an IP-based H.323 or SIP device rather than GR-303 compliant or similar device, and instead of passing it to a Class 5 switch, it could simply send the packets over a packet network (such as the Internet or a private network) to a remote voice gateway at the central office serving the destination. As for the data, it's most commonly carried as packets (these can be nestled in frames or ATM cell traffic) over the Internet or some data network and delivered to its ultimate destination, such as a corporate Intranet, perhaps through an ISP.

Ergo, VoDSL retains its identity as VoDSL only until it reaches the DSLAM, where it becomes either voice traveling over an ATM network or IP (or in some cases, over a Frame Relay network). If IP is used, a telephone call to order a donut and coffee from your downstairs coffee shop must, for example, link to the Internet and may be compelled to travel back and forth across the entire continent just to wind up a few feet from where it originated. Yet, placing a voice call via VoDSL over IP isn't really the same as placing a call over the Internet, which suffers from reliability and quality problems. In most cases the PSTN still handles the voice portion of the VoDSL calls — the signal just travels in a slightly different format among the various pieces of phone equipment. For most VoDSL users, it's difficult to tell the difference between VoDSL (whether it travels over ATM, IP or Frame Relay) and POTS.

Here's an illustrative scenario:

A company's computers and telephones are all tied to an IAD, which has been provided by the VoDSL service provider. The IAD takes the various signals and chops them into digital data packets (ATM or IP format) that can move together in a large bandwidth over twisted pair copper wires to the central office via DSL technology.

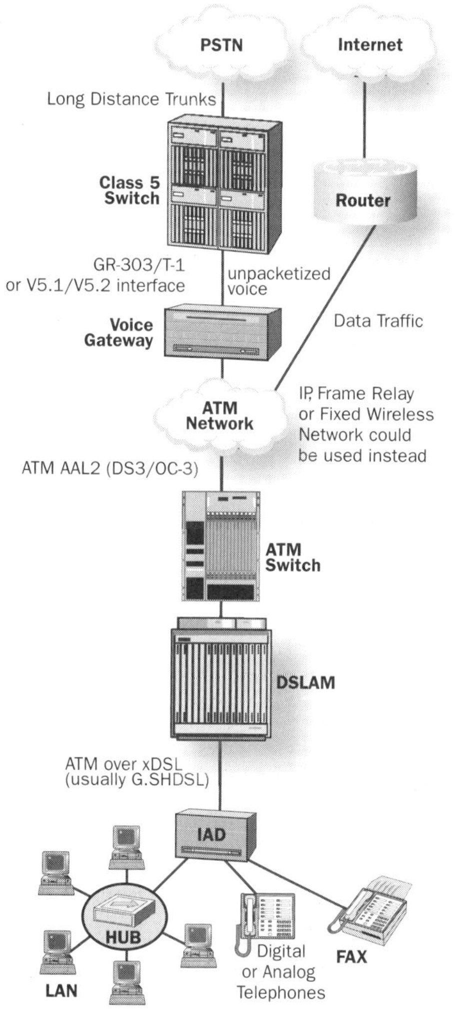

Upstream at the CO, the packets are sorted and routed into the appropriate data or public phone networks. Once they reach their final destination, they're reassembled into their respective voice or data components.

A service provider will generally use its VoDSL technology to maximize the capacity of the PSTN copper plant so as to cram megabytes of voice and data traffic into a single telephone line. But to achieve this, the customer and service provider's equipment must smoothly function in tandem to enable all of this occur, and they must work together to help preserve the quality to certain types of traffic. For example, the equipment can be programmed to give more priority to voice calls or to data as the need dictates. Voice is almost always given priority since it is a real-time application that demands less than 150 milliseconds (ms) of packet latency, less than 100 ms of packet jitter, and less than 1% of packets lost. Modern equipment is sufficiently flexible so that, for example, it could be configured to give priority to multiple telephone calls during the day, but at night, when all is quiet, to give priority to data enabling a network computer to quickly synchronize a database, update a partner's or other division's records, send reports, transmit large software files, etc.

Fig. 4.1. Typical VoDSL Solution.

Decisions, Decisions

VoDSL requires that everyone makes the right decisions, which might begin with what DSL variant to use to provision VoDSL.

But also, on the service provider end, VoDSL service requires a platform of DSL equipment, which is augmented with platform adaptations or additional equipment that can handle the requirements for voice services.

Then, on the customer premise end, VoDSL requires an integrated access device (IAD) in addition to the usual panoply of telephones, a private branch exchange (PBX), key system, fax, modem, and so forth.

The Value Proposition

Delivering voice services over DSL offers a lucrative opportunity for both service providers and their customers. The keys to success are the bundling of both data and voice lines and pricing flexibility.

The Service Provider's Point of View

Why would the service providers want to pursue VoDSL when most already supply T-1 broadband services? Because they want to expand the market to those customers who can't afford a T-1 — a market that represents about 50% of the total business market.

By offering voice on the same infrastructure that already provisions DSL service, the investment made in a data access network can be leveraged to provide a new source of revenue with relatively small additional costs incurred. The payback period for the necessary extra investment in IADs and the voice gateway is well under six months. Thus provisioning VoDSL is very compelling; especially when addressing the SME market which, in most developed countries, has five to ten times the potential of the larger enterprise market. But it's not a free ride; to support the equivalent of POTS quality service for VoDSL there are additional requirements on both the CPE and central office platform.

The service providers see VoDSL offering them various opportunities. First, for their existing DSL customers, it's a relatively simple process. All that's needed is to swap out the SME's DSL ATU-R for an IAD, then connect any existing PBX or key system and the business's LAN to the IAD, and then to the DSL line. Now the service provider is in a position to continue delivering DSL service *and* to offer multiple derived phone circuits — all over one set of copper wires.

Second, there are viable cross-sell opportunities. If the SME doesn't have a PBX or key system, then at the time of the equipment swap the service provider can offer the SME the value-added services normally supplied by a PBX (e.g. call waiting, call forwarding, etc.), a nice perk for the SME and additional revenue for the service provider.

Then there are the additional up-sell opportunities. Service providers can exploit the high bandwidth of the more symmetrical DSL variants, which can support even more telephone lines, while still leaving sufficient bandwidth for data applications.

Yet, from the service provider's perspective (especially data-centric service providers such as ISPs), ADSL variants can also offer considerable opportunities in terms of providing a source of incremental revenue and a way of reducing costs. ISPs typically already have a beach head in the residential and SOHO marketplace, voice service can provide even greater revenue opportunities. By virtue of owning the voice gateways that perform critical circuit-packet/packet-circuit translation, ISPs can become CLECs, communicating directly with ILEC Class 5 switches, and, in some cases, replace CLECs in the voice services revenue stream.

That's not all. VoDSL lets providers further improve efficiency and reduce costs for voice service by replacing edge access and core transport components of the legacy voice network with elements that offer lower capital expense and lower recurring costs.

Another advantage is the ability to bundle services onto a single delivery system. Bundling not only brings incremental revenue but also includes the efficiency of operating one network, and the ability to automate the systems that handle provisioning, billing and maintenance of that network.

The advantages of VoDSL don't stop there. Demand for extra lines is expected to continue to grow as modern households are required to make ever more simultaneous communications to an outside world of databases, services and intelligent appliances.

VoDSL is an ideal alternative to traditional pair gain (or digital leased line) systems. VoDSL is a "data ready" solution. When a customer becomes interested in using high bit rate data access, the operator can rapidly activate the service via "soft" provisioning (software programmed and controlled hardware). This approach not only provides superior value to traditional pair gain systems, but can also reduce the provider's operational expenses associated with spare parts inventory, training, etc.

Because the business case for VoDSL is so strong, many telecom service providers are moving quickly into VoDSL.

The Customer's Point of View

The customers, both business and residential, often under-utilize 50% or more of the bandwidth they currently have available to them. VoDSL, with prioritization of voice traffic, enables the customer to use a single line to connect both their voice and data traffic to the service provider's core network while preserving voice quality.

The SME market will soon be able to buy integrated, richly featured voice/data services in a way previously available to only the larger enterprise. But the large business and enterprise markets are also interested in VoDSL. For instance, one key service is VoDSL for businesses with remote offices. The fact that VoDSL can offer PBX remote functionality — so that branch office phones have the same look and feel as telephones in a main office — can be a compelling business case for VoDSL.

VoDSL can also be seen as an alternative to ISDN for the SOHO, high-end residential and standard residential subscriber markets. Even average residential customer might well find a second line a compelling asset at the desktop, where ADSL is terminated and the PC is located. This second voice line could, for example, be used to connect to a corporate PBX, while data connectivity to the corporate LAN using a PC is provided via ADSL.

Demand for extra lines is expected to continue to grow, as more modern householders find that they must provision ever more simultaneous communication needs. VoDSL in the home has many practical uses. For example, a household member's work extension can also ring at home; there can be additional lines for fax machines, for roommates, and for children's use. Most IADs allow the voice ports to be self-provisioning with the ability to turn them on or off to, for example, limit their use by children.

The economics and tight integration of a VoDSL solution enables the customer to purchase multiple telecommunications services from a single service provider. Benefits to the customer include a single bill with discounts spanning across the entire range of services, a single point of contact for installation, customer service, technical support, not to mention other value-added services.

In Summary

VoDSL offers a number of benefits for both the service provider and the customer.

First, a single DSL-equipped local loop can support integrated voice/data access, with up to 16 (or even 24 or more) voice conversations supported simultaneously.

Second, dynamic bandwidth allocation ensures that circuit usage is maximized at all times, with voice precedence being honored. The cost advantages of a single local loop (comprising a single pair) for both voice and high speed data are clear.

Third, voice features and feature access remain intact, with none of the technical problems associated with signaling and control issues between IP networks and the PSTN that plague pure VoIP. Typically, VoIP gateways are fundamentally LAN devices, connecting to the customer premise LAN and relying on existing non-voice/fax-enabled routers to provide WAN access. On the other hand, ATM-enabled IADs or even their Frame Relay equivalent, the "voice/fax-enabled router" or *Frame Relay Access Device* (FRAD), can connect directly to the WAN as well as legacy telephony equipment.

As VoDSL takes hold, there are still hard decisions that must be made. With the enabling technology continuing to evolve, the process of picking the right equipment is one of the more challenging decisions for service providers deploying VoDSL service. But, perhaps most important, is the choice of whether to use a centralized, circuit-switched (ATM) or a distributed, packet-based (IP) platform. Customers, when looking for a VoDSL service provider, must factor these issues into their decision-making process. We will closely examine these issues in the next chapter, as we look at the four most popular VoDSL network architectures.

CHAPTER FIVE

Four Roads to VoDSL

For VoDSL technology to become accepted by both providers and customers, it must interoperate with existing PSTN equipment. Indeed, VoDSL solutions must integrate seamlessly with the current infrastructure, since providers want and need to continue to offer value-added services such as custom calling and Caller ID and to keep leveraging their huge investment in the many switches housed in central offices. On the customer end the need is to quickly connect the customers' existing analog or ISDN phones, PBXs, fax machines, etc. to the network. This allows the customers to quickly resume their business activities — with the enhanced services and expanded number of channels afforded by VoDSL.

There are four broad forms of VoDSL (each necessitating a different degree of infrastructure replacement) that can provide this compatibility: *Broadband Loop Emulation Service (BLES)* also known simply as *Loop Emulation Service (LES)*; Voice over Multiservice Data Networks (VoMSDN), per-

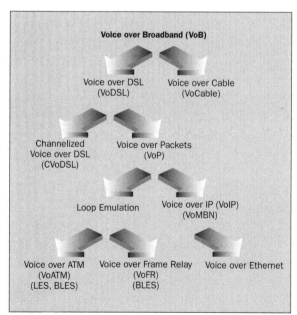

Fig. 5.1. Derived voice taxonomy.

haps better known as *Voice over Multiservice Broadband Networks (VoMBN)*; *Voice over Ethernet*, and *Channelized Voice over DSL (CVoDSL)*.

All the ways of creating "derived" or non-analog voice over broadband are related to each other, as indicated in Figure 5.1.

Although the IP-related VoDSLs are more compatible with so-called "next generation" networks that will supposedly replace the PSTN, nearly all forms of voice over broadband (and all forms of VoDSL) must interoperate in some way with the legacy PSTN since phone calls must always reach their destination, even if the called parties are using analog phones which connect to conventional Class 5 switches in the central office.

Overview of the Four VoDSLs

Having to choose from among four possible VoDSL technologies may at first appear daunting, but each one of these flavors of VoDSL has clear-cut advantages and disadvantages as are indicated in the following overview:

1. Broadband Loop Emulation Service (BLES) is the most popular form of VoDSL. Customers like it because they can simply plug their phones, PCs, fax machines, PBXs and other equipment into an *Integrated Access Device (IAD)*, which turns voice traffic into AAL2 type ATM cells and Internet data traffic into AAL5 type ATM cells. The subscriber's combined voice and data traffic is converted by the IAD into two ATM *Permanent Virtual Connections (PVCs)* — PVCv for voice and PVCd for data — for transport over the local loop copper pair to the DSLAM. The voice and data ATM cells are physically transported by some xDSL (typically ADSL or G.SHDSL) over the access loop to the DSLAM, which is usually the same as the kind formerly used for data-only DSL deployment. The DSLAM aggregates the voice and data traffic received from tens or hundreds of remote IADs. Less sophisticated DSLAMs then simply present these packets to an ATM switch which separates the data traffic (PVCd) from the voice traffic (PVCv); more sophisticated second-generation DSLAMs themselves support ATM QoS, *PNNI (Private Network to Network Interface)* and even a full ATM switch in its entirety. Once the data and voice cells are separated, the data cells then find their way to the Internet via a router or access server connection to an ISP. Meanwhile, the voice cells are sent to a voice gateway, which recodes the multi-voice channels carried in the PVCv, transforming them into the kind of *Pulse Code Modulation (PCM)*, *Time Division Multiplexed (TDM)*, T-1 or E-1 connections acceptable to a Class 5 voice switch. The voice gateway reverses this process for the PCM voice circuits passed from the voice switch towards the VoDSL subscribers. Since the ATM voice virtual circuits are either *Certified Bit Rate (CBR)* or *real-time Variable Bit Rate (rt-VBR)* connections, each voice call is guaranteed bandwidth to ensure full *Quality of Service (QoS)*.

Service providers with lots of legacy equipment love BLES because it enables them to continue to use their existing Class 5 switches in local exchange or end offices and allows them to offer the same Centrex and other normal phone services to customers over the xDSL link. To the customer's telephony equipment and the central office's Class 5 switch, the voice quality and calling features are the same as before, and nothing appears to have changed on the local loop (save for the fact that more voice channels are now available for use). That's why this particular VoDSL service is called "loop emulation" or "loop replacement." (BLES allows customers to use their legacy Frame Relay equipment too, though this is becoming rare.)

2. Voice over Multiservice Broadband Networks (VoMBN) also provides Centrex, custom calling, trunking and data services, but without the need for a Class 5 switch. With VoMBN, the customer still plugs all of his or her telephony equipment into an IAD, which acts like a proxy server and an IP gateway, sending IP packets (usually via ATM AAL5) to a regional multiservice IP or ATM network, which consists of any one of a number of possible combinations of so-called softswitches, call signaling and processing servers and assorted gateway devices. All of these devices and the network collectively replace the Class 5 (and ultimately Class 4) voice-switching network and deliver the typical *Custom Local Area Signaling Service (CLASS)* and Centrex functions familiar to the customer. Of course, the legacy PSTN, its Class 5 switches and the customers who use them don't magically disappear under such a scenario — VoIP calls made to them must traverse a VoIP-to-PSTN media gateway to reach the Class 5 switch on which the called party's local loop is terminated.

Because it's generally based on IP, VoMBN enables service providers to add new value-added services quickly (the customer can in some cases surf to a website and personally provision new services and lines without network operator intervention). VoMBN allocates bandwidth dynamically, so voice calls get priority; while data calls can claim the remaining bandwidth (dynamic resource allocation can increase the link utilization and decrease the need for network buffering).

VoMBN allows for fast and inexpensive provisioning of new customers and new phone "lines" at a customer's site, keeping costs down. The system also scales up much better than pure ATM systems, allowing the sharing of services and facilities among many customers. Indeed, if your business has no phone system but does have a packet data network, VoMBN will allow you to add voice services to it relatively quickly. VoMBN systems can be configured so that the signaling and call control platforms are distributed in the public network and are made redundant, providing for high availability. In general, IP-based VoMBN systems cost less and are more flexible than ATM-based BLES systems. Their only problem (which may now be in the past since a solution seems to have been found) has involved their inability to consistently provide the same toll-call quality of service that can be found with ATM-based VoDSL and the conventional local loop / PSTN.

3. Voice over Ethernet is a method of offering VoDSL where instead of having the IAD convert VoIP packets generated by users on an Ethernet into ATM or Frame Relay cells to travel over a xDSL line, IP-impregnated Ethernet itself can travel to the Internet or enterprise *Virtual Private Network (VPN)* through a common broadband medium, such as a single xDSL line, wireless device or cable modem. Just as BLES emulates the PSTN's local loop and extends the dial tone and calling features of the Class 5 switch over DSL to the customer's IAD, so too can an Ethernet interface transparently extend the bandwidth and features of an IP Wide Area Network over an xDSL circuit to a customer's DSL modem or IAD, and then to the customer's IP phones and other packet devices.

Since the vast majority of all business data traffic is Ethernet, why not leave it as Ethernet rather than experience delays as the Ethernet packets are converted to and from ATM, Frame Relay or even IP? This system provides a workable solution to this query. But, these "voice over IP over Ethernet over DSL" systems are relatively new and do not yet enjoy wide usage. Like other kinds of VoIP systems, quality of service can be something of a problem, since Ethernet treats voice packets the same as data packets, and so additional techniques (such as *DiffServ*, which marks packets so as to prioritize packet delivery) must be used in an effort to give voice packets high priority, ensure their smooth delivery and thus maintain the audio quality of voice calls.

In the case of connecting branch offices together with VPNs, VoIP calls traveling over Ethernet from one branch office to another are kept private by tunneling out through a secure IPSec-type "tunnel" to the customer's ISP and across the Internet to the far end (which is also private customer premise equipment at another branch office or teleworker home). At the far end the IAD, or software running on a PC, sorts out what's voice and what's data using something that prioritizes packet traffic, such as DiffServ. Detected voice packets can then be routed to a VoIP gateway and turned back into PCM/TDM voice channels if that's the way the call needs to terminate.

Ethernet-based systems may become more popular as 1 Gbps and 10 Gbps Ethernet lines are deployed in *Metropolitan Area Networks (MANs)*. A single, access-to-core Ethernet network could be created, though maintaining quality of service with such a "bursty" transmission scheme as Ethernet will always be a challenge. Advanced Ethernet services via an optical Ethernet network (Ethernet over Fiber) are gaining favor because they are relatively inexpensive but they do not yet threaten to completely usurp high quality ATM and Frame Relay services delivered via more conventional ATM / Frame Relay-based networks.

4. Channelized VoDSL (CVoDSL) is the least expensive (and least flexible, from the customer's point of view) form of VoDSL and is targeted at the residential market and perhaps very small businesses. Rather than carrying voice as a packet service (such as ATM), CVoDSL sends voice directly over the xDSL circuit using the old-fashioned telephone network concept of *Time Division Multiplexing (TDM)*. Originally championed by Aware Inc. (Bedford, MA — 781-276-4000, www.aware.com) under the name "Voice enabled DSL," CVoDSL allows for regular analog POTS service (just like non-voice enabled versions of DSL) and, in the case of ADSL, the bandwidth on top of the ADSL framing structure is split into two parallel transport structures: One is reserved for a specific number of derived TDM voice channels traveling over the physical DSL layer while the other is for dynamic ATM data traffic used for Internet access. CVoDSL reserves 64 Kbps of bandwidth for each voice line (just like a pulse code modulated DS-0 timeslot in a T-1), against 80 Kbps allocated under conventional uncompressed packetized broadband loop emulation VoDSL (or 32 and 40 Kbps, respectively, when using codecs enabling compressed voice). Thus, if one sends voice directly over DSL, one doesn't have to lose bandwidth by paying a "cell tax" to packetize voice. However, since each voice call is "nailed up" at full bandwidth, one can't squeeze 16 to 24 lines on a copper local loop, just four voice lines are possible (perhaps eight if some sort of voice compression is incorporated into the system). This is because non-packetized communications can't take advantage of silence suppression or any other bandwidth-saving trick — the voice just stays in constant bit rate format as it is transported over the DSL physical layer.

Service providers may find CVoDSL attractive, however, since it is easy to put into operation — none of the higher layers in the OSI protocol stack are used. But, at the same time CVoDSL systems cannot be assisted by any of the features provided by those layers.

Some proponents of rival systems see the CVoDSL's reintroduction of TDM into DSL local loop lines as a technological step backwards. Ironically, the argument for such a seemingly primitive system stems from the fact that modern, next generation network equipment at the service-provider end of DSL will accept multiple traffic protocols, so there's no longer a reason to use older DSLAMs that understand only ATM and must separate voice AAL2 packets from data AAL5 packets. Why continue, then (so the argument goes) to convert both voice and data into ATM at the customer premises? Even if one makes a voice over IP (VoIP) call on such a system, the rock-solid quality of CVoDSL's TDM over the local loop bolsters the VoIP call's quality of service the same way ATM does in loop emulation or VoMBN systems.

CVoDSL can also save costs in terms of management. ATM-based VoDSL connects users to the central office through *Private Virtual Channels (PVCs)*, with at least one PVC to every customer. CVoDSL, however, allows service providers to connect customers to a DSLAM directly without assigning channels, which somewhat improves scalability.

The biggest obstacle to widespread CVoDSL deployment may be the many deployments of "lite" or partial-rate asymmetric DSL (G.Lite) found in the residential market. Many such ADSL Lite installations average 384 Kbps downstream and 128 Kbps upstream, which means that the upstream bandwidth in particular isn't sufficient to carry multiple 64 Kbps voice channels. The ITU G.SHDSL standard (G.991.2) and ETSI SDSL standard (TS-101-524) have defined a dual-bearer mode which supports simultaneous CVoDSL and ATM transport for applications over symmetric DSL. Symmetric DSLs such as SDSL and G.SHDSL can accommodate more derived voice channels than G.Lite, but the downstream data bandwidth may be lower.

"Digital" Isn't always Digital

As we've seen, "digital" is something of a misnomer when describing Digital Subscriber Line (DSL) technologies. Some of them, such as HDSL and HDSL2, do indeed employ digital signaling techniques such as 2B1Q modulation (these and other similar techniques are the closest approximation to digital signaling that can be achieved by sending electromagnetic signals over a copper wire). Such truly digital DSLs are used to mimic and replace channelized DS-1 transport (T-1 and E-1 services) even to the extent that they can be used with channel banks and *Channel Service Unit / Digital Service Units (CSU/DSUs)* that multiplex digitized voice conversations, chopping up the payload into synchronous 64 Kbps time slots and mapping them to and from the DSL line (such DSLs don't allow for any empty bandwidth at the bottom of the spectrum for analog voice to co-exist with them on the line, so their voice services must be digital *derived voice* services). However, many DSLs, such as ADSL, use the same fundamental analog line coding techniques found in analog modems, albeit over a broader range of frequencies. For example, full-rate ADSL circuitry essentially comprises 256 modems (around 247 of which are actually used) that run simultaneously over different frequency bands, with each "modem" averaging a 32 Kbps transmission capacity. This kind of DSL service is for the most part simply a link or conduit to the public voice and data networks with no substantive switching or routing capabilities of their own, unlike an ATM, IP or Frame Relay network.

Chapter Five

In previous chapters we've looked at ATM and IP "networks." But protocols, such as those defined by the ATM and IP specifications, do not alone provide a network, because the protocols have to run over a physical communications "transport" link. Since DSL technology is as good a physical link as any, ATM and IP protocols can therefore be used to establish rules for transporting voice and data over DSL. Indeed, many times these protocols are used simultaneously in a hierarchical arrangement; for example, to provide a good quality of service, DSL lines can use ATM as the underlying data transport protocol beneath IP. Thus, protocols are usually arranged in layers with the resulting suite called a *protocol stack*.

To save less technically-oriented readers the trouble of reading the next section's more detailed examination of each of these four flavors of VoDSL, here is a short exposition of why you might want to choose one technology over another — or indeed why your service provider may foist a particular technology on you whether you want it or not:

Most of the vendors have adopted either ATM or IP as the transport mechanism for VoDSL. If you want rock-solid voice quality and dependability, the mature and reliable nature of ATM-based systems (BLES) can't be beat. They can be quickly configured since they take advantage of the same Class 5 switch and PSTN infrastructure your business was using before the advent of VoDSL — but now, of course, VoDSL enables more possible channels to the switch. ATM allows for easier service provider billing and maintenance than IP because of ATM's connection-oriented nature. ATM is capable of carrying both voice or voice over IP (VoIP). Indeed, there are myriad of ways to carry voice and data over ATM: Voice can be transported as either Voice over ATM or else as Voice over IP over ATM. IP data can be transported as IP over ATM or IP over PPP over ATM.

ATM-based systems still outnumber IP-based systems (though IP is catching up and will probably surpass ATM) and until recently ATM-based systems have been the preferred choice with most equipment vendors and service providers. Most of the continental network backbones are based on ATM, and so ATM-systems allow for the seamless integration of both the access network and the backbone.

Yet, there are downsides to ATM-based systems. While ATM-based IADs set up quickly and have come down in price thanks to continually advancing technology, the overall ATM environment tends to be more expensive than an IP-based environment. Someone must "pay" for ATM's high quality transmissions and efficient bandwidth management, which is provided by "intelligent" devices built into the underlying network. For it is these devices that make sure that voice cells arrive at their destination with little or no overall delay or variable delay (jitter), regardless of the infrastructure speeds and congestion levels. In the long run, the customer is also indirectly paying for the capital in deploying, and maintaining traditional Class 5 switches that are rigid and outdated and will have to be replaced in time.

Also on the downside is the fact that ATM-based systems such as BLES take the concept of an "access network" literally — they simply replace the analog local loop with multiple xDSL channels for voice and data so as to maintain the same feature set offered by the Class 5 switch that previously connected to the customer premise. The IAD thus acts merely to relay the existing finite number of service options offered by the service provider. Such ATM-based IADs are more focused on deploying or terminating voice-only or voice-and-data services as cheaply as possible, partly in an effort to escape ATM's early reputation as being extremely expensive. Such a simple ATM-based "loop replacement" scenario is okay for the delivery of local calls, but is not ideal for long-distance calls or "on net" calls (calls between two corporate locations over a VPN, for example). It's also an unsuitable (less efficient and more expensive) arrangement for doing things that IP is good at, like handling connectionless and large-scale multipoint to multipoint applications, such as conferencing.

To summarize: ATM-based VoDSL is reliable, has great voice quality, but is more expensive than VoMBN, is not terribly scalable and offers few features in addition to those delivered by the PSTN's local Class 5 switch. It's great for immediately getting a business up-and-running in multiple channel VoDSL, but may cost you in the long run, since you'll probably eventually end up moving to a more IP-centric system.

On the other hand, with the VoMBN model, the Class 5 switch is eliminated and "intelligence" in the network is pushed out to the edge — the customer IAD — and so the IAD is imbued with additional and unique multiservice functionality. Businesses both large and small find this attractive because their requirements change and IP service providers offer services not necessarily found in the PSTN world, so businesses would ideally would like to invest in a single platform that can be upgraded in size and functionality with few or any limits.

IP-based VoDSL systems tend to be less expensive than ATM-based ones, and IP has become the most popular transport protocol in history, thanks to the ascendance of the Internet. Also, with the preponderance of IP traffic at the customer premise and in the core of the network, and the emergence of softswitches for Class 5 switch replacement, the evolution of public networks to a converged IP infrastructure is considered by many to be inevitable.

Note: Nortel offers a VPN-like system where Ethernet, not ATM, is used to transfer IP packets over xDSL, which we list here as our "third flavor" of VoDSL but which is technically a special subset of VoMBN.

"IP-based" VoDSL is often a misnomer, since with most systems the IP packets are generally transferred via ATM AAL5, making them hybrid systems. The problem with IP-based VoDSL is not so much with the DSL technology itself; instead it's more about what happens when the IP packets leave the DSL access network and enter the much-hyped, IP-centric "next-generation" network.

Unfortunately, IP is designed to provide only a "best effort" (not guaranteed) connectionless service. Its variable packets imply a higher latency and are not suitable for transferring voice conversations, since the IP environment offers no inherent QoS guarantees. Various protocols and approaches have been developed to fix this for voice and video, such as Intserv, Diffserv, RSVP, and MPLS but these solutions have not been universally accepted (let alone universally deployed) and so QoS concerns still abound.

This is why many multiservice broadband networks are still relying on a core ATM (or a hybrid ATM / IP) switching structure instead of IP. Thus, although the term "multiservice broadband network" always conjures up the visage of "IP" in the public's mind, MBNs do indeed live up to their "multiservice," name, uniting connectionless IP (Layer 3) and connection-oriented (Layer 2) Frame Relay and ATM technologies.

For example, when Lucent Technologies announced in 2000 that it would supply "multiservice core switches" for Yunnan Telecom's multiservice broadband network backbone in China, the switches delivered under the contract were Lucent GX 550 smart core ATM switches, which are high-capacity switching systems that, because of ATM, provide end-to-end quality of service and the necessary capacity and performance capabilities for simultaneously transporting multiple carrier services such as private line, ATM, Frame Relay, IP and voice. Lucent also deployed at the core and/or edge of Yunnan's network, its multiservice Wide Area Network or WAN switches, the CBX 500, and B-STDX 9000. To tackle any IP quality of service issues, Lucent also supplied Yunnan with their version of MPLS technology, IP Navigator MPLS, enabling a range of IP services from Lucent's multiservice ATM and Frame Relay switches.

So, if you want to be at "the cutting edge" of things and want lots of features, with easy provisioning of new services and users, VoMBN is for you. It appears to be way to the future of communications. Just make sure you sign some sort of Service Level Agreement (SLA) guaranteeing a specified level of voice and data transmission quality.

As for Channelized voice over DSL (CVoDSL), it's inexpensive, reliable and has high voice quality (like ATM or even POTS). At the moment this technology doesn't offer as many voice channels as ATM or IP-based systems, but then it was originally designed for the residential market anyway, so consider it only if you have a large family or SOHO.

The remaining sections of this chapter provide a study of the four VoDSLs in greater detail.

Broadband Loop Emulation Service (BLES)

Voice traffic at the customer premise is handed over to an *Integrated Access Device (IAD)* where it is compressed, echo is removed, and other voice-processing functions are done. After this the voice data is formatted into packets. There still remains, however, the challenge of sending voice encoded in this manner over the PSTN via the central office's Class 5 switch, which is expecting a *Time Division Multiplexed (TDM)* T-1 or E-1 connection from a concentrator or *Digital Loop Carrier (DLC)*, not statistically multiplexed, packetized voice signals from an IAD that resemble the output of a LAN. An equivalent problem at the local exchange is how to extend the services of a Class 5 telephony switch over a packet-based broadband network to the customer.

The solution to this problem is specified in the *Broadband Loop Emulation Service (BLES)* standard that was first defined by the ATM Forum's (www.atmforum.org) reference document entitled "Voice and Multimedia Over ATM - Loop Emulation Service Using AAL2," (AF-VMOA-0145.000 dated July 2000) and by the DSL Forum (www.adsl.org) in their document entitled "Requirements for Voice over DSL Version 1.0" (TR-036) that appeared in August 2000. BLES was the first open standard to promote interoperability for VoDSL. BLES also demands the least overhaul to add VoDSL service to an existing PSTN system consisting of Class 5 Local Exchange switches.

BLES turns the VoDSL packet network link into a sort of invisible extension cord between the output ports of the customer's telephony devices and the corresponding voice ports on the Class 5 switch, tricking the equipment at both the customer premise and central office ends into thinking that they are still communicating over an ordinary analog local loop instead of an xDSL loop running ATM. This allows existing Class 5 Local Exchange switches to provide conventional phone service transparently to VoDSL customers, so voice calls can therefore be set up using the normal DTMF touchtones and call control signals. Aside from standard phone service, the BLES service transparency requirement for the central office includes emulation for residential service, PBX trunks, or off-premises PBX extensions, and support of business services such as Centrex and *Custom Local Area Signaling Service (CLASS)*, which includes items such as Caller ID, Calling ID Blocking, Distinctive Ringing/Call Waiting, Selective Call Forwarding, as well as other custom calling services and analog phone, fax, and modem services. At the customer premises end, BLES networks emulate telephony channels that interoperate with and support the full functionality of all existing Telcordia and ETSI-compliant POTS and ISDN telephony devices, such as analog phones, ISDN terminals, dial-up modems, fax machines, PBXs, key systems, TTY terminals, and Point-of-Sale (POS) devices. All of these devices must be able to operate with the same service parameters as under the PSTN for end-to-end delay, echo cancellation, dial tone delay, hook-flash signaling, call teardown delay, etc.

In order for the BLES architecture to emulate the local loop, two functional concepts are added to a conventional DSL system architecture to adapt it to the PSTN: 1) the access network or *Customer Premise Interworking Function (CP-IWF)*, and 2) the *Central Office-Interworking Function (CO-IWF)*. "Interworking" means that these devices interwork with both voice and data. Basically, the CP-IWF and CO-IWF are just fancy names for signaling and bearer channel translation devices that must be situated in the customer premise and at the central office. Their job is to make sure that the existing equipment at both ends of the local loop will continue to communicate with each other just as if there's still an ordinary subscriber local loop connecting them.

Such IWF equipment can be standalone devices or integrated into the traditional equipment. The first generation of such devices tended to be standalone since no one really wanted to replace a lot of equipment at either end of the loop.

At the customer premises, all telephony devices are plugged into the IAD which performs the CP-IWF, providing DSL modem and telephony interfaces, data interfaces such as USB, Firewire or 100Base-T Ethernet as well as associated bridging / routing functionality (moreover, the development and availability of digital signal processor integrated circuits incorporating BLES capability have even made possible the inexpensive integration of the IAD's CP-IWF functionality with a DSL modem).

When a phone connected to the IAD goes off hook, the IAD notifies the voice gateway, which in turns uses GR-303 to negotiate with the Class 5 switch to use a DS-0 channel. This DS-0 is just one of many DS-0s which are physically connected between the Class 5 switch and the voice gateway. Once the DS-0 is granted by the Class 5 switch, the voice gateway cross-connects the DS-0 to the ATM virtual circuit, and the user of the phone at the IAD hears the dial tone generated by the Class 5 switch.

The IAD digitizes and packetizes the voice signals, places them on the same line with data packets from the PCs, and sends them all along with certain signaling events over the loop to the central office. In the case of VoDSL, the interface between the IAD and the DSLAM or ATM-based edge switch is generally a flexible, symmetri-

Note: In a non-VoDSL DSL system, the DSLAM doesn't need this level of intelligence since a splitter is used to separate analog voice from data packets; in a VoDSL facility equipped with an older, less sophisticated DSLAM, the packets are simply aggregated from subscribers and forwarded to an ATM switch, which has the intelligence to identify and separate the voice packets from the data packets.

cal DSL such as G.SHDSL. When the packets reach the central office, the DSLAM must separate voice packets from the data packets. Once the voice packets have been extracted, they are sent over an ATM circuit (usually with a DS-3 capacity, around 45

The GR-303, TR-008, V5.1 and V5.2 Standards

Bellcore (now called Telcordia) produced the first General Recommendation 303 (GR-303) specification in 1995, but the first document describing it was called TR-303, and a few people still refer to the standard by this older name. An older related specification is TR-008 (Bellcore TR-TSY-000008 Issue 2, Revision 1, September 1993) which describes a digital interface between the old SLC-96 digital loop carrier system and a local digital switch.

Although many people think of the GR-303 standard as a T-1 interface to a Class 5 switch, the GR-303 specification actually describes an access protocol that runs between the Class 5 switches — which the standard refers to as *Local Digital Switches (LDSs)* or *Integrated Digital Terminals (IDTs)* — and the *Access Equipment* (also called *Remote Digital Terminals or RDTs)* which provide network access for the subscribers. In particular, GR-303 encompasses the functionality of a *Digital Loop Carrier (DLC)* system which concentrates analog telephone traffic and feeds it to the Class 5 switch as T-1s. In American National Standards Institute (ANSI) markets, primarily North America, DLCs terminate and aggregate analog subscriber loops, converting them to G.711 DS-0 channels, multiplexing them into T-1s and sending them to the Class 5 switch. In the case of VoDSL, the GR-303 protocol is used by the voice gateway to communicate with the Class 5 switch to dynamically negotiate usage of voice circuits (DS-0s) that are connected between the Class 5 switch and the voice gateway. GR-303 is required to dynamically negotiate DS-0 usage because there are typically more DS-0s at the edge of the network than there are DS-0s between the Class 5 switch and voice gateway. Since not all phones are in use at all times, GR-303 allows the maximum usage of expensive Class 5 DS-0s. The ratio of DS-0s at the edge of the network to DS-0s connected directly to the Class 5 switch is known as oversubscription. A common oversubscription ratio is 4:1.

Under GR-303, T-1 circuits exiting the switch go directly to the RDT equipment, without the need for additional equipment in the central office. GR-303 runs on dedicated redundant control channels also known as *Timeslot Management Channels (TMCs)* and *Embedded Operation Channels (EOCs)*. TMC data links are for timeslot allocation and deallocation during the assignment and management operations on the LDS interface (signal-

Mbps) to a voice gateway (performing the CPO-IWF function) that translates the voice packets back into a T-1 or E-1 form palatable to the Class 5 switch's existing GR-303 or V5.x interface. Meanwhile, the data packets are sent to an ISP's router and

ing bits are used to indicate call control for such things as call setup and tear down). The EOC leads back to the RDT equipment for remote management operations such as maintenance and alarm surveillance.

The resulting T-1 circuits are configured for *Extended Superframe Format (ESF)* framing, and usually have *Bipolar with 8 Zero Substitution (B8ZS)* encoding enabled to accommodate the "ones density requirement" in the public network. This involves inserting one of two special violation codes for strings of eight consecutive zero voltage states, the intentional bipolar violation codes being inserted in bit positions 4 and 7 of the datastream line encoding. The first two T-1 circuits each carry the TMC and EOC for redundancy. The EOC is carried in timeslot 12 of the first and second T-1 circuits, and the TMC is carried in timeslot 24 of the first and second T-1 circuits.

GR-303 allows for expandability from two to 28 T-1 circuits that can carry up to 668 channels simultaneously. When concentration / oversubscription is taken into account, thousands of subscriber channels can be handled. Subscriber lines can even be ISDN circuits (both BRI and PRI) and multiple *Interface Groups (IGs)* are supported, so that the remote equipment can simultaneously interface to multiple switches.

GR-303 had its foundation in the old "slick" SLC-96 mode 2 specification, but the two standards differ in several respects: GR-303 is expandable, whereas SLC-96 is fixed at two T-1 circuits and 96 subscriber channels. The GR-303 protocols emanate directly from the switch, whereas SLC-96 needs additional equipment in the CO. GR-303 also has a comprehensive EOC which allows an operating company to do Operation, Administration, Maintenance and Provisioning (OAM&P) remotely, whereas SLC-96 is limited in its capabilities. Finally, GR-303 has continual redundancy, whereas SLC-96 has an optional back up scheme

In European Telecommunications Standards Institute (ETSI) markets, V5.1 and V5.2 are specifications that define the communication interface between the Access Network (AN) and the Local Exchange (LE) switch, particularly for a DLC system that handles E-1 circuits. E-1 standards use HDB3 (High Density Bipolar 3) line coding instead of B8ZS. Both the V5.x and GR-303 interfaces were designed to provide for on-demand concentration, digitization and multiplexing of voice traffic to minimize the number of interfaces leading to the local exchange's Class 5 switch.

onto the Internet or private corporate data network.

The BLES concept is not only used with xDSL to connect a customer's premises and a service provider or service node (such as a Class 5 switch) attached to the PSTN, but is also used with other transport techniques such as broadband wireless and *Hybrid Fiber Coax (HFC)*, a system where optical fiber is used for backbone distribution and terminates in a neighborhood remote unit where an optoelectric conversion takes place, allowing the signal to pass on to coax cables which carry the data the last leg to the individual business, residence, dormitory room, etc.

ATM or Frame Relay for Loop Emulation?

When creating a VoDSL system based upon broadband loop emulation, the first enormous question to be answered is — what kind of packet protocol should be used to transport voice traffic over the DSL service between the customer premise and the central office?

The broadband loop emulation service recommendation allows for two possibilities: One part of the specification defines ATM, an approach that's generally most cost-effective where the service provider already offers traditional voice services over an existing switching network, and wishes to offer revenue-producing, churn-reducing voice services that use the existing infrastructure. The other transport environment that can be used is Frame Relay, based on the Frame Relay Forum standard "Voice Over Frame Relay Implementation Agreement" (FRF.11).

While ATM has the highest QoS reputation, Frame Relay also has been used for VoDSL since many customers have long owned this inexpensive equipment. As illustration: Frame Relay has often been used for low cost LAN-to-LAN connections, tying together multi-protocol networks and devices, and connecting PBXs and telephony devices in enterprise branch offices using voice-enabled *Frame Relay Access Devices (FRADs)*.

In theory *Voice over Frame Relay (VoFR)* can be an efficient transport protocol on the access network, since it can integrate voice, data, and fax over a single packetized link. Unfortunately, Frame Relay is a Layer 2 protocol and as such does not guarantee end-to-end frame/packet delivery, although Frame Relay does have a primitive traffic control / service level agreement parameter called the *Committed Information Rate (CIR)*. Securing high CIRs from a carrier can be expensive; they are usually set well below the possible speed of the link since traffic can "burst" above the CIR. Any packets that are in excess of the CIR are marked as "Discard Eligible" — congestion control is achieved by simply discarding frames / packets, and during periods of high network congestion, these packets will be disposed of immediately. Terminal equipment can detect this condition and retransmit frames/packets accordingly, but there is no specific provision to support the notion of "priority" traffic (voice is a real-time phenomenon demanding high priority packet delivery).

Modern IADs are more sophisticated than old, single-service IADs, such as those that handle merely a voice-to-Frame Relay conversion. Instead, the newer, more sophisticated IADs act as multiservice concentrators, which funnel in data from your LAN and voice from your PBX (over either T-1 or analog links) and transmit them as either ATM cells or frame relay packets. Many of the high-end devices allow you to use ATM and still keep your existing outdated FRADs, since they support such ATM-Frame Relay internetworking standards as FRF92.08 (the "Frame Relay Network-to-Network Interface Implementation" that describes how to map Frame Relay onto ATM).

Still, even under the ATM transport environment, BLES was originally defined to use a *Frame-based User-Network Interface (FUNI)* as its framing option. This is a frame format for access to ATM networks, very much like Frame Relay but with a few additional bits reserved for mapping into the ATM control bits in the cell format. Both FUNI and the Frame Relay format can pass through a frame switch.

ATM-based VoDSL

If a DSLAM is not located near a switch (for example, if the Class 5 switch is owned / leased by a CLEC and the DSLAM by a carrier, or *vice versa*) then the DSLAM must communicate over a high bandwidth, robust, high quality network to reach the voice gateway and the Class 5 switch. ATM has been used for years for universal trunking of voice and data in the long haul network backbone, since network operators did not relish the thought of maintaining separate backbone networks for circuit switching, Frame Relay, ATM, IP, X.25, etc. And, since nearly all of the first generation of DSLAMs used ATM transport (before the rise in popularity of the Internet and IP), it made sense that early VoDSL installations would be based upon a form of ATM somewhat similar to what is used for voice over ATM in the backbone or long haul part of the network. Thus, the most popular form of VoDSL could actually be termed Voice-over-ATM-over-DSL.

In order for voice traffic to be carried over an ATM network, it must be adapted to ATM transport, meaning that, among other things, user application packet streams must be inserted into the 44 or 48-byte payloads of 53-byte ATM cells, then extracted again and reassembled at the destination. Such processes are taken care of by the ATM Adaptation Layer (AAL). As we saw in Chapter 2, several AAL types have been standardized by the ITU-T for different kinds of voice and data applications: AAL1, AAL2, AAL3/4 and AAL5. The ATM version of the broadband loop emulation service is based largely on the later versions of the ATM Forum's Voice Telephony over ATM (VToA) standard, which recommends ATM Adaptation Layer 2 (AAL2), a protocol extension specifically designed to provide real-time voice service, and an underlying infrastructure that supports the QoS functions necessary for customer satisfaction.

Prior to the appearance of DSL, older techniques for sending what was then called Voice over ATM (VoATM or VoA) relied on ATM Adaptation Layer 1 (AAL1), which employs byte-interleaved multiplexing, a form of TDM over ATM. In 1995 the ATM Forum came up with the term *Circuit Emulation Service (CES)* to describe AAL1 virtual circuit trunk connections that emulate the characteristics of the standard uncompressed 64 Kbps DS-0 digital voice channels (detailed in their recommendations AF-VTOA-0078 CES and AF-VTOA-0085.000).

CES allows a whole T-1 or T-3 to be mapped statically to an ATM *Virtual Channel Connection (VCC)*. Thus, CES is based on converting T or E carrier circuits into a stream of ATM cells. AAL1-based CES can be run in two modes: unstructured mode, where the whole T-1 can be mapped to an ATM VCC; and structured mode, wherein customers may manually select particular individual DS-0s of a T-1 / T-3 span to transmit. The emulation reproduces a real-time, constant bit-rate, dedicated bandwidth *Pulse Code Modulation (PCM)* TDM circuit over an ATM network — for this reason network providers often refer to this service as *Constant Bit Rate (CBR)* service. The AAL1 service class thus supports the highest quality VoATM service, which most resembles the kind of service found in the circuit-switched PSTN (AAL1 illustrates how ATM, although based upon cells, can nevertheless blend both circuit and packet communication mechanisms).

Moreover, AAL1 generally uses ATM to multiplex data and voice on top of an optical fiber link such as a SONET / SDH path; AAL1 requires a link that supports isochronous network clocking (such as SONET) because it relies on timing synchronization between the source and the destination. This further adds resiliency, for SONET / SDH can recover from transmission errors in just 50 microseconds (and from a loop cut in under 50 milliseconds) — this is particularly important for voice if an entire T-1 or E-1 circuit containing many multiplexed voice calls or 64 Kbps circuit data traffic are sent from one clocked PBX to another clocked PBX, since the rate of sending and receiving T-1 / E-1 frames must remain exact, or callers could suddenly find themselves talking over the wrong voice channels.

Of course, when voice of any sort is transported across a network, it is important to synchronize the data that is transmitted from the speaker to the listener. This can be achieved by such standard mechanisms as *Adaptive Clocking* and *Synchronous Residual Time Stamping (SRTS)*, a clock recovery technique in which difference signals between source timing and a network reference timing signal are transmitted to allow reconstruction of the source timing at the destination. In SONET/SDH networks, one node, called the clock master, provides a time reference to the other, called the slave. Somewhere in the network there is at least one extremely accurate primary reference clock, with long term accuracy of one part in 100,000,000,000. This node, the accuracy of which is called *stratum 1*, provides the reference clock to secondary nodes with *stratum 2* accuracy, and these in turn provide a time reference to *stratum 3* nodes. This time synchronization hierarchy is crucial for the proper functioning of the network as a whole. However, these mechanisms are effective only for master-slave environments or point-to-point communication. Owing to the availability of the global timing standards, it is easier and more practical to adopt an externally synchronized model in multi-point services. In an externally synchronized model, each node is synchronized to a single external clock source.

AAL1 can handle synchronization and cell delay jitter (resulting from unequal packet delay times) along with cell loss and wrong cell insertion. The ATM layer takes care of the traffic management and switching functions necessary to bring data efficiency up to acceptable levels, but it does so at a high cost, with significant complexity, limited scalability and considerable inefficiency owing to ATM-related overhead. Each 53-byte ATM cell has a 5-byte header and 48-byte payload. The use of AAL1 takes up one additional byte for the overhead out of the available 48 bytes for payload. Consequently, there are 6 bytes (5-byte header + 1) of overhead for every 47 bytes of payload. This additional bandwidth wasted on overhead processing is often referred to as a "cell tax," which in AAL1's case is about 11.5%. Thus, in order for AAL1 to carry a standard 64 Kbps DS-0 digital voice channel, 71.3 Kbps of bandwidth must be used in the network. To help conserve bandwidth normally wasted by AAL1-based voice over ATM trunking, an AAL1 *Dynamic Bandwidth Circuit Emulation Service (DBCES)* was developed that still uses substantial 64 Kbps certified bit-rate DS-0 circuits, but allows the pipe's capacity to vary in 64 Kbps increments as a function of the number of active voice circuits. (The approved specification appeared in the ATM Forum document AF-VTOA-0085.000 entitled "(DBCES) Dynamic Bandwidth Utilization in 64 KBPS Time Slot Trunking Over ATM-Using CES" published in July 1997.)

Almost at the same time DBCES appeared, the ATM Forum unveiled its initial specification for Voice and Telephony over ATM (VTOA) which defined how one could do TDM / ATM internetworking over a public or private network, at last making possible end-to-end VoA calls using AAL1 CES.

Still, AAL1 l had too much going against it in terms of cost and efficiency. For these reasons, when it came time to pick an ATM service class to handle the voice part of VoDSL, AAL1 had fallen out of favor, having been used primarily not for individual voice connections but for trunking between core switches only, in particular as a private line service supporting business enterprise networks, capable of moving entire T-1 / T-3 transport circuits transparently over the ATM network. AAL1 CES allowed enterprises to interconnect their existing private branch exchanges (PBXs) transparently and/or to interconnect their LANs.

AAL2 for Voice Packets over Loop Emulation-based VoDSL

Although it's possible to use AAL1 to transport voice over DSL (some IADs offer AAL1 as an option), all DSL and ATM standards bodies now favor the more efficient AAL2 — which is sometimes called Voice over AAL2 (VoAAL2). Whereas AAL1 was designed to handle wasteful, uncompressed (G.711) voice applications, AAL2 was designed to reduce the packet delay incurred in highly compressed voice applications while maintaining the bandwidth efficiency gains by utilizing all the payload available in each ATM cell.

The DSL Forum and the ATM Forum both developed specifications based on the BLES architecture, which uses AAL2 for voice bearer channels and signaling. The DSL Forum's original reference document is TR-036 "Requirements for Voice over DSL," dated August 2000, and the ATM Forum reference document is "Voice and Multimedia Over ATM - Loop Emulation Service Using AAL2," AF-VMOA-0145.000, dated July 2000. Any exceptions in compliance to AF-VMOA-0145.000 are documented in the following superseding interface specifications: DSL Forum's document TR-039, "Addendum to TR-036 Annex A; Requirements For Voice Over DSL, Version 1.1," dated March 2001, and "OpenVoB D1 Model Implementation Guide" (VoDSL GR-303 LES-CAS Interoperability Guide), dated March 5, 2001. The latter document defines an architecture by the OpenVoB or Open Voice over Broadband consortium (www.openvob.org) which was first posited in an earlier document entitled "Voice Over Broadband: Implementing Standards Based, Phone-to-Phone Interoperability," published in August 2000. The OpenVoB consortium has formulated several voice over broadband architectures, some similar to that described by the DSL Forum and based on additional standards existing at that time. In the document the OpenVoB refers to the ATM Forum's AF-VMOA-0145.000 document mentioned previously, as well as CableLab's PacketCable Specification 1.0, and residential media gateway control protocols, such as MGCP and Megaco / ITU H.248.

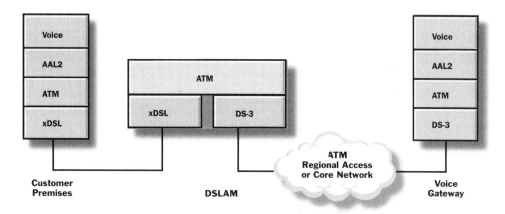

Fig 5.2. Protocol stacks for VoATM AAL2.

AAL2 consists of a *Common Part Sublayer (CPS)* and a *Service Specific Convergence Sublayer (SSCS)*. The ITU-T's AAL2 CPS recommendation (I.363.2) was approved in September 1997, while the AAL2 SSCS consists of two component standards: Segmentation and Reassembly (I.366.1) approved in June 1998, and Trunking (I.366.2) approved in February 1999.

The AAL2 protocol's features include the following:

• Definition of the packet header for voice packets that can be used for other real-time media, since it includes a way of identifying media content, i.e. compressed voice, uncompressed voice, etc., as well as a means to mix encoding schemes in the same virtual circuit. This allows AAL2 to do things such as changing encoding on-the-fly if a fax is detected signaling on the line.

• Support for the transfer of short or variable length packets containing uncompressed G.711 PCM or compressed G.726 *Adaptive Pulse Code Modulation (ADPCM)* voice payload, signaling control information, etc. The PCM voice encoder is the same one used by the PSTN to encode your analog voice calls into digital ones, except under VoDSL the IAD does the digitization right at the customer premise, just like an ISDN phone would.

• Allows for both standard uncompressed G.711 PCM and compressed G.726 ADPCM voice coders to exist on the same virtual circuit to improve bandwidth efficiency. According to the BLES parent document TR-036, if no specific voice compression codec is specified, voice channels default to standard G.711 "compression" (which is a standard 64 Kbps digitized channel and is not really compressed at all). ADPCM compressed voice has a default 32 Kbps bandwidth. So-called encoding delay occurs as a result of the encoding of an analog signal to a digital form. Under the

G.711 standard, PCM-encoded voice samples are transmitted at the rate of 64 Kbps, which means that it takes around 6 milliseconds to fill the entire 48-byte payload of an ATM packet. Encoding delay is directly proportional to the level of voice compression employed; greater compression implies more processing time and hence greater delay in packet delivery. The encoding delay is heavily dependent upon the level of voice compression because an ATM packet's length is fixed. The delay in packetization can be reduced either by partially filling packets or by multiplexing several voice calls into a single ATM *Virtual Circuit Connection (VCC)*. The voice channels must support both mu-law and A-law encoding and they must be selectable as befitting the country or local operating environment. Both PCM and ADPCM offer voice quality equivalent to that of a POTS line. Both G.711 PCM and G.726 ADPCM have *MOS (Mean Opinion Score)* ratings of 4.0, which means that the perceived voice quality is indistinguishable from that of a POTS line.

- The ability to accept and transport fax and modem transmissions with little or no problem, since the G.711 and/or G.726 encoding of the channels it transports also provide nearly transparent fax and modem support. G.711 passes both high- and low-bit-rate fax calls without alteration, while G.726 passes low-bit-rate fax calls and high-bit-rate fax calls that are preceded by a 2100 Hz echo canceller disabler tone,

The Mean Opinion Score

For assessing speech quality over telephony, the ITU recommended rating scale for both listening-only and conversation tests is the five-point Mean Opinion Score (MOS) category scale, which ranges as follows:

1. bad
2. poor
3. fair
4. good
5. excellent

Listening-only tests can also be assessed via the listening effort scale, which supposedly measures the effort required to understand the meaning of sentences. It progresses like this:

1. no meaning understood with any feasible effort
2. considerable effort required
3. moderate effort required
4. attention necessary; no appreciable effort required
5. complete relaxation possible; no effort required

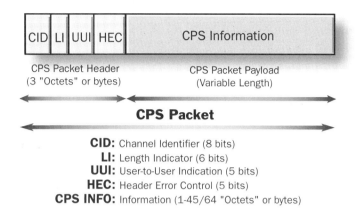

Fig. 5.3. AAL2 CPS Packet Format.

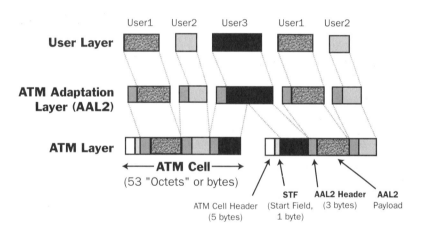

Fig. 5.4. AAL2 protocol in action.

which passes as an in-band signal. If G.726 is active and this tone is detected, system logic can be incorporated that allows falling back to G.711, which passes both high- and low-bit-rate fax calls. The same scenario is true for modems.

• Allows packets with variable inter-arrival times to accommodate packets from different media sources, e.g., different rate codecs or silence suppression descriptors.

• Allows packets to cross ATM cell boundaries to more efficiently pack the payload of each ATM cell, thus reducing the "cell-tax." Large AAL2 packets or datagrams of information can be easily divided into small 53-byte ATM cells and then reconstructed at the destination. This so-called *Segmentation And Reassembly (SAR)* function, a process also encountered in SMDS and X.25 networks, employs a 3-byte *Common Part Sublayer (CPS)* header and a 1-byte *Start Field (STF)*. An 8-bit *Length Indicator (LI)* field in the CPS header identifies the length of the packet payload while a 5-bit *User-to-User Indication (UUI)* field is responsible for linking user services or groups of services with the *Service Specific Convergence Sublayer (SSCS)* and the 5-bit *Header Error Control (HEC)* field. To avoid audio distortion, packet reconstruction must be done in real-time, which also means that packets carrying voice traffic should be transmitted from the origin to the destination in real time. If there is a delay in the transmission of packets, the SAR function might not have any data to process. This is called an *under-run*. An under-run results in annoying gaps in conversation. These gaps can be prevented by accumulating arriving cells in a buffer queue prior to packet construction by the SAR function. To avoid an under-run event, the buffer size must exceed the maximum predicted delay. Each packet that arrives at the emulated circuit's line rate traverses through the buffer, which results in a buffer delay that is a function of the size of the buffer. Therefore, to minimize the network delay, the *Cell Delay Variation (CDV)* needs to be tightly controlled. The VoDSL architecture references several voice quality standards such as G.114, which specifies end-to-end one-way delay for an international connection of less than 150 milliseconds as acceptable for most user applications.

- Minimization of the jitter of partially filled cells and guaranteed real time response.
- Allows multiple connections to be multiplexed on one virtual channel. Because of the statistical multiplexing nature of a packet-based ATM network, the circuit paths in an ATM network truly exist only when there is actual traffic flowing through them; hence the term virtual circuits. Virtual circuits can either be established on a permanent basis via provisioning (Permanent Virtual Circuits or PVCs) or on demand via end user signaling (Switched Virtual Circuits or SVCs). AAL2's introduction of the concept of a "channel" within a virtual circuit thus introduces a third level of multiplexing / switching in the ATM VPI / VCI hierarchy, which we looked at in Chapter 2. An AAL2 virtual circuit can be subdivided into 255 channels, with each one identified by a unique 8-bit *Channel Identifier (CID)* contained in a field found in the 3-byte CPS header. If a voice packet is shorter than an AAL2 ATM cell payload, then rather than waiting for more packets from the same voice call to fill up the payload, one ATM cell (and hence one ATM virtual circuit) can be shared among several voice calls.
- Supports services such as real-time packetized voice that do not have a constant data transmission speed (the data flow tends to be "bursty" in nature) but do have some quality of service requirements similar to constant bit rate services; hence the term used to describe AAL2 is "real time variable bit rate service." (Note, however,

that the full BLES specification recommends but does not require the AAL2 configuration to support variable bit rates.) AAL2 is more efficient than AAL1 because it allows the network to allocate bandwidth dynamically on the DSL service between the demands of voice and data services. The BLES standard stipulates that voice traffic cannot compromise the customer's use of broadband for data traffic; to keep data traffic flowing, BLES-based VoDSL systems transmitting over AAL2 support a dynamic bandwidth allocation scheme whereby the bandwidth is reserved for active voice channels while unused voice channel bandwidth is allocated to data traffic. Therefore, if at a certain moment no voice services are in use, then all of the bandwidth can be dedicated to data services. This is totally unlike what happens when AAL1 with PVCs are used (AAL1 emulates a circuit-switched channel right down to its wasteful shortcoming of reserving the bandwidth needed for voice in case it should suddenly be needed).

• The ability to modulate the voice quality to conform to network conditions and/or the service being provided.

• Since systems transmitting over AAL2 can provide *Variable Bit Rate (VBR)* voice services, cells not containing voice packets can be allocated to data packets. This permits the system to augment data traffic bandwidth on a cell-by-cell basis via *silence suppression*, which removes the need to packetize the intervals of silence in a phone conversation (pauses when no one is speaking) and instead inserts data into what would have been a packet stream of silence (another "gap" relates to the fact voice communication is essentially half-duplex, which means that one person is silent while the other speaks). Normal phone conversations contain silent gaps which can range from about 40% to 78% of the total conversation time, so silence suppression alone can reclaim over half of the bandwidth allocated for voice traffic (the actual amount of silence in a two-way conversation statistically varies according to various demographic factors including age, gender, nationality, language, and even social status). Note, however, that in order to provide its many bandwidth allocation features, AAL2 also imposes between 17% and 35.5% of extra overhead, depending upon the length of the voice packets. This is not so bad, since it is offset by AAL2's silence suppression capabilities. Additionally, AAL2 inserts special *Silence Indicator Cells (SIDs)* at the beginning of each silent gap which cues the far end receiver into not only reproducing the silence's duration for the recipient but also generating background "comfort noise." (Human listeners, having long been acclimated to the "background noise" found in the PSTN, find truly silent gaps in conversation somewhat eerie and are either disturbed by them or mistakenly believe that the connection has been dropped and the line is dead.) The level of such *Comfort Noise Generation (CNG)* is specified by the near-end equipment, with some vocoders such as G.723.1 and G.729 sending spectrum data of the near-end silence to aid CNG production at the far-end.

Control Signaling over AAL2

Long ago it was realized that, for large-scale deployment of AAL2 based voice, a standard way had to be devised to carry not just voice over AAL2 but the call signaling information too: dialed digits, the on-hook / off-hook status of the call, in-band tones, and signaling messages used for routing and control. A way had to be found to embed a management channel between the CO-IWF and the CP-IWF. Fortunately, the ITU-T recommendation for AAL2 trunking (I.366.2, or "SSCS for Narrowband Services over AAL2") was developed to enable an AAL2 virtual circuit to provide the same functionality associated with a PSTN line. This also serves as the basis for the Loop Emulation Service for AAL2 specification (though there are a few differences between AAL2 trunking and loop emulation, as we shall see). A complementary ITU-T specification, I.366.1, was also introduced that specifies how to carry out-of-band signaling information over AAL2 channels.

The combination of I.366.1 and I.366.2 working in concert with the AAL2 Common Part Sublayer allows AAL2 to carry just about any kind of media or signaling payload such as compressed / uncompressed voice, silence insertion descriptors, circuit mode digital data, narrowband ISDN messages (such as Frame Mode Data), dialed digits and tones, demodulated fax signals, alarms, local management information, etc.

Media information and signaling control messages are carried over AAL2 via two types of packet formats defined by I.366.2 trunking and referred to as Type 1 and Type 3 respectively. Type 1 packets are the default format used on an AAL2 virtual circuit. Type 1 packets are not protected in terms of any kind of error correction coding or checking such as the *Cyclical Redundancy Check (CRC)* and are not retransmitted if they are lost in transmission over the network. The Type 1 cell's information payload is typically used for raw data transfer, i.e. fax, image, audio, etc. Type 3 packets, however, are not only protected with CRC-

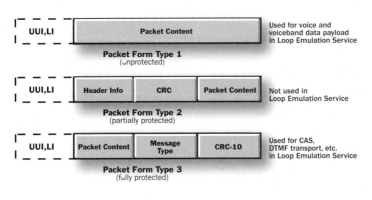

Fig. 5.5. I.366.2 packet formats.

10, but are also transmitted three times in succession (achieving triple redundancy) which ensures that the control information reaches the destination. This packet format is ideal for the transport of control data, i.e. dialed digits, signaling bits and fax control information where incorrect communication results in mis-routed connections and/or interoperable connections (note that if a voice codec is used that is transparent to DTMF tones, then a dialed digit will not be sent three times as Type 3 packets).

One might wonder whatever happened to Type 2 packets. There does exist a Type 2 packet format defined under I.366.2 trunking — it is a "partially protected" format that is used in trunking but not in the loop emulation service.

Information is transmitted in the form of primary and secondary streams. Information such as compressed voice, fax and circuit mode data are classified as primary stream, and only one of them can be active at a time per channel. However, information such as dialing digits, signaling bits, alarms and frame mode data are considered to be secondary stream types and may be sent in the same channel with a primary stream.

CAS and CCS

BLES-based facility signaling can be divided into two basic categories: Those that use older bit-oriented *Channel Associated Signaling (CAS)* and those that use the more modern, message-based *Common Channel Signaling (CCS)* methods. CAS is an "in-band" signaling method which means that no separate signaling channel is needed. CAS is used by North American standard PSTN switch interfaces, e.g. GR-303.

In particular, CAS bits are normally inserted directly into T-1 / E-1 channels by PBXs. Line states are represented by 4-bit codeword. These four CAS bits, called ABCD bits, indicate the signaling state of each voice channel; 16 states are possible, their meanings are defined by the GR-303 standard (examples of bit configurations for upstream line states include 0101 for on-hook and 1111 for off-hook, while downstream bits could be 0000 for ringing and 0101 to signify an idle line). The most common CAS is carried in the same T-1 or E-1 frame as the voice signals, but not enough bits are used so as to interfere with speech. T-1 bits are actually "robbed" for this purpose, dedicating specific bit positions (and thus bandwidth) in the voice-band information to permit the continuous transmission of the ABCD bits (the seventh bit position in frames 6, 12, 18 and 12 of a 24-frame extended superframe format), while E-1 goes so far as to devote an entire timeslot to carry the four bits for each of the 30 other channels. In both forms of this kind of bit-based signaling, the commands are continually repeated without acknowledgment.

In the case of BLES using permanent virtual circuits, these CAS ABCD bits are carried in signaling packets on the same AAL2 channel (CID) as the voice with which they are associated. Whenever ABCD bit state changes occur, a Type 3 packet is sent with triple redundancy and CRC-10 error correction. CAS is very popular for simple end-to-end switching between enterprise PBXs over permanent virtual circuits. Because bit-based signaling is essentially a one-way communications mechanism, it lacks optimal robustness and provides only minimal capabilities for acknowledgements, retransmission, and problem logging.

As for CCS, this consists not of bits but of whole digital messages that travel on a path separate from the call with which they are associated ("out-of-band" signaling). The public ATM network carries these messages to support the PSTN's SS7 network, which is a form of CCS that's a separate signaling network used to control the voice circuit network. By offloading signaling functions to a separate overlay "metanetwork" of separate high-speed links, SS7 frees up TDM network capacity and increases network efficiency. Entire dedicated timeslots of a T-1/E-1 circuit are used as the data channel to transmit CCS messages.

Interestingly, whereas SS7 is used only for signaling between network elements, the ISDN D channel extends the concept of out-of-band signaling from the switch all the way to the subscriber interface; ISDN service uses signaling that is conveyed between the user station and the local switch over a separate digital channel called the D channel. The voice or data which comprise the call is carried on one or more B channels.

CCS messages are essentially addressed signaling messages for individual trunk circuits and/or database-related services between signaling points in the intelligent network. Because of this network complexity CCS data traffic from one site requires switching to be delivered to two or more endpoints, so the signaling channels must be terminated and interpreted at the ATM switch in order for the correct information to be passed to the correct endpoint.

CCS is also used by standard switch interfaces outside North America, e.g. V5.1, V5.2.

Digital CCS messaging is a much more robust and flexible communications mechanism for telephony signaling because it provides an extensible message set for implementation of an essentially infinite range of functionality, as opposed the mere 16 possible states of the older CAS technology. Also unlike CAS, CCS allows for bi-directional messaging which allows for explicit acknowledgments and end-to-end negotiation. The bi-directional acknowledgment capability allows a network to handle so-called "glare" issues in which calls initiated from both ends of the line attempt to seize the trunk simultaneously. While one-way, bit-based signaling requires cumbersome system workarounds to avoid collisions, the built-in two-way intelligence of digital message-based signaling can completely resolve such situations through auto-negotiation within the signaling channel.

This inherent ability of CCS to do bi-directional communication, along with its nearly infinite message extensibility, enables the message-based SS7 system to do everything we have come to expect in the *Intelligent Network (IN)*: efficiently manage a large range of critical telephony functions such as automatic callback, calling number delivery, network message service, and network automatic callback distribution. It also allows SS7 to be used in the implementation of advanced revenue-enhancing services such as Caller-ID, "one number" services, number portability, etc. Since the public network uses a separate signaling path for CCS / SS7, when ATM carries these messages it must also use a channel separate from voice. CCS for all voice channels can be carried over AAL2 using channel identifier CID = 8. However, since these signaling messages can be considered as data traffic, one does not need the high quality, constant bit-rate or real time variable bit rate capabilities of AAL2, and so in the case of ATM trunking with CCS signaling, the CCS messages can be sent as AAL5 cells, with each CCS channel having its own permanent virtual circuit. Also, if CAS is used with SVCs, CAS bits can be converted into CCS messages which can be forwarded across the ATM backbone, either via AAL2 using channel 8 (CID=8) or else via AAL5 in which case the messages occupy an entire separate AAL5 channel.

Subscriber Profiles

No all-encompassing provisioning standard yet exists for IADs (and therefore each IAD has different capabilities in this area). Yet, once installed, all IADs, in addition to doing some kind of self-configuration to interface with the appropriate network elements at the customer premise, must also download what are called *service subscriber profiles* required from the gateway or service-creation back-office system before voice communications can commence.

A profile is simply a set of entries, usually a list of capabilities of the available voice encoder algorithms, also called "voice codecs" or "vocoder" algorithms (such as G.729, generic PCM G.711, etc.). Each entry in the list consists of a UUI along with the packet size used by the algorithm. Both sender and receiver must agree on what profile to use, unless the type of voice codec to be used is already configured in advance. By transmitting the profile index, both sender and receiver can convey the necessary information about the supported vocoders. It is even possible to change a vocoder in mid-call without the intervention of any external procedure. This capability is useful since the voice quality is modulated to adapt to the particular service being provided and/or network conditions.

The Loop Emulation Service can use any of the ITU-T I.366.2 pre-defined profiles, though there is mandatory support for ITU-T profile #1, which is a generic PCM encoded 64 Kbps channel carried in 40-byte packets. Additional profiles defined in the LES spec for optimum performance include those for ADPCM 32 Kbps, silence removal, and 44-byte packets for maximum bandwidth efficiency. One profile for broadband loop emulation that achieved popularity early on was Profile #7 (defined in document VTOA 0113).

LES defines two modes of CP-IWF operation: *Independent Mode* wherein the CP-IWF (the IAD) determines profile entry to use on its own, and *Master/slave Mode* wherein the CP-IWF sets its profile entry to match that chosen by the CO-IWF. While AAL2 permits the transmitter under loop emulation to determine which profile entry (voice encoding) to use, it is not always appropriate to let the customer's IAD determine its own profile entry since the choice of entry impacts bandwidth usage and the service provider may want to be in control of this.

AAL2 PVCs and SVCs

We've seen how *Switched Virtual Circuits (SVCs)* need little provisioning when compared to *Permanent Virtual Circuits (PVCs)* — only the ATM addresses of the end points and routing tables in the network switches need to be provisioned, whereupon the end user devices can then establish virtual circuits automatically between themselves whenever they are needed, setting them up and "tearing them down" after use. However, when applied to voice transport over DSL links, AAL2-based loop emulation was originally defined using PVCs, which are based on packet-interleaved multiplexing.

Using AAL2 with PVCs also meant that the voice gateways tended to be unintelligent devices, sometimes taking an AAL2 PVC directly from the IAD and converting each voice call into a standard G.711 64 Kbps digitized channel and multiplexing them in a T-1 type of format for the GR-303 interface on the Class 5 switch. In such a model the gateway acts as a remote terminal with most of the intelligence residing on the Class 5 switch (as was the case previously with an analog local loop), so when

Note: More often a DSLAM is first used to provision separate PVCs to and from the CPE in order to provide a broad differentiation among media types; the DSLAM then aggregates them into one or more PVCs to the voice gateway, or perhaps sends them across a regional ATM network to yet another aggregation device, that aggregates all PVCs from all subscribers via the DSLAMs, so that subscriber traffic is carried on fewer PVCs.

the user picks up an analog phone that is connected to the IAD, the remote Class 5 switch provides the dial tone and DTMF functions.

I say "most" (not all) of the intelligence still resides in the switch because AAL2's variable bit-rate (VBR) service capability specifies a method for packing multiple voice calls into a single PVC to optimize bandwidth utilization, and for this method to work effectively, it must be understood by the DSLAM as well as the CPE and gateways. To add this type of intelligence while still maintaining basic (and ever-increasing) capacity and speed requirements, DSLAMs must take on some of the characteristics of ATM switches, moving functionality that was previously handled higher up in the network closer to the edge. In particular, DSLAMs will increasingly have to "understand" QoS issues since all the traffic goes through the DSLAM, and as more involved QoS considerations enter the picture DSLAM technology must become increasingly aware of ATM Layer 2.

Still, the AAL2 PVC loop emulation model keeps the carrier in control of the architecture. The IADs can be made simple and inexpensive, since under this model they don't do much other than packetize the voice and data and send them to the DSLAM (it also means they can't offer very many service options). Also, with the right equipment in place, PVCs can be provisioned in the carrier's network using the traditional highly structured service order and support structure collectively called *Operations Support Systems (OSSs)*, which allow carriers to manage their VoDSL customers the same way they do customers connected to digital loop carriers, directly from their Class 5 voice switch. This enables the carriers to, among other things, activate or deactivate customer VoDSL phone lines the same way as traditional phone lines (unfortunately, like conventional PSTN solutions, it's not a very scalable system).

In the next phase of VoDSL evolution, however, the technology will extend its reach to provide connectivity not just to local switches, but also to long distance voice switches and PBXs at networked corporate sites. Achieving such new functionality demands some additions and modifications to the original architecture. The first requirement is implementing SVCs, not necessarily to replace PVCs but to augment them, which enables the future build-out of advanced services.

Formerly known as Accelerated Networks, Occam Networks (Santa Barbara, CA — 805-692-2900, www.occamnetworks.com) was one of the first manufacturers of IADs, DSLAMs and voice gateways to offer the choice of using SVCs instead of just PVCs to simplify network deployment and provide substantial flexibility in the routing of voice calls. Interestingly, devices from Occam Networks can set up both AAL1 Certified Bit Rate (CBR) and AAL2 real time Variable Bit Rate (rt-VBR) PVCs and SVCs. They call their technology Switched Voice over DSL (Switched VoDSL). Instead of hard-wiring the subscriber to a local Class 5 switch port using GR-303 signaling, Switched VoDSL allows subscribers to dynamically connect to a local Class 5 switch, a long distance Class 4 "tandem" switch, emerging softswitches, or directly to other corporate locations. The IADs tie into a carrier's SS7 intelligent network policy server in order to determine call routing instructions. Occam Networks has also designed each of its products to be interoperable with third-party equipment, so that a service provider can choose to work with different vendors for specific parts of the network. The company has submitted a Switched VoDSL proposal to the DSL Forum to make it an industry standard.

And, provided that the IADs served by a central office are all "intelligent," are software programmable (the so-called "soft IADs" discussed in an upcoming chapter), and have all successfully loaded the same Loop Emulation Services-compliant software, telcos can manage all the IADs running such software regardless of the IADs manufacturer. Critical tasks such as bulk software and configuration download, flow-through provisioning of large numbers of IADs, off-hours scheduling of maintenance windows, remote troubleshooting of IADs, and detailed performance measuring across the access network can be carried out without a truck roll. All of these capabilities enable service providers to rapidly serve subscriber requests and facilitate network maintenance of many customers.

It seems that the future trend for ATM VoDSL customers will be the use of SVCs so that they can configure connections themselves instead of having the network operator do it. Customers will be able to instantly order and drop phone and data lines so, for example, customers could access a corporate network during the day for teleworking, and then switch over to their own Internet connection in the evening for surfing, game playing, downloading multimedia files, or some other form of entertainment.

Loop Emulation Service vs. AAL2 Trunking

Although similar, broadband loop emulation using AAL2 and AAL2 trunking between core network switches are not exactly the same. Let's look at the several differences.

In AAL2 trunking, the ingress gateway is capable of sending and receiving signaling messages that allow it to establish connection paths across the ATM backbone. In the case of the loop emulation service, however, the customer's IAD cannot make any routing decisions on its own; it simply forwards the signaling messages from the customers telephony equipment transparently through the DSLAM and voice gateway, and on to the Class 5 switch, which then routes the call.

AAL2 data units can overlap between two ATM cells. The loop emulation service also allows for a "CPS Lite" alternative in which each ATM cell only contains a sin-

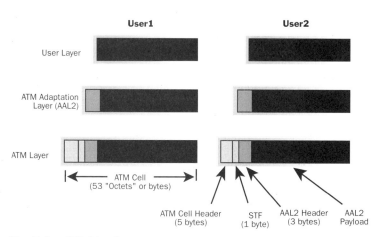

Fig. 5.6. CPS Lite Option in the Loop Emulation Service.

gle AAL2 data packet, which always follows a "Start Field." Under the Lite option, if the payload is less than 44 bytes, then the remainder of the cell is padded with zeros. This process (which is the equivalent to setting the "CU_timer" defined in I.363.2 to a value of zero) simplifies deployment, decreases customer premises IAD cost and can even optimize performance for a small number of voice channels on a virtual channel connection but also lowers overall DSL bandwidth utilization. Most high quality IADs can support full-featured AAL2.

AAL2 trunking is based on peer-to-peer messaging between two gateways at either end of an ATM backbone network. The loop emulation service is more of an asymmetric affair, with the network-based voice gateway controlling the customer premises IAD.

AAL2 channel management is different in loop emulation and trunking. Channel management specifies procedures for allocating voice and signaling channels to AAL2 channels (CIDs) and specifies procedures for activating and de-activating AAL2 channels. There are two ways of activating AAL2 channels, implicitly and explicitly.

At the customer end, the AAL2 channels are allocated implicitly, wherein customer premises ports are statically assigned to CIDs within these ranges:

Analog POTS phones have a CID range of 16 to 127
ISDN D-channels have a CID range of 128 to 159
ISDN B-channels have a CID range of 160 to 223

Under such circumstances the CO-IWF (network gateway) activates a particular CID by commencing AAL2 transmissions on that CID, and the CP-IWF (IAD) considers a CID active if it is receiving AAL2 packets on that CID.

At the network end, AAL2 channels can be assigned explicitly. This simply requires the network gateway to use the common signaling channel 8 (CID = 8) to send messages that dynamically allocate customer premises ports to specific CIDs. The messages thus both "explictly" assign CP-IWF ports to AAL2 CIDs, and activate the CIDs themselves. Conversely, other messages can explicitly de-assign and de-activate the AAL2 channels. Both of these processes require the support of a special protocol called the *Emulated Loop Control Protocol (ELCP)*. ELCP is interesting in that it can perform dynamic CID (channel ID) allocation, which makes possible a certain degree of call concentration. Concentration is accomplished by configuring several user ports in the voice gateway and IAD and by using fewer CID channels at any time than the number of user ports. This cannot be done in implicit or CAS signaling where the CIDs are statically defined to specific user ports. Such a concentration capability comes in handy when a customer is using a low bandwidth DSL, such as an ADSL line.

The AAL2-based loop emulation service reserves an *Embedded Operations Channel (EOC)*, CID = 9, which is more sophisticated than the EOC of the GR-303 standard, being instead a take-off on the *Interim Link Management Interface (ILMI)*, an ATM Forum-defined interim specification for network management functions between an end user and a public or private network and between a public network and a private network. ILMI is based on a limited subset of the *Simple Network Management Protocol (SNMP)*. Operating at the OSI Application layer, SNMP is the most common method by which network management applications can query a management agent using a supported *MIB (Management Information Base)* and is the basis for most network management software. SNMP defines a set of six simple query operations that can be performed on network hardware and specifies how the communication of the queries and information is handled. SNMP is a standard, not an API; thus it is not tied to any specific language or platform (HP OpenView and Cisco Systems IOSoftware also use SNMP). AAL2's EOC can monitor and manage downstream equipment via SNMP messages. Under AA2 loop emulation, the voice gateway in the network runs SNMP management applica-

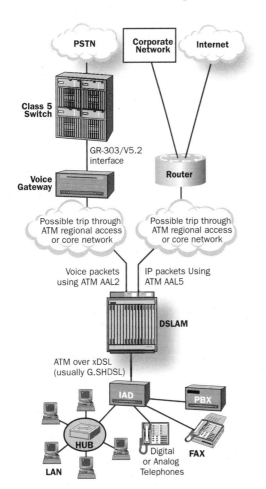

Fig. 5.7. Broadband Loop Emulation ATM VoDSL.

tion software which communicates over the EOC with the SNMP agent installed in the IAD. This allows the voice gateway to query the status of and otherwise manage the customer premises IAD over the EOC, with support for remote configuration, inventory retrieval, alarms and performance reporting.

AAL5 for IP Data Packets

Our extensive discussion of ATM-based loop emulation for VoDSL has until now focused on how voice and signaling messages are sent using AAL2. But Internet access is a prime motivator for a customer to order any kind of DSL service, so let's look at how these data packets are handled.

ATM-based VoDSL systems handle voice and data cells in different ways and on separate virtual circuits. Voice packets and signaling are sent using AAL2, while "simple" data packets, which tend to be IP packets, are segmented into ATM cells using AAL5 (because of technological limitations, two different AAL types such as AAL2 and AAL5 are never used on the same PVC). AAL5 thus allows for what one could call "IP over ATM over VoDSL." In some systems, IP data also can be transported as IP over PPP over ATM.

In major metropolitan areas, it is common for large central offices to have multiple DSLAMs, which can lead to challenges regarding data aggregation of the IP packets traveling over AAL5. Many service providers initially employed a simple optical SONIC Add/Drop Multiplexer (ADM) to bridge broadband data and voice traffic onto the service provider's SONET rings for backhaul to an ISP's point of presence (POP). An ADM can insert or extract DS-1, DS-2, and DS-3 channels or SONET signals into/from a SONET bit stream, without having to go through the onerous process of demultiplexing and remultiplexing the whole signal, which is required in the traditional T

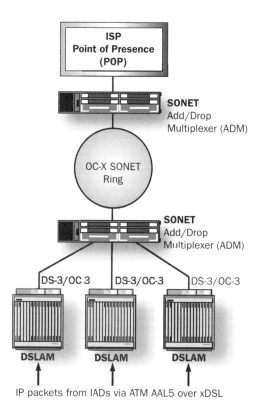

Fig. 5.8. Inefficient xDSL data packet aggregation and transport.

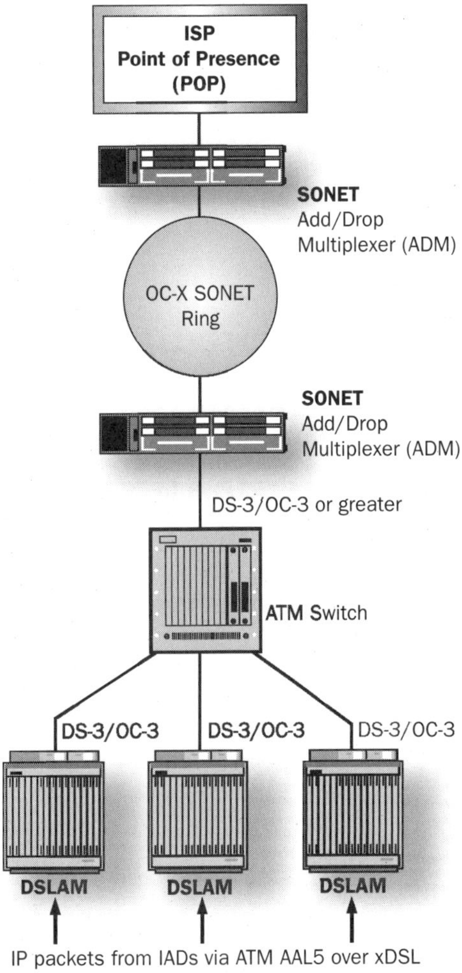

and E carrier systems. Each DSLAM in the CO is connected to an optical SONET ADM via an individual DS-3 or in some cases an Optical Carrier-3 (OC-3) connection. The SONET ADM maps each DS-3 into a 54 Mbps channel and each OC-3 into a 155 Mbps channel on the SONET ring regardless of the actual utilization, which can be as little as 15% to 20%. Each DS-3/OC-3 from the DSLAMs in the CO is mapped individually since traffic on a SONET network is channelized and thus SONET doesn't allow sharing of the allocated bandwidth. This results in SONET facilities that are greatly underutilized.

By utilizing only a small percentage of these large pipes from the SONET network, service providers dramatically increase the cost of their broadband network, since new core transport facilities must be added as broadband deployments grow, even though current facilities are underutilized. One way to fix this type of problem is to simply expand the capacity of the fiber network. One vendor in this field General Bandwidth (Austin, TX — 972-372-5000, www.genband.com), estimates that service providers increased their spending on SONET networks by over 15% in 2000 alone, an expenditure totaling $8.4 billion. Thus, there are great potential savings incurred by service providers aggregating DSLAMs in a DSL network efficiently.

Fig. 5.9. DSLAM aggregation with ATM switch.

As DSLAMs achieve greater capacity, some solutions exist to better handle the SONET pipe, such as the OC-48 WaveDirect SONET ADM card from Alidian Networks (Milpitas, CA — 408-487-9700, www.alidian.com). Unlike restrictive point-to-point implementations using so-called "SONET MUX" cards, the multi-port SONET WD-ADM modules from Alidian flexibly consolidate traffic around a fiber ring, sharing the available timeslots of an OC-48 aggregate signal for full bandwidth utilization. Two OC-48 ADM module types are available — one with seven OC-3 / OC-12 tributary ports and

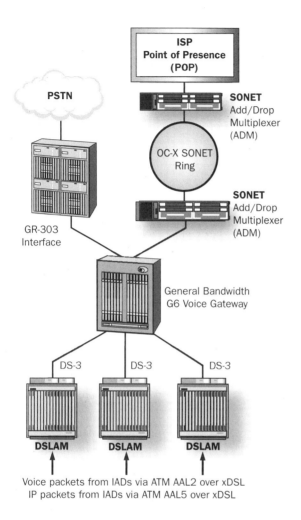

Fig. 5.10. Integrated voice and aggregation achieved by using the General Bandwidth G6.

another with a single OC-48 tributary port. The single-port WD-ADM module consolidates traffic from multiple seven-port SONET WD-ADM cards into a single, high-speed OC-48 signal connected to large SONET ADM systems. The OC-48 signal, transmitted at a precise *Dense Wave Division Multiplexing (DWDM)* wavelength, allows co-existence with legacy SONET and other multi-service wavelengths in a fiber-optic ring. The net result is highly efficient TDM transport at lower costs over a reduced number of 2.5 Gbps wavelengths.

Another possible solution to solve the SONET bandwidth utilization problem is to introduce an ATM switch to aggregate and groom the traffic from the access network. The switch concentrates the data and voice traffic before it reaches the SONET ADM, which enables improved utilization on the SONET ring. But this is also an expensive solution. It adds not just per port hardware costs but also considerable "soft" costs associated with the management and provisioning of the additional element in the network. This increased cost is not offset by new service creation, and as a result it directly impacts the benefits of increased utilization of the core network.

Another way of dealing with the problem is to introduce a voice gateway into the network that also has the capability to act as an aggregator. Essentially, service providers want to add packetized voice to their DSL network to better leverage their DSL infrastructure and more cost effectively provision voice services, and they need to aggregate and groom the traffic originating from these networks prior to handing off the traffic to the core network. By introducing a new network element that is both a voice gateway and an aggregator, the cost of the aggregation solution is lowered (no ATM switch is needed), allowing service providers to achieve proper utilization of their network resources and get VoDSL for "free."

One such integrated product is the G6 from General Bandwidth, a multi-application platform capable of providing carrier-grade, toll-quality, voice over broadband services while simultaneously enabling aggregation of traffic from DSL and cable networks, next generation Digital Loop Carriers and Passive Optical Networks. This is an easy task for the G6 since it has a massive 38.4 Gbps backplane, expandable to 60 Gbps, which also enables service providers to leverage the G6 for multiple applications within the central office.

This combination of AAL5 for IP data and AAL2 for uncompressed or compressed voice payload bearer channels and signaling comprises the whole Broadband Loop Emulation Service (BLES). BLES was the first and continues to be the most popular form of VoDSL since service providers can use their existing Class 5 switches in local exchanges or end offices. As DSLAM technology improves, BLES can be made even more economical, since, for example, a voice gateway can be integrated into a DSLAM, so a network operator need no longer buy an ATM switch for each central office and reserve space for a separate voice gateway.

Voice over Multiservice Broadband Networks (VoMBN)

For all its initial success, broadband loop emulation was looked upon by some experts as a mere interim measure, capable of providing VoDSL until the development and deployment of more sophisticated "next-gen" networks based on IP. This idea is plausible because after examining the loop emulation architecture in detail, one might be tempted to ask, "can't you just get rid of the expensive, semi-obsolete Class 5 switch and pipe packets directly from the DSLAM into a long-distance data network?"

Yes, it is possible. The next approach to VoDSL we examine is Voice over Multiservice Broadband Networks (VoMBN), which provides voice service (custom calling and Centrex services, as well as trunking for customer premises equipment), along with data services over DSL from the regional broadband network without the need for a Class 5 switch. This is not as strange as it seems, especially when one considers that the carriers run both voice and data trunks over the same optical fiber backbone. One can get a hint of how this can be done even in a broadband loop emulation system — a hobbyist could load some Internet phone software on his or her PC, then make an IP-based Internet telephony call over the AAL5 "data" packets, creating a "voice-over-IP-over-ATM-over-DSL" system. Indeed, VoMBN often uses AAL5 to offer end-to-end VoIP service which bypasses the Class 5 switch.

ATM vs. IP

But since we've previously examined in detail ATM AAL2 for voice, one might wonder why there is such a hubbub over IP and the Internet. Couldn't our all-packet network be pure ATM instead of bringing IP into the picture? Couldn't the whole network just be made a series of ATM switches? After all, ATM is already used in the core and can support voice, video and data transport over a common infrastructure. (IP advocates would similarly argue for an all IP network.) Indeed, instead of sending the AAL2 voice packets to a voice gateway and Class 5 switch, the packets can be seamlessly sent over a public or private ATM network, with the voice packets remaining in AAL2 format. It's an attractive idea, since one of the reasons ATM AAL2 was originally chosen for transporting voice over DSL was because of the way that ATM in the local loop adroitly handles the "large data packet induced jitter problem" that occurs when a small voice packet gets queued up behind a 1500 byte data packet on a low speed DSL link. So long as the ATM AAL2 virtual circuits are run as rt-VBR, both delay and jitter are minimized, owing to the small ATM cell size of 53-bytes. In fact, there are some ISPs that use ATM on their backbone (UUNET for example), since, aside from ATM's high quality of service, ATM enthusiasts will tell you that it's easier to manage a core network consisting of ATM virtual circuits than a network of IP routing tables. In the event of circuit failure, one can reroute an entire interface or group of routes by re-directing the virtual circuit across alternate capacity. (IP proponents will counter that circuit management is easier with IP, noting that an ATM mesh array capable of handling a huge multioffice enterprise or an ISP the size of AOL, for example, would require many more ATM circuits than one would need with IP-over-SONET.) Also, the low latency characteristics of the multiplexing ATM packet architecture make it easier to provide Service Level Agreements (SLAs) between service providers and customers. ATM is designed to transmit diverse traffic types and distinct classes of service are available.

The problem comes when attempting to provide flexible, differentiated services and supporting the increasing complexity of ATM switches that supply such services. It's easier and cheaper to devise and offer enhanced IP services.

Also, until recently, provisioning under ATM was troublesome because there was no virtual circuit aggregation through the DSLAMs when using VoATM — DSL providers with ATM networks used ATM switches to aggregate subscriber traffic from their DSLAMs at OSI Layer 2. Such Layer 2 aggregation becomes inefficient as the DSL network scales, since at least one virtual circuit (generally a permanent virtual circuit, or PVC) must be kept open for each subscriber IAD all the way back to the service provider's point of presence in the serving area. It wasn't until the 2001-2002 time frame that subscriber management systems appeared that could provide Layer 3 aggregation. For example, in 2002 the VantEdge Broadband Services Concentrator (a DSLAM) was introduced by Copper Mountain Networks (Palo Alto, CA — 858-410-7305, www.coppermountain.com). It can aggregate subscriber virtual circuits into single service tunnels rather than provision a virtual circuit for each subscriber connection, thus solving the ATM switch "PVC exhaust" problem right at the central office. The number of virtual circuits handled by upstream ATM switches can now be reduced by two to three orders of magnitude, enabling providers to use these switches to their full switch-fabric and port-bandwidth capacities (another similar solution was the Cisco IP DSL switch, which we will examine shortly). The VantEdge also eliminates the need for metro-based subscriber aggregation gateways and reduces the need for additional ATM switching equipment.

But what if the DSL network / system expands to point where another Class 5 switch is needed? Why would any service provider want to invest in more seemingly obsolete Class 5 switch technology? It is possible to buy a new broadband-enabled Class 5 switch (which is basically a kind of centralized softswitch), but this just underscores the fact that VoDSL, relying upon ATM-based loop emulation, suffers from some of the same limitations of traditional voice networks, in particular the constraints imposed by the traditional network's inherent centralization of call control and services creation. (Note that the VantEdge device described above supports a migration path to IP.)

Also, ATM standards are quite complex, and ATM is not the best option for carrying Internet data, especially because of its "cell tax" — ATM's basic cell design principle of using a small, fixed size cell results in about a 10% header overhead. The 53-byte cell size (48 bytes of payload) of ATM was a compromise between two small proposed cell sizes of 32 and 64 bytes, both championed to avoid the need for echo cancellation. Since time scales with the data rate, the DSLAMs and ATM switches must make quick decisions about ATM cells during the short time it can scrutinize them (for example, 681 nanoseconds at an OC-12c rate). It was not until the end of the 1990s that superfast ATM chips appeared for host-network interface packet segmentation and reassembly. Moreover, network interface memory design and buffering strategies would have been neater and easier to put into service if the cell payload length had been a power of two (64 bytes, for example). This is because processors use binary arithmetic, gulping bits in powers of two (8, 16, 32 bits) which makes binary division easy — division by powers of two reduces to shifting all the 32 or 64 bits a certain number of places to the right. Also, network caches use memory chips that

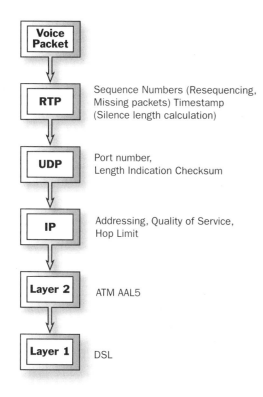

Fig. 5.11. Transporting a Voice Packet via IP.

store these bytes in segments that have boundaries that are powers of two — for example, 1 Kilobyte isn't 1000 bytes, it's 1024 or 2 x 2 x 2 x 2 x 2 x 2 x 2 x 2 x 2 x 2 bytes. Therefore, code running on a processor that deals with cell streams and addresses memory runs faster if everything is packaged in chunks that equal powers of two.

One could argue, of course, that if VoMBN achieves high quality of service VoDSL by sending voice over IP over ATM AAL5, then flexibility gained by using IP has a price as great (or greater) than the ATM cell tax, depending upon the degree to which voice and packets headers are compressed. This is because voice packets are placed in one type of protocol packet, then those are diced up and inserted into the packets of a different protocol. The practice of sending data of one protocol within the transmission of a different protocol is called *encapsulation*. An encapsulated protocol is sent within another protocol just as a gift-wrapped box can be shipped across the country while enclosed within a Federal Express box. There are so many layers of protocol involved in running "voice over IP over ATM over DSL" that efficiency decreases as the system maps and encapsulates IP packets successively into the *Real-time Transport Protocol (RTP)*, then the *User Datagram Protocol (UDP)*, then IP and finally ATM cells. One ends up with combined packet headers that are quite large, which leads to high overhead.

For example, imagine we're going to transport a 40-byte, 10 millisecond G.726 voice packet and we'd like to know what the overhead is going to be. First, we add 40 bytes for the combined IP / UDP / RTP header that are included with the data in each VoIP packet, even for small payloads, which means that two ATM AAL5 cells are needed to transmit each VoIP packet — the first cell would have 48 bytes of the IP datagram, while the second would have 32 bytes of the IP datagram plus the 8-byte AAL5 trailer at the end of the cell. The remaining 8 bytes between the segment of the voice packet and the AAL5 trailer would be padded yielding an overhead of 98.8% .

One might as well have gone with direct VoATM, which encapsulates the voice coder output directly in a single ATM cell. One big problem is that, since ATM cells carry a fixed 48-byte payload (minus 1- or 2-byte AAL headers), if the transported data does not fill the cell, then the additional payload space is padded.

However, VoIP does support a configurable number of codec samples per packet, and ATM payload efficiency increases to 94% when VoIP is properly tuned. For example, five samples of G.729 at 10 bytes each, along with 40 byes of IP / RTP / UDP header, yield 90 bytes of data in two ATM cells. Using a supposedly bandwidth saving G.723.1 codec with VoIP, however, results in an ATM payload efficiency of only 73%. Because G.723.1 uses a 30 millisecond frame, placing multiple frames into a packet to improve the ATM payload efficiency is not practical because of additional packetizing delay.

There are some other "tricks" that can be used to shave some precious milliseconds off of packet transmission times, such as compressing RTP headers and disabling UDP checksums. Still, some experts may consider these to be stopgap measures and not reliable solutions to serious delay problems that may appear when ubiquitous residential and business VoIP finally appears. Still, every millisecond counts, and those few milliseconds can add up when applied over a hundred or a thousand termination points.

Also, keep in mind that these conversion processes must preserve the voice packet format integrity end-to-end to eliminate any additional latency and transcoding impairments to voice quality. Achieving this demands sophisticated interworking between VoDSL and VoIP signaling protocols, which can result in the considerable consumption of processing overhead and bandwidth.

The fact remains that until recently ATM has served as a sort of crutch to compensate for IP's deficiencies. The primary reason we're all not using end-to-end VoIP on a pure IP network today centers on the problem of Quality of Service (QoS). When IP voice and data packets are mixed together and sent across a large IP network (such as the Internet), real time voice and video quality cannot be guaranteed with the same confidence as with ATM or conventional circuit-switched connections, even though there are

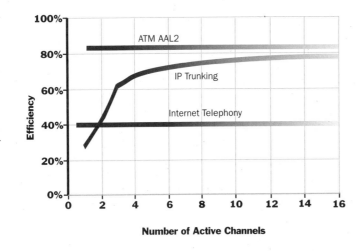

Fig. 5.12. Bandwidth efficiency of Voice-over-IP and AAL2 on ATM-based DSL (44-byte voice packets).

various techniques that are undergoing refinement that may solve the problem — these include *Differentiated Services (DiffServ), Multi-Protocol Label Switching (MPLS)*, and even the simple overprovisioning of bandwidth.

To conserve bandwidth, VoIP networks often use high-compression coding algorithms, such as G.723.1/A and G.729A/B, which reduces the impact of voice on data traffic. But this approach has voice quality penalties. The G.711 A-law/mu-law and G.726 ADPCM speech coding algorithms specified by ATM-based BLES VoDSL provide better perceived speech quality than the high-compression voice coders. Also, because G.711 and G.726 are sample-based algorithms that can fully utilize any cell or packet size, they enjoy much lower delay times than G.729A/B and G.723.1/A, which operate on 10 millisecond and 30 millisecond fixed frame sizes, respectively.

Also, a dropped packet in a BLES system will generally not degrade voice quality as much as in a VoIP system, because each uncompressed packet carries a shorter interval of speech. One packet of BLES corresponds to 5.5 ms of voice, whereas one packet of G.723.1/A or G.729A/B corresponds to 30 or 10 ms of voice, respectively.

By offering toll-quality voice and telephony service transparency while maintaining a high data traffic bandwidth, ATM-based BLES steers clear of the many current sources of customer dissatisfaction with packetized voice over broadband. The service transparency also allows customers and telephony service providers to continue using existing PSTN equipment, which ultimately keeps down the cost of finally switching over to a packetized network. So, BLES still does have some powerful arguments in its favor.

Furthermore, in order for VoIP users to enjoy the same services they were accustomed to in the PSTN world (voice messaging, call waiting, caller ID, etc.) a major enhancement to the world's communications infrastructure must take place, with various new kinds of gateway, signaling and switching devices that can handle not only VoIP but legacy circuit-switched PSTN calls too. Such equipment has taken years to be perfected and deployed.

Still, most experts feel that both the future information highway and telecom byways will be based mostly on IP / Internet technology simply because IP is ubiquitous, offers universal addressing, and universal connectivity. The trend in the telecom industry is to push intelligence to the network's edge or even to the endpoints themselves, and IP is a good way to do that. Also, IP systems scale better than do either ATM or Ethernet systems. IP makes it easier to assign individual lines and circuits, so it's also easier to do network management. And from the viewpoint of developers and integrators, IP offers a uniform platform for applications and services development.

At the customer premises end in the late 1980s and 1990s, "computer telephony" often involved physically connecting a PBX with a computer via something as primitive as a serial cable, then running some "middleware" and application software that worked "most of the time." The rise of IP leveled the playing field and made the process of integrating disparate communications devices far easier than ever before, so much so that the original reason for adopting IP — free long distance "toll call bypass" — was almost totally forgotten. (I say "almost" because there are still Internet telephony service providers and Internet telephony exchange carriers who make a profit by doing end-runs around international tariffs using VoIP connections between foreign countries that employ both the Internet and private managed IP networks.)

Although the hype surrounding the idea of an all-IP converged network continues unabated in some quarters, service providers and vendors have realized that the most optimistic migration to IP will be a long, three-phase evolution: ATM, a hybrid ATM / IP network and then possibly an all-IP network. The world has now entered the "ATM/IP hybrid" phase, but the 2001-2002 economic collapse of the telecom sector underscores the fact that the entire communications infrastructure cannot be completely transformed overnight. In early 2002, the VoDSL industry was shaken by the demise of Jetstream Communications, which had taken a conservative approach by manufacturing ATM-based VoDSL gateways under the theory that the Regional Bell Operating Companies (RBOCs) had relatively new ATM networks as well as many Class 5 switches which they weren't going to dispose of immediately. Some looked upon the cause of Jetstream's failure as a result of backing older ATM / Class 5 technology instead of IP. In actuality, Jetstream had been the leading VoDSL gateway manufacturer, accounting for approximately a third of all VoDSL gateway revenue during 2001 — the real cause for its downfall was a tremendous plunge in all telecom capital expenditures, particularly those by incumbent local exchange carriers.

Carriers are not only looking for a way to accrue incremental revenue, but also to reduce capital expenditures and operational expenditures. Thus, the transformation of the network will involve a "build-as-it-grows" approach with support for multiple network architectures. Although GTE Internetworking (now part of Verizon) began building a new national fiber backbone in the late 1990s with separate circuits for voice (ATM and Frame Relay), even then their plan was to ultimately converge everything to an all-IP network, using high-speed routers to deliver prioritized IP traffic directly into a SONET multiplexer for transport across the backbone. The router examines the IP packets and can deliver both class of service and quality of service, is able to mix real-time traffic like voice, video and multimedia and can essentially create an "express lane" through the network while mixing that traffic with lower-priority data, all at the

IP layer. A similar piecemeal migration will occur in the local loop.

Moreover, the "big controversy" over whether the future belongs to ATM or IP-based VoDSL systems is muted somewhat when the argument is brought down to the level of the access network. At one time, what kind of transport packet protocol one chose (IP, ATM or even Frame Relay) would have had a huge effect on the actual implementation of VoDSL system architecture and infrastructure, particularly in terms of the speed at which providers could deploy the service, the size of the investment they would have to make in the infrastructure, and the cost and complexity of ongoing operations and management.

The way things have been evolving, ATM-based VoDSL systems send voice cells from a customer premise IAD to a DSLAM, then to an ATM-based voice gateway, and then to a Class 5 switch, while VoIP-based VoDSL systems tend to send voice packets from an IAD to a DSLAM, then to an IP router and softswitch or Internet-based voice gateway, which ultimately brings it back from the Internet to the PSTN and thence to the called party.

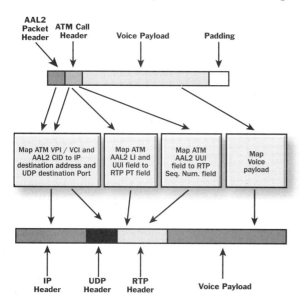

Fig. 5.13. Mapping of AAL2 Packet Voice to Voice over IP.

But consider that, with current technology, an ATM AAL2-based (not IP) VoMBN access network can deal with the voice traffic in any number of ways once it reaches the DSLAM. Instead of the usual "loop emulation" route from the DSLAM to the voice gateway and Class 5 switch, the voice traffic can be routed over a different kind of packet network, such as an IP network (the voice packets must be extracted from AAL2 and mapped into RTP, then UDP, then IP, as shown in figure 5.13). The new generation of softswitches and gateways can provide efficient conversion, not only between the PSTN format (PCM on TDM) and the packet network format (typically compressed and packetized voice), but also between packet network voice protocols, particularly VoIP and ATM-based VoDSL.

Or, in the spirit of an VoMBN, the voice packets can remain in the AAL2 format and be seamlessly transmitted over a public or private ATM network. (ATM enthusiasts dream that IP is a passing fad and that the whole telecom world will one day come to its senses and completely switch over to ATM technology.) Since ATM is a switched technology, ATM virtual circuits do not have to terminate at the central office voice gateway, and can be switched through ATM networks at either end of the local loop. At the customer premise end an ATM LAN could be used, while at the central office the ATM cells can be taken off of the DSL link and be sent directly over a long haul ATM network. If a call is made from such a system to somebody having a conventional phone system, the voice traffic would travel from the ATM network through a voice gateway (which decodes the packets back into PCM TDM) and would enter the Class 5 switch through a long distance trunk interface instead of a GR-303 line side interface as would be the case in loop emulation.

In short, the cost of doing protocol translation is decreasing, as is the cost of both ATM and IP routing equipment. The entire core switch-router product line from Marconi plc (London, UK — +44-(0)-20-7493-8484, www.marconi.com) has the ability to switch and route IP, MPLS (Label Edge Router and Label Switch Router) and ATM simultaneously, and scales from 2.5 Gbps to 480 Gbps. For example, Marconi's BXR-48000 is a core router with packet-only interfaces, a distributed forwarding table, central route processing and a payload agnostic fabric, making it ideal for multiservice IP cores. With the flexibility afforded by Marconi equipment, service providers can adjust the role of Marconi's switch-routers in their network to accommodate their timeframe for evolution to a multiservice IP network. Also added to this mix are DSLAMs, which are becoming highly flexible and are turning into intelligent multiservice platforms, and other devices such as General Bandwidth's G6 Voice Gateway / Packet Telephony Migration Platform, which can handle both ATM and IP access aggregation, as well as VoATM and VoIP support to connect either to Class 5 switches via GR-303 / TR-08 or else to softswitches. Also, Occam Networks' innovative Ethernet- and IP-based Broadband Loop Carrier (BLC) is a complete loop carrier that combines the functionality of a DLC, DSLAM, and media gateway to deliver traditional and packet voice, as well as a variety of broadband and IP services from a single, converged all-packet voice and data access network.

As such technological achievements continue, one can see that the IP *versus* ATM controversy may ultimately lose much of its acrimony. Any given VoDSL deployment will likely depend less on the relative popularity of IP or ATM, and more on the immediate needs of both the customers and service providers, as well as whatever the most cost-effective infrastructure is currently available.

Moreover, when even higher capacity *Very High bandwidth DSL (VDSL)* technology finally appears, it will not be backward-compatible with any other flavor of DSL, so carriers won't be concerned whether their embedded base of xDSL users are served by ATM- or IP-based equipment when selecting VDSL products. Most early VDSL proposals used ATM as the Layer 2 transport technology, but there are large cost savings that can be realized by using Ethernet as the transport over the loop.

At the moment, however, regardless of the end-user architecture (IP or ATM), xDSL services of all types (including VoDSL) will tend to use an ATM backbone network connection over a wide area network for real-time voice and video traffic. Until end-to-end VoIP parameters such as delay, jitter and packet loss are "bounded" once and for all at acceptable levels, near toll-quality voice calls will demand the use of private ATM backbones rather than the Internet. Because of its rich QoS support and extensive traffic-management capabilities, ATM enhances IP in DSL access networks, ensuring that IP voice calls will receive sufficient network resources to guarantee circuit switched performance levels, thus assisting the transition to a converged IP network. This is why a company such as Cisco, for example, promoted at the end of the 1990s an architecture it called "IP+ATM" (which led to what they now call IP DSL switches) and many other companies are working toward a similar end — to add IP routing capabilities on top of a platform that is, in most other regards, ATM-centric. These efforts have all emphasized or have completely taken the form of QoS initiatives, no doubt to help bolster IP's Achilles' heel. MPLS and DiffServ have appeared prominently among DSLAM vendors' product upgrades (DiffServ QoS parameters map directly to ATM QoS parameters), indicating that QoS — specifically, the need to differentiate among individual packet streams within a given virtual circuit and to handle them accordingly by assigning priorities based on the differing delivery requirements demanded by particular applications — will help drive the movement toward IP.

MPLS, however, does much more to affect how a network is engineered than the basic class of service parameters defined within ATM. Called by some "ATM without fixed length packets," MLPS uses very small standardized headers having labels which contain routing information specifying the route of the packets from source to destination. (MLPS can theoretically be applied to any type of packet to force constant-routing of a packet stream.) By way of the labels it attaches to packet streams as they enter the network backbone, the protocol essentially allows IP-style routing to occur over ATM, which means that packets define their routes through the network dynamically, based on their destination, as opposed to requiring hop-by-hop SVCs as normally is the case with ATM. The end result is an infrastructure that is faster, more scalable, and less expensive, like IP, but which can nonetheless effectively differentiate among the various types of traffic it is carrying, just like ATM. Whether MPLS turns out to be effective on a large scale remains to be seen.

UDP

UDP is described in the Internet Engineering Task Force document RFC 768. UDP is part of the TCP/IP suite of protocols which literally make the Internet work. UDP is the lesser known brother of the more famous TCP. Both of these host-to-host protocols run on top of IP networks, but most packet traffic on the Internet is TCP. A TCP connection lets one process send a stream of data and ensures that the other end receives the exact same stream. TCP is used for many applications such as e-mail transport and Web browsing. It's also used by Telnet for remote interactive logins and by *FTP (File Transfer Protocol)* which uses two TCP connections (a control connection and a data connection) to transfer files across the Internet. UDP is called an "unreliable connectionless" packet protocol for network transport, while TCP, a connection-oriented protocol, is described as a "reliable transport." This is because UDP offers few error recovery services, with no attempt made to detect dropped or out-of-sequence UDP packets, whereas implementations of TCP make use of sophisticated error-correcting and retransmission techniques to prevent data loss. Indeed, TCP is often used to detect lost/duplicated IP packets and fix things up since IP itself, while responsible for getting packets from the source endpoint to the destination endpoint, is a best-effort protocol that doesn't guarantee delivery of packets.

UDP is a connectionless protocol in that it establishes no official communications "session" (setting up connections and then tearing them down when finished) — it simply allows datagrams to travel directly from one host to another host's particular port (a port is like a mailbox at the receiving end — sending information to them via UDP is efficient since there is no handshaking process between the two hosts; the system assumes that an application is always ready and waiting at the destination to receive the packets coming through the port). Since UDP allows applications to send IP datagrams to other applications without having to initially establish a connection and then release it later, communication speed is quite fast, albeit at the possibility of delivering lost or out-of-sequence data. Another difference between TCP and UDP is that TCP allows the transmission of data streams of essentially unlimited size, whereas UDP is limited to about 65,507 bytes.

But since UDP does so little, it enjoys high performance, unlike TCP, which uses substantially more operating system resources than does UDP and is too slow to support real-time applications. Moreover, there are 20 bytes of header overhead in every TCP segment, whereas UDP only has 8 bytes of overhead, so more users can run client application software connecting to a server over UDP than TCP. Without such a burden of overhead, UDP thus offers high scalability and so can be found in heavily-used environments where servers must

Chapter Five

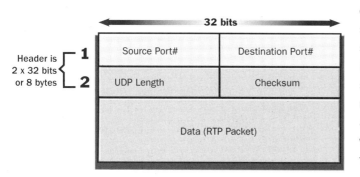

Fig. 5.14. UDP segment.

continually handle many simultaneous clients. Four such environments are instant messaging systems such as ICQ, voice-over-IP, multicast communications applications such as network audio and video conferencing, and online game servers (hosting games such as Quake, Baldur's Gate, Diablo, Star Trek, etc.) which depend more on data delivery being prompt rather than ordered and reliable.

Like all packets, UDP datagrams consist of a UDP header and some data. The UDP header contains the following fields:

- Source port (16 bits): Port number of the sender.
- Destination port (16 bits): Port number of the intended recipient. UDP software uses this number to demultiplex a datagram to the appropriate higher-layer software (e.g. a specific connection).
- Length (16 bits): Length of the entire UDP datagram, including header and data.
- Checksum (16 bits): Checksum of entire datagram (including data). The checksum field includes a 12-byte "pseudo-header" that's not actually part of the UDP datagram itself. The information in the pseudo header comes from the IP datagram header:
- IP source address (4 bytes): Sending machine.
- IP destination address (4 bytes): Destination machine.
- UDP Length (2 bytes): Length of UDP datagram, as given by the lengths in the IP header.
- Protocol (1 byte): Protocol field of the IP header.
- Zero (1 byte): One byte padding containing zero.

Each application running over TCP or UDP distinguishes itself from other applications by reserving and using a 16-bit port number. Destination and source port numbers are placed in the UDP and TCP headers by the originator of the packet before it is given to IP, and the destination port number allows the packet to be delivered to the intended recipient at the destination system. TCP or UDP examines the port number in each received frame and uses it to determine which server gets the data. Both TCP and UDP have their own similar set of port numbers.

Thus, at a higher level, it becomes clear that the real motive for introducing IP routing functionality closer to the edge of DSL networks, and accompanying it with an effective QoS mechanism, is to create a more direct interface to an Internet-based infrastructure and to the many differentiated services that this can make possible. While ATM will likely still be used as a common transport method in DSL lines, a blended network based on MPLS can more easily provision services such as IP VPNs, while remaining agnostic to the underlying transport layer. Such a network architecture will facilitate a more transparent and standards-based deployment of voice-over-IP in the local loop, by tying subscribers more tightly into Internet-based applications and still effectively managing QoS.

Since VoIP quality of service hasn't been fully worked out, it makes perfect sense that, at least in the access part of the network (the local loop), the IP voice traffic of VoMBN is transported over a DSL line using the reasonably robust ATM AAL5 (although it is possible to send data via the even more robust AAL2 normally reserved for ATM voice). For the moment, then, both IP and ATM must rely on each other to varying degrees.

The Particulars of VoMBN

At the customer premises, VoMBN equipment is similar to that for the Broadband Loop Emulation Service — it's the now-familiar box called an IAD. On the network side, however, a regional multiservice IP or ATM network, consisting of softswitches, call signaling and processing servers and various gateways and gateway controlling devices, replaces the Class 5 voice-switching network (we'll examine each of these components in detail shortly). Used together as a distributed switching system, these components perform call control and routing for subscriber telephony traffic placed between packet voice endpoints or between packet and circuit-switched subscribers. Once in place, such an architecture makes it easy for a service provider to add voice services to existing data networks.

Transporting a Voice Packet over VoMBN using IP and ATM

Under IP-based VoMBN, the customer IAD must at minimum act as an IP gateway. The IAD must first digitize the customer's voice channels or else simply accept pre-coded and packetized voice from an IP-PBX, IP phone or similar device. The voice packets are then mapped by the IAD into *Real-time Transport Protocol (RTP)* which carries the actual media; it "timestamps" real-time packet data such as voice and video, giving it a higher delivery priority than ordinary "connectionless" data (during silent intervals, the timestamp can be used to determine the duration of silence). It also generates sequence numbers that allows packets received out of order to be resequenced into the proper order, at the correct time, and synchronized between streams. RTP is supported by RTCP *(Real-time Transport Control Protocol)*, which is a control channel that carries status and control information (QoS, synchronization information, participant identification, session control) via the *Transmission Control Protocol (TCP)* so that the quality of data delivery can be monitored, even if large multicast networks are involved. RTCP packets are sent at intervals to the same IP address as the RTP session but on a different UDP port number. Participants in a call (even a multiparty call) can leave the call by sending a "BYTE" RTCP packet.

Actual UDP port selection for RTP sessions depends on what type of call signaling is used. In an H.323 environment, H.225.0 signaling determines which UDP ports will carry the RTP traffic. In a SIP environment, a SIP message sender indicates via a session description the UDP ports on which it will receive the RTP stream.

After the voice packets are mapped into RTP, they are then mapped into UDP, which lies at the next level below in the protocol stack.

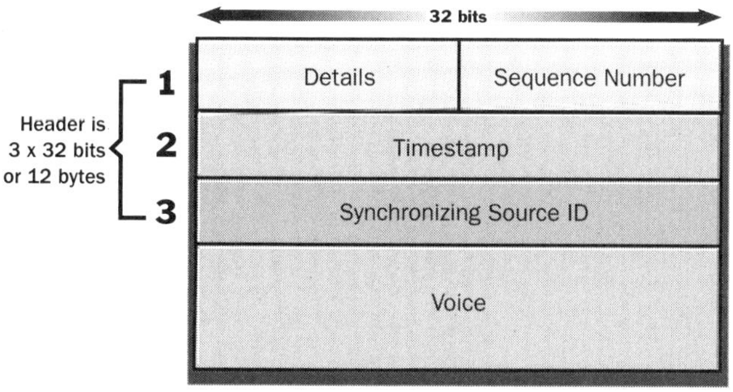

Traditionally, UDP, not TCP has been used as a transport layer protocol for such real-time applications. One might wonder why anyone would want to use UDP in any way for voice transport, since it follows the datagram concept of having neither control over the order in which packets arrive at the destination nor how long it takes them to get there, and both of these factors are important *vis a vis* voice quality. Ironically, this apparent weakness is UDP's strength. UDP is so simple, such a "raw" network interface, that UDP can meet the demands of delay-sensitive real-time applications so long as they can implement their own flow control and retransmission schemes. This is why in VoIP applications RTP packets are nestled in the UDP packets: RTP enables the receiver to put the packets back into the correct order and not wait too long for packets that have either lost their way or are taking too long to arrive. Voice doesn't need every single packet to be accounted for, but a natural, continuous flow of many packets in the correct order is necessary to acceptably reproduce speech.

Fig. 5.15. RTP Segment Structure.

The pseudo header provides additional validation that a datagram has been safely delivered. To see why this is appropriate, recall that because UDP is a transport protocol it really deals with transport addresses. Transport addresses should uniquely specify a service regardless of what machine actually provides that service.

So, aside from the multiplexing/demultiplexing function and some light error checking, UDP adds nothing to IP. In fact, UDP "talks" almost directly with IP. UDP takes messages from an application process, attaches source and destination port number fields for the multiplexing/demultiplexing service, adds two other fields of minor importance, and passes the resulting "segment" to the network layer (the IP layer does not care if a packet comes from UDP or TCP, since everything after the IP header is simply considered data). The network layer encapsulates the segment into an IP datagram and then makes a best-effort attempt to deliver the segment to the receiving host. If the segment arrives at the receiving host, UDP uses the port numbers and the IP source and destination addresses to deliver the data in the segment to the correct application process.

IP traffic is likely to be transported over ATM using AAL5 to get over the DSL line. Techniques for achieving this include encapsulation methods such as "Classical IP over ATM" (documented in RFC 1483/1577) and MPLS with the aid of Point-to-Point Protocol over ATM (PPPoA) and Point-to-Point Protocol over Ethernet (PPPoE). Most IADs may already use these methods to send IP data traffic over ATM over the xDSL access line, so there's no additional cost for transporting voice packets in a similar manner.

Like all of the AAL protocols, AAL5 accepts data from the upper-layers (i.e., USER layers) of the end user and prepares it for handoff in 48-byte segments to the ATM Layer of the ATM protocol stack. The ATM Layer then provides each segment with an appropriate header and hands the 53-byte ATM cell down to the Physical Layer for transmission over the ATM network. AAL5 is the most popular of the AAL protocols since it supports both connection-oriented and connectionless data. AAL5 was originally called the *Simple and Efficient Adaptation Layer (SEAL)* because the SAR sublayer simply accepts the *Convergence Sublayer Protocol Data Unit (CS-PDU)* and segments it into 48-octet (byte) *Segmentation and Reassembly Protocol Data Units (SAR-PDUs)* without reserving any bytes in each cell. AAL5 makes the assumption that error recovery is performed by the higher layers, so that all 48 bytes of the payload may be used to carry data.

Ironically, it is possible to send voice directly over AAL5, but this can only be done end-to-end: from one ATM desktop to another. There is no standard for interworking AAL5 voice into the PSTN for switching to non-ATM destinations. However, AAL5 is the easiest AAL to work with, since it doesn't necessarily distinguish voice from data traffic, and is a perfect fit for IP, since most VoIP systems do not as yet mark or prioritize voice packets over data packets. In an IP-based VoDSL environment, then, we tend to send voice indirectly over AAL5, thus once again taking advantage of ATM's robust nature.

VoMBN and the "Next Generation" Network

As the reader may recall, the point of performing all of these protocol gymnastics with VoMBN is so that the customer doesn't have to connect to a Class 5 switch at the central office. Instead, they rely on less costly devices to perform the same functions that comprise the *Next Generation Network (NGN)*.

In a traditional circuit-switched network, Class 5 and Class 4 switches concentrate all network activity, performing all of the call control functions required in a voice network, including switching, signaling and services functions. In a genuine next-generation model, the core features of traditional Class 5 switches (such as the Lucent 5ESS and Nortel DMS-100) and the GR-303 gateway are "decoupled" or "decomposed," broken out into several discreet functional units, each typically implemented in its own standalone box. These boxes can be distributed throughout the carrier network at optimum points, and properly sized for each point of presence. Also, by separating the call control layer from the service and transport layers in this way, and

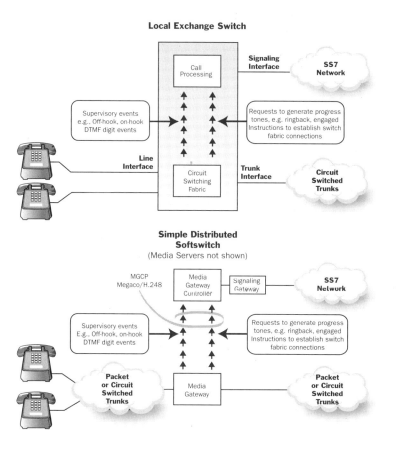

Fig. 5.16. Comparison between a Local Exchange Switch and a Softswitch.

using a common API across different vendor products, the carriers, service providers and developers will now enjoy the freedom to mix and match best-of-breed products and applications.

There are five main components in a next-generation VoIP network

- The "Softswitch" which encompasses the workings of all the components.
- The Media Gateway Controller (MGC) is responsible for mediating call control (including call setup, tear-down), between the signaling gateway and media gateway and controlling access from the IP world to/from the PSTN.
- The IP / Internet Services Layer (IP/ISL) where media servers and other devices provide "value added" or "enhanced" services to the network.
- The Signaling Gateway (SG) is responsible for interfacing to the SS7 network (just like a Class 5 switch would) and passing signaling messages to the IP nodes. An SG can relay, translate, or terminate SS7 signaling.
- The Media Gateway (MG) handles the actual transport of the voice traffic, and is responsible for packetization (and depacketization) of voice traffic as it travels back-and-forth between the circuit-switched PSTN and the packet network on its way towards the destination.

When VoIP was first formulated by the ITU in the 1990s, everything was based on the H.323 standard, a suite of protocols which carries its own nomenclature. In the H.323 VoIP environment a media gateway controller is referred to as a "gatekeeper," while the signaling gateway and media gateway are known singularly as a "gateway."

A second standard, roughly similar though newer than H.323, has gained widespread recognition — the IETF model, which relies on the following protocols:

- **Stream Control Transmission Protocol** (SCTP): Along with its adaption layers, SCTP is a better replacement for TCP that makes up the "signal transport" or Sigtran protocol suite, which is typically used to carry SS7 messages back and forth between the signaling gateway and the media gateway controller.
- **The Media Gateway Control (Megaco)**: A control protocol also defined by the ITU as H.248. This is the control protocol between a media gateway controller and the media gateways (an earlier version by the IETF called MGCP is still in use).
- **The Session Initiation Protocol (SIP):** This is a call control protocol running between media gateway controllers or between media gateway controllers and IADs or SIP based phones (SIP being a competitor and successor to H.323 and other protocols). The MGCP and Megaco / H.248 protocols act as "glue," allowing the distributed softswitch architecture to work as a whole in mirroring the functions of the PSTN's SS7 architecture and thus emulate the functionality of central office switches found in circuit-switched networks. This results in a sort of "IP Central Office" that

Chapter Five

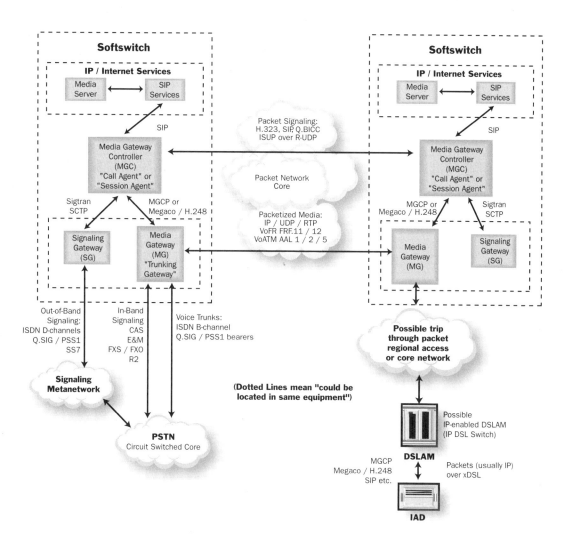

Fig. 5.17. Integrated PSTN and Internet Telephony.

relies on these protocols as control protocols to deliver services across the network via the media gateways. However, these two protocols differ from SIP's (and the Internet's) distributed service model. MGCP and Megaco / H.248 pre-supposes the existence of switching hardware and "dumb" terminal endpoints, while SIP abstracts the signaling layer from the network and pre-supposes "smart" terminals.

• **The Real Time Protocol (RTP):** This is specified by the IETF as the same RTP used by H.323 for carrying packetized media.

Our discussion of the NGN and its relation to VoDSL focuses on the model as illustrated in Figure 5.17. Let's now examine the "next generation" architecture in greater detail, since it is said to be a blueprint of the world's future communications infrastructure...

Softswitches are confused with Media Gateway Controllers

The media gateway controller has also been called a softswitch (along with call agent, and session agent). This is something of a misnomer, for although a media gateway controller performs many of the softswitch's "intelligent" functions, a true "softswitch" has actually become more of an abstract entity than a real physical "box," one that arises from the interactions of the other four components listed above. If we place the media gateway controller, media gateway and signaling gateway in one box, then we have a real, physical softswitch. But we can disperse these components around the world as individual, standalone boxes, and collectively they could still be termed a "softswitch." Of all the components, the media gateway controller has most of the intelligence and so tends to be mistakenly referred to as a softswitch. This is all possible because the components of a softswitch — the media gateway controller, media gateway and signaling gateways — are logical (not physical) entities themselves. In physical implementations, the functions may be kept distinct, even far apart, or else combined in single network nodes.

In North American PSTN networks, the signaling (SS7) and media/voice bearer links are carried on physically separate links. Thus, it makes sense to have physically distinct signaling and media gateways. In European PSTN networks, signaling and voice channels tend to share physical links. Thus, in a European "next gen" network it would make sense for the signaling gateway and media gateway to be co-located.

In any case, the overall "softswitch" acts as the intelligence in the network; as an example, it is the database that contains names and numbers, connects parties together and performs the billing. If enough components are grouped in a single box to form a real physical "softswitch," then the media gateway controller takes the form of software running on the softswitch.

With the distributed model, off-net traffic may be handed off to the PSTN at multi-

Note: Clever network architects can design an environment where the softswitch, although residing in an IP domain, can simply reach out and use the databases already existing on the SS7 network rather than maintaining a duplicate database of its own.

ple locations, with the call agent determining the optimum trunk gateway and routing for the call. On-net voice bearer traffic between packet voice endpoints is forwarded by distributed routing elements, including the IAD, IP-aware DSLAM (Cisco's version of this is called the IP DSL switch), and routers located in the core of the packet network.

But whether the softswitch is an abstract collection of dispersed components or a centralized real box, the softswitch communicates through the media gateway, DSLAM and over the xDSL line with the VoDSL subscriber's IAD, which itself contains signaling intelligence. The softswitch and IAD use a signaling protocol such as MGCP, Megaco/H.248, H.323 or the Session Internet Protocol (SIP) to provision the individual voice ports, dial numbers, indicate progress tones, and initiate features like caller ID and call transfer. The IAD becomes a SIP client (for example) delivering line-side features like call waiting and call transfer, while the media gateway controllers and other softswitch components in carrier regional offices act as servers, controlling all signaling, feature administration, billing and call routing for both line-side features and network trunk interactions.

The IP / Internet Services Layer

The next component, the IP / Internet Services Layer, is connected to the media gateway controller but also works with the media gateway. In this layer one finds media servers, devices that are generally responsible for "value-added" services on a voice network, such as voice mail, automated speech recognition (ASR), text-to-speech (TTI), call logging, faxing, multi-port conferencing, video, and interactive voice response (IVR). They also enable voice browsers and wireless web access which provide users with access to web content through voice commands, etc.

Like the overall softswitch, the media server is a packet-switched (IP) entity. All of the features supported could reside on one server or be split into separate servers (i.e. the ASR Server, the fax server, etc.) within the IP network. The media server features which present the biggest challenge to implement are conferencing (because different vocoders must run on the bridge) and video (because it involves the compression of data).

The IP / Internet Services Layer also contains servers or other devices handling SIP presence services, SIP proxy servers, etc.

Signaling Gateway

The conversion of the PSTN's out-of-band signaling is handled in conjunction with the signaling gateway, which connects to the North American SS7 network (or an equivalent signaling metanetwork in other locales).

The signaling gateway terminates the SS7 network call control data governing calls that come from the circuit-switched telephone network; it also manages the signaling and control interfaces between packet networks and the more conventional circuit-switched networks. The signaling gateway thus has the greatest concentration of switching intelligence, and so this is where manufacturers spend time and effort ensuring fault tolerance because it's going to control a lot of other equipment. (Unlike standalone media gateway controllers, which don't have to be as fault tolerant — data can be re-routed to another box in case of hardware failure).

For interworking with PSTN, IP networks need to transport signaling such as Q.931 or SS7 ISDN User Part (SS7 ISUP) call set-up protocol messages between the signaling gateway and other IP nodes, in particular the media gateway controller. The signaling gateway might even handle the transport of signaling ("backhaul") from a media gateway to a media gateway controller. As mentioned previously, the signaling gateway communicates with the media gateway controller via SCTP which enables *Sigtran*. Sigtran is defined in the IETF document RFC 2719 as "an architecture framework for transport of message-based signaling protocols over IP networks." Thus, Sigtran is a protocol stack for the transport of conventional circuit switched network signaling over an IP transport protocol. The protocol stack resides in the signaling gateways directly at the interface between the two networks.

The signaling gateway also generally connects to a remote database called the *IP-enabled Service Control Point (IP-SCP)*. This exists entirely within the IP network, but is addressable from the SS7 network via the signaling gateway. This is the IP version of the kind of SCP existing within the SS7 network that supplies the translation and routing data needed to deliver advanced network services. For example, the SCP translates an 800-IN-WATS number to the required routing number. It is separated from the actual switch (or softswitch), which makes the process of introducing new services on the network much easier.

Media Gateway

The media gateway controller also connects to and controls the media gateway. Media gateways (or "media translation gateways") translate between networks that follow differing communications standards. In most schematic diagrams detailing softswitches, the media gateway is depicted as a "trunking media gateway" that sits between the packet-switched and traditional circuit-switched domains, taking the voice streams on TDM bearer circuits from the traditional telephony environment and converting ("transcoding") them to packets for the VoIP (or VoATM) environment and *vice versa*. A TDM (PSTN) interface sits on one side (e.g. T-1, T-3, OC-3, etc.) of the media gateway, and a packet (WAN) interface on the other (e.g. IP, ATM, Frame Relay, etc.). A standardized protocol such as the *Media Gateway Controller Protocol (MGCP)* may be used for sending call-setup, tear-down, and related commands from the media gateway controller to the media gateway. MGCP will probably be ultimately superseded by a newer standard, jointly agreed to by the ITU and the IETF, called H.248 by the ITU, and Megaco by the IETF (we shall refer to it as *Megaco / H.248).*

It would be a mistake to categorize media gateways as mere "dumb" machines (like a digital cross-connect or Ethernet-to-token ring bridge), simply connecting two different end-points together as instructed by the external intelligence of the softswitch / media gateway controller. It is true that the protocols used to control them, MGCP and Megaco / H.248, derive their technological philosophy from the telco engineering world and suffer somewhat from the more traditional telephony paradigm of a nearly one-way interface between (or the "intra-domain control" of) "less intelligent" gateways and "more intelligent" switches. However, it is also true that media gateway technology has been at the center of the "convergence" phenomenon that places voice traffic onto packetized networks (since they essentially reduce the need for multiple networks, converging all of their transport media to a single IP-based network), and they are capable of many functions in their own right.

Note: A similar rapid evolution happened to DSLAMs, which were once little more than simple boxes routing traffic as it came through but then became more sophisticated devices that could handle various types of DSL and enable many value-added services to be delivered to customers.

Media gateways adapt the packetized traffic (using compression and echo cancellation), create and attach IP headers, set up media paths and send the packets on their way through the network according to the media gateway's instructions. For example, packets arriving from the IP network are received at the gateway interface, input to a jitter buffer for "dejittering" (creating a smooth and fairly constant playout by adding sufficient delay, i.e. transforming unacceptable jitter into delay, which is not desired but less objectionable than jitter) and re-ordering the packets in case any were received out of sequence (this is quite likely in an IP network where packets do not all take the same route to their destination). The packets can then be handed off to the voice codec which decodes the packets and decompresses the speech into normal audio before sending it to the TDM network. Usually the codec will implement a *Packet Loss Concealment (PLC)* algorithm or suite of algorithms to mask the effects of lost voice energy by adding comfort noise, replaying the last packet, or simply interpolate what the lost voice segment would have sounded like.

Lost packets are a major problem on IP networks. Lost packets may disappear while traversing the network or be the result of "jitter buffer discards," which occur when there is excessive delay in receipt of the packet, i.e. the delay time of the packet is greater than the delay added by the jitter buffering. In the PSTN-to-packet network direction, applications in the media gateway process the voice channel data (each being a typical pulse code modulated 64 Kbps channel) using a board containing some *Digital Signal Processors (DSPs)* to transcode (and compress if need be) the voice into IP packets that can be transmitted over the IP network.

Telchemy (Suwanee, GA — 770-614-6944, www.telchemy.com) is a company that thinks voice quality assessment should also be a key media gateway function, which they have implemented via their VQmon technology. After all, aside from actually listening to sound, how else would a customer know if the network is delivering packets "appropriately" or if the jitter buffer is properly "tuned" or if the PLC actually works? Something is needed to inform the customer and/or service provider if the calls are good (toll quality) or bad.

Telchemy also notes that data quality should also be included in media gateway technology. Gateway vendors rarely tell their carrier customers that their subscribers may not be able to use their products for modems, so fax and dial-up must use other kinds of access points. Conventional packet-loss concealment and error-recovery techniques for voice does not work for data. Concealing a lost packet from a high-speed modem transferring a software program can lead to disastrous results. Modems are much more sensitive to delays and jitter than voice, and as such, present a problem and will fall back to slower rates or even disconnect when encountering network conditions that do not adversely affect voice calls. As with Voice over IP (VoIP) and Fax over IP (FoIP), there is much effort going into the creation of techniques and standards to address these challenges.

To resolve these issues, the gateway must detect the type of call and apply the necessary processing resources to the call stream to handle it effectively. Some gateways try to carry modem traffic over packet networks by establishing a transparent VoIP G.711 (64 Kbps full-duplex) channel that carries the raw modem samples over the network between two media gateways. In such a scenario, the voice-related Echo-Canceller (ECAN), Voice Activity Detector (VAD) and Comfort Noise Generation (CNG) mechanisms are disabled. This approach wastes bandwidth and doesn't handle packet congestion very well, and any resulting change in packet delay, packet loss, or jitter behavior will cause a modem to retrain, or disconnect.

In regular voice-enabled (managed) IP networks, where QoS conditions of 1% packet loss are common, modem throughput is expected to be very low when modern modem protocols (such as V.34, V.90 and V.92) are used. Again, the reason is that modems do not tolerate delay and jitter, and fall back to slower rates. Data-modem calls thus get dropped and faxes get messy or fail to transmit entirely. Therefore, just as every voice gateway has voice-specific stream processing that depends on the characteristics of the human ear and IP networks, the more capable gateways that offer robust transport of data-modem and fax streams use modem-specific processing for data modems and fax-specific stream processing for fax calls.

While these gateway vendors employ the ITU T.38 fax over IP recommendation in their equipment for handling fax calls, *Modem-over-IP (V.MoIP)*, the comparable standard for modems, is still undergoing development and ratification by several standards bodies. So, certain important interoperability standards have yet to materialize, a reminder that the "next generation" network technology (and the industry resulting from that technology) is still in its infancy.

For the purposes of our discussion, the salient point is that the gateway does a lot more than interconnect two methods of voice transport. As we've seen, codecs, packet loss concealment and jitter buffer technology are key attributes or parts of the media gateway function.

Our discussion of media gateways has until now centered on "trunking" media gateways. The IETF now recognizes various other kinds of media gateways. For example, if a customer has a local VoIP network running on the company LAN or VPN, yet needs to connect to the PSTN over conventional TDM T-1 / E-1 lines, then an *access gateway* would be installed. "Access gateway" has become a more generalized term for any mediation point that can connect traditional user network interfaces (such as ISDN or analog services) to a VoIP or VoATM network. The access gateway has the ability to pass the TDM call signaling onward to a softswitch/media gateway controller. One specialized form of access gateway interconnects two distant wireless networks (such as two IS-41-based wireless networks) via an IP-based network that is used instead of SS7 to transport short messages between the two wireless networks. Since such short messages are not time-critical, the TCP/IP network can be used to inexpensively interconnect the widely-separated networks.

Another kind of large-scale access gateway similar to a media gateway is the *VoIP gateway* or *Hop-on Hop-Off (HO-HO)* server installed at various locations around the world by service providers offering low-cost long-distance services. The budget carrier places its VoIP Gateway equipment in an ILEC central office, and passes voice-traffic onto the Internet for no-cost transport of the voice traffic to a similar VoIP gateway in the local area of the long distance party. A calling party usually reaches the low-cost VoIP gateway services by dialing an access number. One such service is "10-10-321" which, as its name implies, requires the customer to dial 10-10-321 + 1 + area code + destination party's telephone number.

Another specialized form of the access gateway is the highly popular *Network Access Server (NAS)*, a computer server used by an Independent Service Provider (ISP) to provide customers with Internet access. The NAS has interfaces to both the local PSTN service provider and the Internet backbone, and terminates modem calls or *High level Data Link Control (HDLC)* connections (HDLC is an ITU-TSS link layer protocol standard for point-to-point and multipoint communications) to provide connected customers with Internet access. The server receives a dial-up call from customer devices (such as a PC) wishing to access the Internet, performs user authentication and authorization via a user name and a password, then allows packets to flow to other hosts (computers or IP-enabled devices) on the Internet. An early example of NAS technology was the Cisco AS5300 server, which essentially combined a series of dial modem banks and a router. With the coming of xDSL modems and VoDSL, early DSLAMs containing banks of xDSL modems took over many of the NAS functions, and, in conjunction with IADs, established ATM PVCs for voice (to voice gateways) and ISP routers (to the Internet backbone). This meant that service providers needed to provision the subscriber in several different places (in the DSLAM and any other aggregation devices) before they could offer services.

With the transition to more advanced, IP-centric, distributed next generation softswitch architectures and VoMBN, it was decided to do something about the ungainly process of creating many virtual circuits from the gateways to the IADs. Cisco took the initiative by introducing the Cisco 6000 series *IP DSL switches* which introduce the functionality of IP in traditionally Layer 2 DSLAMs. Such IP-enabled DSLAMs combine DSLAM and large-scale aggregation functionality into the same device, aggregating the subscriber virtual circuits by using common access encapsulation methods. Also, these IP DSL switches can act as a so-called *Provider Edge (PE)* device for MPLS cores and can offer VPN services. An IP DSLAM acting as a PE router can associate the incoming subscriber traffic to their corresponding VPNs. A PE typically is situated at the edge of the service provider's network and communicates with various routers or IADs in the customers' networks and executes functions that were previously handled by an external device other than the DSLAMs.

Service providers not wishing to take advantage of the IP functionality in the IP DSL switch can continue using the current ATM architecture, in which case the DSLAM will act as a traditional ATM switch. As networks migrate to the next generation model, most service providers will probably want to use the IP DSL switch's IP and ATM functionality (which Cisco calls their IP+ATM services architecture) and will thus have the flexibility to provision a subscriber on one network object (the IP DSL switch), aggregate subscriber traffic from various regions and forward it to their final destination using MPLS VPN, IP routing or tunneling. The IP DSL switch can provide IP multicast services — therefore, for multi-tenant dwelling unit applications such as apartments or hotels, an IP DSL switch in a basement can be connected to interactive video gaming servers and video on demand servers. A professional park with small offices can now offer server backups, web hosting, video training and other IT-based services, and the IP DSL switch can be used to bring web-caching to the edge of the network.

At the other, smaller, end of the gateway spectrum, residences or SOHOs would use a *Residential Gateway* (essentially another name for an IAD in the VoDSL world) to connect up to 10 or so analog telephony devices to an IP network over a broadband xDSL line employing ATM, Frame Relay or Ethernet to move the IP packets.

Distributed vs. Centralized Next-Gen Networks

Things do get confusing for those trying to understand the softswitch when equipment manufacturers decide to centralize or "unify" the next generation architecture to varying degrees by "bundling" two or more of these functional concepts together in a single product.

Since the VoDSL IAD must interoperate with both transport gear and the softswitch, some experts have predicted that the industry may reverse the trend toward distributed processing and collapse softswitch functionality (at least in part) into the media gateway itself. IP-based gateway vendors such as TollBridge Technologies (Santa Clara, CA — 408-585-2100, www.tollbridgetech.com) believe this gives them an advantage when it comes to creating new services like Caller ID with Web-Push technology or voice/data Follow-Me services that use DHCP for dynamic IP addresses, in a sort of local number portability scheme. Then there's the ICS2000 gateway from Convergent Networks (Lowell, MA — 978-323-3300, www.convergentnetworks. com), which has so many features it's referred to as a broadband switch.

Possible configurations of softswitch components lumped together include a media gateway controller + media gateway; a signaling gateway + a media gateway; a media gateway controller + media server; a media gateway + media server; or even bundling everything together into a single box that can be called a softswitch. Since the softswitch components reside on a so-called "open" architecture as opposed to its traditional counterparts, the PBX and Class 5 switch, they tend to run on a workstation or rackmount PC rather than on dedicated hardware. One can add functions simply by

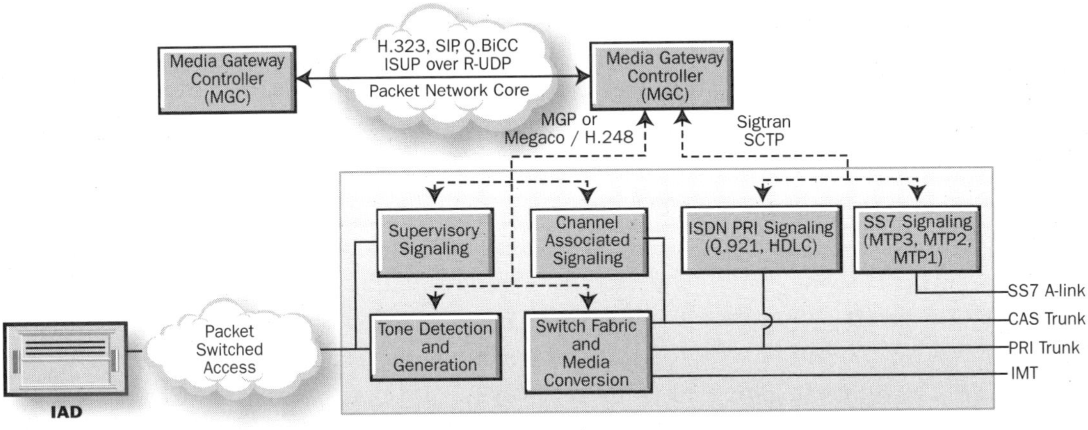

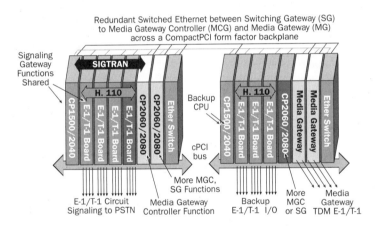

Fig. 5.18. Media Gateway for packet-to-circuit access with integrated signaling gateway.

Fig. 5.19. Sun Microsystems' design for a complete softswich in a rudundant, eight-slot CompactPCI chassis. Plug-in resource boards enable the media gateway controller to be easily combined with the signaling controller and the media gateways on the same backplane.

plugging in a different set of resource boards into the PC backplane. Hence, softswitch component boundaries probably will never be clearly defined and both vendors and service providers will continue to experiment with different approaches.

One totally integrated softswitch architecture can be found in the Converged Local Exchange (CLX) from Gluon Networks, Inc. (Petaluma, CA — 707-285-4001, www.gluonnetworks.com), a compact and powerful next generation broadband enabled Class 5 local switch. It integrates virtually all of the network elements and functions found in a central office environment into a single compact system, including voice and DSL physical access, voice and data switching and routing, circuit-to-packet interworking, signaling and transport.

The CLX brings softswitch distributed intelligence to a more efficient point of presence — the physical interconnect layer of the end office — rather than higher in the net-

Fig. 5.20. Several examples of the Gluon Networks CLX, shown here in a central office setting. The model at right has the front cover removed to show the system's inner workings.

work cloud, which enhances the service provisioning process, simplifies operations and maintenance, and improves survivability (lifeline service) in emergency standalone situations.

In addition to complying with new federally mandated capabilities such as CALEA and LNP, the CLX fully supports revenue generating legacy Class 5 services and advanced services while providing the carrier with a simple transition plan to the all-packet infrastructure of tomorrow. For instance, the CLX supports from 50 to 50,000 lines, and allows network migration to one consisting entirely of packets.

Voice-over-Ethernet-over-DSL

Voice has been sent over Ethernet for many years, originally for Intranet communications over an office LAN. More recently, 3Com offers its NBX system which runs a voice over Ethernet protocol and can handle up to 100 users on one box (the voice over Ethernet protocol used is a 3Com protocol and is not an open standard).

Since Ethernet can be sent over DSL, IP packets containing voice can be sent via this method instead of relying on ATM as the underlying transport protocol. Such Ethernet-based systems typically use the *Point-to-Point Protocol over Ethernet (PPPoE)*, a simple protocol popular among service providers for facilitating residential broadband Internet access by, among other things, obtaining a dynamic IP address from an ISP as opposed to providing a dedicated static IP address. PPPoE takes standard PPP packets (containing IP packets) and inserts them into Ethernet packets. PPPoE provides the ability to connect a network of hosts over a simple bridging access device (modem, router or IAD) to a remote *Access Concentrator (AC)*. Under such an architecture, each host uses its own PPP stack and the user's interface is unaffected. When connecting with the outside world, access control, billing and type of service can be

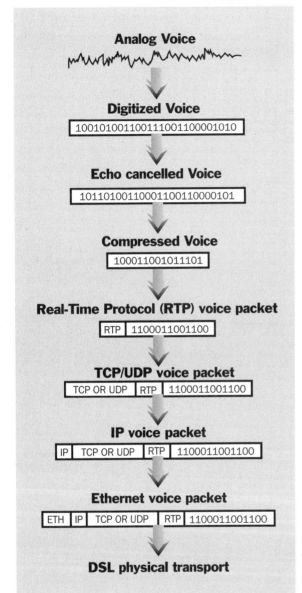

Fig 5.21. Shown here are some of the basic steps necessary to convert analog voice into an Ethernet packet ready for physical transport by xDSL. Other protocol encapsulations are possible and are not pictured, such as PPPoE.

performed on a per-user, rather than a per-site, basis.

For example, the Business Communications Manager (BCM) from Nortel Networks Corporation (Brampton, Ontario, Canada — 905-863-0000, www.nortelnetworks.com) delivers an integrated communications system to multi- and single-site businesses and is capable of delivering both IP-enabled and pure-IP solutions. The Business Communications Manager integrates key system/ PBX capabilities, VoIP gateway functions, and quality of service data routing features in a single system.

Nortel's 2002 release of the BCM includes support for PPPoE which allows the system, working as a router, to use a DSL or cable modem connection to acquire a dynamic IP address with the customer's ISP. When that occurs, the BCM thinks it's dealing with just a different IP-based Wide Area Network (WAN) connection, even though the underlying transport is an Ethernet connection into the DSL modem instead of ATM or Frame Relay. And any of the devices behind the BCM (such as IP phones) using Nortel's address translation capabilities can communicate out of the customer premises, or even PCs on a LAN behind the BCM can send data traffic out onto the service provider's network. Ethernet systems treat IP voice packets literally as just another form of data.

Since the voice packets are not differentiated from the data packets, one wonders how quality of service can be maintained, since Ethernet tends to be "bursty" in nature? PPPoE can encapsulate non-IP protocols. Any protocol which can be encapsulated by PPP can be sent via PPPoE, but PPPoE can be annoying in that the maximum Ethernet frame allowed is 1518 bytes long. 14 bytes are taken by the header, and 4 by the frame-check sequence, leaving 1500 bytes for the payload. For this reason, the *Maximum Transmission Unit (MTU)* of an Ethernet interface is usually 1500 bytes. This is the largest IP datagram which can be transmitted over the interface without fragmentation. PPPoE adds another six bytes of overhead, and the PPP protocol field consumes two bytes, leaving 1492 bytes for the IP datagram. The MTU of PPPoE interfaces is therefore 1492 bytes.

Nortel's BCM can compensate for this. As opposed to the Voice over DSL mechanisms where you actually are channelizing or identifying to the far end of the connection that this is specific voice information, the BCM treats a voice packet as just another piece of data, but the BCM marks the voice packets with a particular *DiffServ Code Point (DSCP)* field, formerly called the *Type of Service (TOS)* field.

In most cases the BCM would be deployed in a scenario resembling voice networking PBX-extension running over a VPN, where your business can route voice and data packets back and forth to branch offices over its own virtual network without having to use a telco for expensive voice calls. The calls are transparent to the transport mechanism, which simply establishes a packet stream connection among all of your offices over which your business can send any kind of packet. The VoIP packets traveling over Ethernet / xDSL and to an ISP can be made secure enough to travel over the Internet by tunneling out of the main office through a secure IPSec type tunnel, then across the Internet to the far end (a branch office) where the company's CPE can sort out what's voice and what's data by examining the DSCP for each packet.

A BCM in a San Francisco office can thus send all data to a New York City office and the NYC BCM, if it could speak, would tell us: "okay these are data packets, but I see a DSCP that tells me this is a voice packet, so I'm going to route this packet to my VoIP gateway and turn it back into PCM/TDM voice if that's the way the call needs to terminate."

Essentially, Nortel's system simply creates a PPP session within the Ethernet layer for transporting IP packets, some of which are voice and some of which are data. You don't have to terminate your call on a particular DSLAM to be able to interpret the way you're doing VoDSL, since you're working with a more abstracted system, one running on OSI Layer 3 as opposed to Layer 2. As long as a VPN connection exists, one can simply send voice packets without worrying what the underlying hardware is.

Interestingly, Ethernet has moved from the LAN out into the WAN world with the RBOCs offering *Metropolitan Area Network (MAN)* Ethernet access and some inter-exchange carriers are offering inter-state access. Ethernet has gone from a humble file and printer-sharing protocol to a full-blown WAN service, competing with ATM and SONET. Gigabit Ethernet services that will let customers connect geographically dispersed offices at speeds of up to 10 Gbps will become a dominant networking technology for both LANs and WANs.

For example, Occam Networks' Broadband Loop Carriers combine the functionality of a DLC, DSLAM, and media gateway to allow service providers to create Ethernet- and IP-centric access networks. They have an interesting feature called Ethernet Protection Switching (EPS), an Occam-developed technology that provides Ethernet with the reliability and resiliency of SONET for "five nines" (99.999%) network reliability. Occam offers an end-to-end BLC system that includes the BLC 1200 Remote Terminal, BLC 1240 Central Office Terminal and a line of remote cabinets for deployment flexibility.

Channelized VoDSL (CVoDSL)

This least expensive form of VoDSL is designed for the residential and SOHO markets. The CVoDSL approach eliminates the need for packetization of voice traffic over the copper loop into upper layer protocols, but gobbles up bandwidth and thus offers a fewer total number of available channels than a system based on packet transmission.

CVoDSL is singular among voice over DSL solutions in that it transports voice within the fundamental physical layer, rather than packetizing voice into ATM and IP packets before sending it over the copper phone line. This places it technologically at the first layer of the OSI Reference Model. For this reason, CVoDSL requires less processing by both the home and central office equipment and contributes almost no delay, or latency, to the access network. In ADSL systems, this allows the transport of derived voice channels while maintaining both POTS and standard-compliant full-rate or G.Lite DSL data access. In its simplest form, CVoDSL reserves well-defined physical layer "channels" of DSL bandwidth over the local loop to deliver standard 64 Kbps PCM DS-0s from the DSL CPE to the next-generation access equipment. The access equipment then transmits the voice DS-0s directly to the circuit switch via PCM.

CVoDSL operation on DSL implementations were not standardized prior to full-rate ADSL, but these chipsets are still capable of working within a standards-compliant ADSL system. SDSL and G.SHDSL do have standard-compliant mechanisms that enable the use of the CVoDSL method.

CVoDSL access equipment could alternatively packetize the DS-0s into ATM or IP for transport to a media gateway or packet switch. The result is a simple, flexible, cost-effective method to enable next-generation equipment with derived voice functionality.

CVoDSL reserves a channel of upstream and downstream DSL bandwidth for each voice line, while allocating the remaining bandwidth to other applications such as surfing the web. When using PCM voice with no compression, each voice channel consumes 64 Kbps of bandwidth in each direction without overhead. ADPCM voice compression can also be used to reduce bandwidth consumption to 32 Kbps per voice line. The voice bandwidth can be dynamically allocated, so that when the voice lines are not in use, the bandwidth is utilized for data traffic. Multiple voice lines can be active simultaneously, depending upon available bandwidth.

CVoDSL offers residential and small office customers true toll-quality voice like that of standard POTS phone call, plus all of the additional features such as call waiting that this group of customers value.

Requirements for successful deployment of VoDSL are presented in DSL Forum's TR-0391, the goal of which is to ensure that a subscriber will not perceive any degradation in voice service as the result of the service being delivered by means of VoDSL.

CVoDSL makes efficient use of bandwidth, conserving the upstream bandwidth. It delivers voice and signaling like a POTS call, making it easier to integrate CVoDSL into existing legacy networks. Although at first glance CVoDSL may appear primitive or a throwback to earlier technology, it is in fact extremely versatile in its implementation, fitting seamlessly into a number of network architectures.

Other, competing VoDSLs do have a head-start on CVoDSL, so it should be interesting to see how it fares.

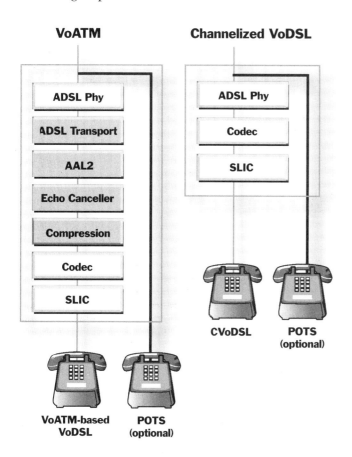

Fig. 5.22. CVoDSL has a less complicated architecture and therefore offers cost savings over VoATM-based VoDSL.

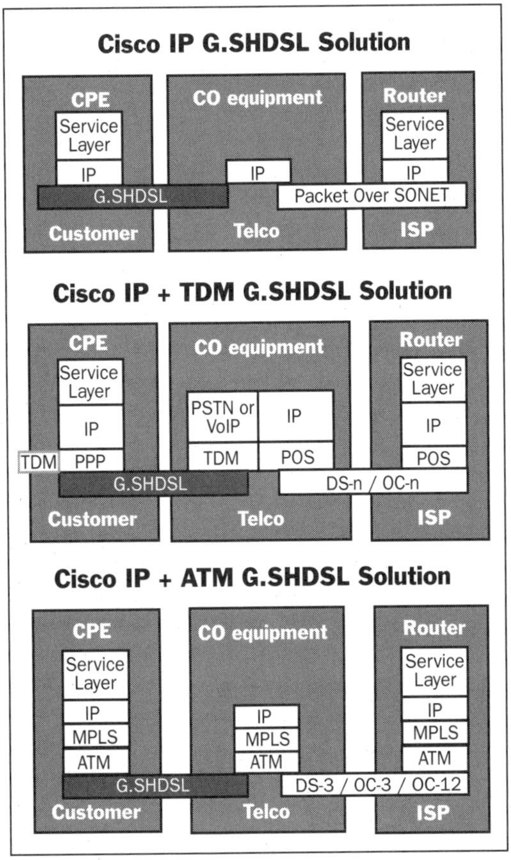

In Summary

If we look at each one of the four roads to VoDSL in terms of the OSI Reference Model, we see that voice can make its way over DSL at different levels in the protocol stack. Be it voice directly over DSL, or voice over ATM over DSL or voice over IP over DSL or voice over Ethernet over DSL.

Most forms of VoDSL will rely heavily on the ITU's G.SHDSL protocol, the first internationally recognized xDSL implementation. G.SHDSL is also perhaps the most flexible symmetrical DSL that takes into account ATM, IP and channelized transport of voice and data (for example, it has provisions for mapping ATM cells directly into the DSL channel).

Which VoDSL road is for you? Before deciding, perhaps you should read chapter 7, which discusses the considerations that should go into choosing a VoDSL service provider.

Fig. 5.23. Cisco's access network solutions demonstrate G.SHDSL's versatility. These three architectures, IP, IP+TDM and IP+ATM, can use G.SHDSL in the local loop.

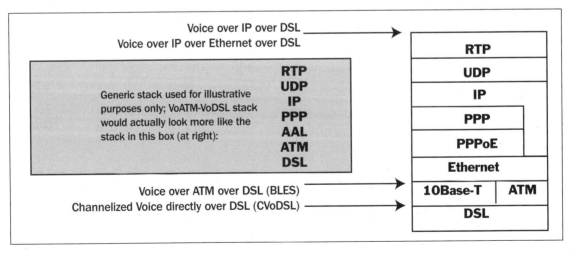

Fig. 5.24. Voice entering the stack.

CHAPTER SIX

Customer Premise Equipment

DSL, rather than traditional T-1/E-1 or ISDN lines, is quickly becoming the telecommunications access method of choice for small / midsize enterprises (SMEs). These same companies find DSL more appealing than T-1/E-1 (or ISDN) for merging their data and voice communications systems. Voice-over-DSL not only represents the next step in telecommunications, it also represents a tremendous evolution in the way service providers deliver telecommunications services to their business and residential customers. With VoDSL, customers can realize lower voice and data service costs, while retaining the high-quality voice and high-speed data connections they want and expect. This allows both the customer and service provider to maximize their equipment investments.

Combined voice/data hookups offer a simple broadband solution for many VoDSL customers. The target market for VoDSL (residential, SOHO and SMEs) rarely needs more than a few voice lines in addition to their data feed. For example, one of the really interesting opportunities in VoDSL applications is with key systems. Customers with key systems typically have somewhere in the range of four to 20 POTS lines, and the opportunity is there for a service bundle that integrates with such systems.

Another selling point that the targeted market finds appealing is that they can still use many of their current telecommunications devices, such as analog telephones, IP telephones, fax machines, PBXs, and so on.

Indeed, Bizfon (Salem, NH — 603-870-9400, www.bizfon.com) has added VoDSL capability to its Model 680, a smart little phone switch that has up to six outside lines and eight extensions. Its features include eight private voicemail boxes, five-party conferencing, auto attendant, automatic fax detection, and internal and external call forwarding. You can also dial into the box from anywhere and get a local dial tone.

However, service providers find themselves caught in a Catch-22 situation as they scramble to deliver more competitive differentiated services. They ponder how to reduce their overall costs while delivering new cutting edge services that can differentiate them from their competitors. In answer to this dilemma and to create competitive differentiated services so as to grow market share and reduce customer churn, service providers are offering integrated service bundles which may include local and long distance voice and Internet access, web hosting, back-up services and the like.

The IAD - The Enabler

Most DSL hookups have been concentrated in residential and small business areas where a simple piece of CPE, commonly referred to as a DSL modem (technically speaking, a kind of bridge), is sufficient. However, with the recent advances in DSL-related technology, especially VoDSL, more SMEs are entering into the converged voice-and-data marketplace. Essential to the uptake of VoDSL is the *Integrated Access Device (IAD)*. The most sophisticated versions of these complex devices combine the functionality of several other devices (such as a modem and a router) and enable service vendors to integrate voice, data, and Internet services onto a single network connection. Although the IAD, like the simple "DSL modem" is typically located on the customer's premises, it's a much more sophisticated piece of equipment. Fundamentally, the job of the IAD is to take a data connection and let you connect standard phone sets into it, ideally with multiple lines. Unlike the plain "DSL modem" all IADs can, at a minimum, inte-

Fig. 6.1. These CopperCom MXRs are IADs that can combine data and voice traffic into packets onto a single DSL line. POTS phones, fax machines, key systems, PBXs, PCs, and Ethernet hubs plug into the MXR for telephony and high-speed Internet access. Shown here is the MXR 300 (bottom) for the small business market that supports eight voice lines and data. At the top is the MXR 500 that supports 16 voice lines and data. All MXR models are compatible with industry leading DSLAMs. Not shown is the MXR 100, a basic four port fixed configuration device designed for the upscale residence.

grate multiple traffic types over a single access line through packetization and multiplexing of voice and data traffic streams so they can flow over the DSL line to the service provider's DSLAM and into the PSTN and/or Internet.

The IAD concept isn't new. Similar devices have been around for a few years. Traditionally, IADs have been used with existing PBX or key systems and predate DSL. The industry started off with very basic TDM-based, T-1/E-1 type IADs that took a T-1/E-1's worth of bandwidth and divided it up for voice and data applications. Now these vendors are producing DSL-based IADs, and even going a step further, to IP-based IADs for true next-generation, end-to-end IP infrastructures.

An IAD is what processes and routes traffic from the customer's premise. The IAD that provisions VoDSL typically possesses the following features and functions:
- A platform that's reliable, resilient, and scalable.
- A channel bank.
- A multiplexer for both voice and data.
- Routing capabilities.
- Ability to transmit not only frame relay/PPP (Point-to-Point Protocol) data but also use the DHCP protocol.
- A firewall.
- Remote management capabilities.

When considering a VoDSL service provider, find one that offers an IAD that can, at a minimum:
- Support cost-effective analog voice.
- Provide built-in IP routing.
- Provide an integrated Nx64 V.35 port.
- Support a full feature set as required by today's telecom network.
- Is highly integrated, i.e. TDM and ATM capable.

Differing IAD Architectures

An integrated access device or IAD is a compact, scalable access device that provides a platform to allow the combination of multiple network access functions. The IAD should provide support for between 2 and 24 analog or digital voice ports over a single DSL line. These ports can connect to any analog telephone, fax machine, analog modem, or key telephone system, with some IADs offering support for analog or digital PBX systems. The IAD should also support data connections to the PC via USB or Ethernet, and since it's connected to a LAN it should handle such network protocols as Ethernet, Token Ring, SNA, IPX, ATM25 and AppleTalk, along with all and sundry bridging, encapsulation and routing protocols. Some IADs will be able to support customer-located routers or Frame Relay devices via V.35 connections. The IAD should have compatibility with existing telephone equipment, including fax machines, analog telephones, modems, and PBX and key systems to ensure ease of migration.

Since the IAD can aggregate multiple services such as voice, data, and Internet access, a single access line into the customer's premise can replace multiple access lines that would otherwise be required to provision those different services.

The customer will typically find one of three fundamental types of IADs being offered by the VoDSL service provider market:

First is the *basic IAD*. This device performs the normal functions of a multiplexer (such as Coppercom's MXR 100); and when it is deployed in a customer's existing legacy network its primary function is to aggregate voice and data traffic emerging from the customer's network and to channelize that traffic onto a single network connection. For voice services, the IAD operates as a channel bank and will front-end an existing PBX or key system to aggregate voice traffic through either a foreign exchange station interface to connect to an analog key system or a digital system

IAD Defined

While the industry experts seem to agree that IADs are positioned to play a large role in the future of telecommunications, these same experts are less clear in their definition of an IAD.

In the past, IADs were defined as boxes with TDM architectures and backplanes that did TDM (not statistical) multiplexing. They were originally essentially just channel banks; i.e. devices that aggregate a company's 64 Kbps DS-0 voice streams and multiplex them into a T-1 line, or, conversely, splitting a T-1 into POTs lines for the company phones and providing battery and analog call supervision.

An IAD can do that and more. It can, depending on the amount of voice traffic, also aggregate the correct number of DS-0s for data traffic and provide a router interface for them or do the switching itself using a simple built-in router. An IAD has interfaces for analog telephony devices (e.g., FXS, FXO, E&M), as well as digital ones (e.g., CAS, CCS, Q.931, and Q.2931). It can take this and data traffic (such as Ethernet data) and send it all across multiple types of uplinks, such as the following:

- TDM just like a channel bank (TDM multiplexing).
- ATM where the voice is converted by the IAD to Voice over ATM (such as AAL1 or AAL2) and sent over a DSL uplink (or a T-1 if DSL is unavailable). To achieve good voice

Chapter Six

cross-connect interface to connect to a digital PBX.

On the data side, the basic IAD typically provides a serial port and a Nx64k interface for connection into the legacy data network, which normally will consist of bridges, hubs, routers, Frame Relay access devices and so forth.

Second is what I call an *office-in-a-box*. This IAD provides an all-in-one voice and data solution by supplementing the basic IAD features with integrated routing and "edge" or "DMZ" functions, such as Internet firewalling, Network Address Translation (NAT), the Dynamic Host Configuration Protocol (DHCP) service and other advanced protocol functionality.

The all-in-one office device is available in the market today. However, most of these fall short of some key carrier-class attributes such as:
- High reliability.
- Resilience (Telcordia NEBS standards).
- A robust operating system (e.g. real-time OS).
- Remote management capability with the ability to easily integrate into the service providers' higher end OS system.

---

quality, DSL lines often use ATM as the underlying data-transport protocol beneath TCP/IP.
- Packet where everything is converted to IP including the voice (VoIP) and the uplink is DSL, ATM AAL5 / DSL or Ethernet / DSL.

Many IADs provide voice compression, enabling more voice channels to fit in the same amount of available bandwidth. Moreover, some IAD's are now providing advanced functionality such as SIP, the Media Gateway Control Protocol (MGCP) and Megaco / H.248 to allow phone services to be controlled / delivered through softswitches. And, as previously pointed out, the IAD also offers many opportunities to service providers who want to bundle services.

In the old days, experts considered ATM access devices to be a separate product line. Now, since ATM is often used with VoDSL to achieve quality of service, all of those devices and product lines plus the new broadband devices, such as the DSL IAD, are commonly lumped together under the IAD umbrella.

At the same time, there's a move to collapse the functionality of multiple devices into a single product — the IAD — for ease of network management, lower capital costs and more efficient use of space within COs or other equipment sites.

For the purposes of this book, an IAD is a service provider-managed device that sits at the customer premises and aggregates traffic — in most cases both voice and data — and has an internal router that sends voice to the circuit-switched network and data to the cell- or packet-switched network.

Finally, there's the *next-gen IAD*. This device takes the office-in-the-box to the next level by providing the ability to gracefully migrate to Voice over Packet services. It's the ideal customer network edge device. The next-gen IAD can be deployed seamlessly into the customer's existing legacy network where it can offer a migration path to packet-based voice and data (ATM/IP) services via software download. Such an IAD can not only function as a customer premise equipment gateway, but can also be responsible for the translation of all legacy voice and data services to the new class of packet-based services and *vice-versa*. The customer can, therefore, retain all legacy network configurations including telephone numbers.

The most advanced IAD products offer the ability for the customer to set up a virtual private network (VPN) that can accommodate teleworkers and business partners, and engender a decentralization of an organization for security purposes. Such IADs are produced by Occam Networks (Santa Barbara, CA —805-692-2900, www.occam-networks.com), Merlot Communications (Bethel, CT — 203-730-1791, www.merlot-com.com) and Nortel (Research Triangle Park, NC — 800-466-7835, www.nortel.com). Advanced IADs should also be able to offer PBX- or Centrex-type voice capabilities, such as those from Merlot and VINA Technologies Ltd. (Newark, CA — 510-492-0800, www.vina-tech.com).

Today's state-of-the-art, next-gen IADs not only extend the economic and management benefits of convergence to the customer, but they also allow for cost-effective optimization of the local loop. These next-gen devices move the service demarcation point into the customer's network, thus distributing integration functions and enhancing network scalability for both the customer and the service provider, positioning the service provider to furnish a wide range of services, which enables customers to buy integrated solutions without the need to manage their own networks.

Value-Added Services

The IAD concept is still in its infancy. Most IADs only possess the ability to combine voice and data traffic before sending it on toward the public networks via xDSL. But many IAD vendors know that these devices can and should be able to do much more. The IADs on the drawing boards, as prototypes, and in trials have the ability to be the key service nodes through which service providers can offer a variety of voice, video and data applications.

The vendors envision the IAD moving from the transport level into the application level with an IAD platform that provides web access that acts as an e-mail server and has the ability to access any service. The IAD of the future will be not only be accessible through the typical PC, but also e-mail protected and safeguarded from over subscription.

With the industry-wide movement to take the intelligence out of the core network — switches and SS7 — and deliver it closer to the customer, future IADs at the customer's premise will deliver more services, give customers more control of these services and will do all of this over a lower-cost infrastructure. (The service provider only has to install IADs at its customers' premises, rather than invest in more expensive network-based equipment and then hope customers will sign up for service.)

As a baseline, these devices must be able to support dial tone and high-speed Internet access, but beyond that are higher margin services such as Frame Relay, VPN and security services. The question is, "where do you put these capabilities?" The least expensive place to put them and the place with the best security is inside the IAD.

The capabilities built into the IAD allow the service provider to effectively offer a portfolio of differentiated services, ranging from a voice and data service bundle to a fully managed family of value-added services. For instance, a value-added voice service may include the ability to deliver local Centrex services, obviating the need for the customer to have a standalone telephone system or to dedicate Centrex trunk lines for every telephone extension. Or perhaps the value-added services will be data-related and will include bundled Internet services with e-mail, virtual private networking, and even web-serving support. Some IADs even allow customer-based management for the feature-rich value-added services offered. For example, for a local Centrex service the customer could manage and custom-configure specific service features (e.g. turn 3-way conference calling on or off or institute a call blocking service when the situation dictates).

VPNs

As more service providers enter the VPN space, it will be possible for companies to extend the PBX functionality at their headquarters to remote locations because signaling can be securely sent out and synchronized between sites, and the IAD can support the ability to add such new services. Cisco Systems Inc. (San Jose, CA — 408-526-4000, www.cisco.com) is integrating VPNs into its IADs. (The company already offers VPN capabilities in some of its other edge devices.)

Among the vendors offering Centrex-like functionality are:

The aforementioned VINA with its Business Office Xchange (BOX), which allows small offices to have the equivalent of Centrex services because the device provides dial tone and can route calls.

Marconi plc (London, UK — +44-(0)-20-7493-8484, www.marconi.com) has added the Megaco/H.248 signaling stack to the Mariposa Technology product line (Marconi acquired Mariposa Technology). This allows for more intelligent conversations between end user and carrier equipment in order to interpret a phone number and determine where to route the call, so not all calls have to go through the CO.

Currently in the Centrex world, if phone A wants to talk to phone B on the same premises, it has to go all the way to the CO. But a "smart" IAD with an H.248 signaling stack would recognize that and, as a result, the company could recover some of the bandwidth on its DSL line used for unnecessarily routing calls to the CO.

Many IAD vendors want to support voice VPNs for office-to-office communications. In that scenario, the IAD could use the the Session Initiation Protocol (SIP), the Media Gateway Control Protocol (MGCP) or Megaco/H.248 signaling to point to a device, such as a call agent or a softswitch, that stores phone numbers and could operate a dial plan enabling customers to set up their own dial plans. A dial plan might specify that, for example: "any call preceded by the number 8 is long distance and any call starting with 9 is local." In fact, that's exactly what Smart Bandwidth On Command does from WorldCom Inc. (Clinton, MS — 800-465-7187, www.wcom.com).

Of course, there are many more enhanced services beyond VPN and PBX/Centrex services that the IAD can help carriers bring to market.

Another new feature that could be added to IADs in the future is the addition of wireless handset interfaces, which would allow for wireless LANs and in-building wireless telephony. Such applications would make sense for end users such as retailers whose salespeople use wireless handsets to communicate with other staff within the store environment.

Future IADs will support MPEG or a similar video protocol to enable devices to more easily access stored video on servers. Running MPEG on an IAD might be overkill, but there is a small (but growing) market for it. The idea here, as in much of the rest of the network, is combining multiple devices — in this case an IAD and a videoconferencing control unit — into a single device.

Although some IADs are evolving to include new features and functionality, souped-up IADs won't make sense for all applications. There will be two kinds of IADs in the next-generation network. One is a dumb device which I've dubbed the *thin IAD*, where the intelligence resides in the older centralized network. The thin IAD shouldn't cost more than $200 so it definitely makes sense in the residential and SOHO market space, where there is always a sensitivity to price.

Then there's the *thick IAD*, which is laden with PBX functionality and lots of other bells and whistles, as dictated by the target market. These devices will run anywhere from $500 upward to $3000, depending on the features and functions provided.

Meet the Vendors

Many TDM IAD vendors are expanding into the DSL IAD market. TDM IAD vendors include such names as Cisco Systems Inc. (San Jose, CA — 408-526-4000, www.cisco.com), Clarent Corp. (Redwood City, CA — 650-306-7511, www.clarent.com), Carrier Access Corp. (Boulder, CO — 800-442-5455, www.carrieraccess.com) and VINA Technologies (Newark, CA — 510-492-0800, www.vina-tech.com).

Also numerous ATM IAD vendors are moving into the DSL IAD market place. This group includes Lucent Technologies Inc. (Basking Ridge, NJ — 908-719-7657, www.lucent.com), Marconi plc (London, UK — +44-(0)-20-7493-8484, www.marconi.com), and Nortel Networks Corp. (Research Triangle Park, NC — 800-466-7835, www.nortel.com).

DSL IAD vendors — at least the ones that support both data and multiline voice — include such companies as Occam Networks (Santa Barbara, CA —805-692-2900, www.occamnetworks.com), CopperCom (Santa Clara, CA — 408-987-8500, www.coppercom.com), Jetstream Communications Inc. (Note: Jetstream, at one time the leading VoDSL equipment vendor, ceased business operations on April 12, 2002), Polycom (Milpitas, CA — 408-526-9000, www.polycom.com) and TollBridge Technologies Inc. (Santa Clara, CA — 408-585-2100, www.tollbridgetech.com).

Take Cisco, one of the most well-known vendors, as an example. The company has sold TDM IAD products for a number of years. But Cisco, realizing that ATM technology is making its way into edge devices like IADs to enable carriers to deliver QoS and VoIP and VoATM alternatives to SMEs, is growing its IAD product line to accommodate many eventualities — VoDSL capable IADs fall within this category.

Extensive price variations exist among IAD vendors; and service providers often give customers deep discounts on the IAD as an incentive to sign on to their services.

According to Research First Consulting (Birmingham, AL — 205-995-8866, www.researchfirst.com), a market research and consulting firm covering the telecommunications industry, in their 2001 report entitled *Enriching the Broadband Offering: the Small Business Voice over DSL Opportunity*, SME decision makers want to buy all of their communications equipment from a single provider. They don't want to buy this equipment from a Staples or Office Depot. Nor do they want to mix and match and deal with evolving standards. They are quite satisfied with leasing their IADs from service providers, just as they would other upscale data products.

There are a good number of vendors offering credible IADs. To give the reader an idea of the different IADs available, I give a small sampling of the typical IAD vendors who currently are offering IADs to the VoDSL service provider. Like the tail wagging the dog, many times the IAD will be the determining factor in choosing your VoDSL service provider.

3Com Corp. (Santa Clara, CA — 408-326-5000, www.3com.com) has rolled out a series of products. Its OfficeConnect equipment allows VoDSL service providers to offer the SME market alternatives in voice telephony that were previously available only to large enterprises. For instance, the OfficeConnect base unit boasts a converged voice and data platform or a data platform that can be easily upgraded to a converged system.

ADTRAN's (Huntsville, AL — 256-963-8000, www.adtran.com) offerings are the Total Access 604 and 608, which offer a variety of different network interfaces. These four- and eight-port IADs can be configured to handle a variety of broadband types, including ADSL, and G.SHDSL. ADTRAN's Total Access 850 is an expandable device that can handle up to 24 voice ports.

Carrier Access' (Boulder, CO — 800-442-5455, www.carrieraccess.com) Adit 105/205 family of products deliver up to eight or 16 voice lines, including an integrated IP router. The box can be used indoors or out, and it includes remote management, optional battery backup, and a lifeline feature that automatically switches calls to a separate POTS line if things go wrong.

CopperCom's (Boca Raton, FL — 561-322-4000, www.coppercom.com) MXR 500 line supports 16 voice lines and can meet voice and data transport requirements in both residential and small businesses.

Although Jetstream Communications went out of business in 2002, there were many of their IAD-402's put into operation, each of which delivers of up to eight telephone lines and high-speed Internet over an ADSL interface.

Netopia's (Alameda, CA — 510-814-5100, www.netopia.com) 4752 IAD works with SDSL on up to eight voice lines, and includes a built-in firewall, built-in VPN features, support for standard Centrex features such as three-way calling and caller ID, and an integrated DHCP service to enable automated IP address assignment. According to Netopia, their IAD works with a large variety of existing DSLAM equipment.

Polycom (Milpitas, CA — 408-526-9000, www.polycom.com) has several lines of IADs. The NetEngine 6104 and 6108 devices are ADSL IADs with four and eight voice ports, respectively. They work with a variety of DSLAMs and voice gateways, and include TFTP (Trivial File Transfer Protocol — a simple version of FTP protocol that has no security features) for software upgrades, IP filtering, and a variety of calling features. The NetEngine 6300-4 and 6300-8 IADs work with SDSL, a popular DSL flavor for the SME market.

Symmetricom (San Jose, CA - 408-433-0910, www.symmetricon.com) was one of the first out of the gate with a G.SHDSL product, GoWide IAD for G.SHDSL. The produce offers up to eight access lines and supports VPNs with burstable data rate up to 10 Mbps.

Interoperability

It is risky to buy CPE equipment on your own and then shop around for a provider whose DSL Access Multiplexers (DSLAMs) are compatible, mainly due to the lack of interoperability within the VoDSL service provider community. The IAD offered by the service provider will typically be the only IAD that will work *easily* with the service provider's equipment.

That's not to say that some interoperability among other IADs isn't possible. Yet, even when a certain IAD (say the NetEngine 6300-8) is on the service provider's list, don't assume it will work with another service provider's equipment even if that provider also lists the NetEngine 6300-8 as its IAD of choice. Service providers use different line codings on their networks that can prevent the same make and model of an IAD from the same vendor from working in different situations.

The best hope for mix-and-match interoperability, with the accompanying potential for more competitive pricing and features, are products built to the G.SHDSL Symmetric High-Bit-Rate DSL or SHDSL standard. Occam Networks, Cisco, Lucent and Symmetricom are among the vendors who claim G.SHDSL support for certain product lines.

Getting to Know Your IAD

Once you sign on with a provider, that company largely makes the IAD decision for you. If a customer already has DSL service onboard, the CPE being used will normally fall into two main categories (neither of which can be used for VoDSL service): bridges and routers. While it's possible to find a VoDSL-capable IAD in a plain vanilla DSL installation, they are rare.

DSL bridges (a DSL modem is technically a bridge) are essentially unintelligent pipes to the DSL network, with a single Ethernet port coming in and a DSL port going out. They're only used in a business or residential network environment that can itself provide sufficient routing and security for effective DSL configuration.

DSL routers are essentially standard network routers with an added DSL card inside the box. They do, however, have some differentiating features, such as a sufficient number of Ethernet ports for a customer's specific needs; flexibility in assigning IP addresses with features, i.e. NAT and DHCP; firewalls; VPN support; POTS lines (which provide backup dial-up service) and some even offer Voice-over-DSL (VoDSL) functionality.

VoDSL IADs take over the router functions. They are also flexible: Many of the higher-end IADs are upgradable, either via software switches or with the addition of new boards. And because it is important to provide support off site (both to save costs and to expedite fixes), IADs come with software for remote diagnostics and provisioning over FTP and telnet.

Voice capabilities on the DSL side usually include interoperability with standard Centrex features, including caller ID, call waiting, stutter dial tone, and call forwarding. It is also helpful if the IAD can recognize when a fax call comes in, so that it can deal with the special compression scheme used by fax machines.

Other distinguishing features include interoperability with a wide variety of DSLAM equipment and gateways, backup features, and the number of DSL formats supported. And, as is vital with always-on broadband connections, most IADs include security functionality such as built-in firewalls and a secure tunneling protocol (such as IPsec) for establishing a VPN.

In fact, the next-gen IADs and other advanced hardware have enabled service providers to target a new, smaller-sized customer: home offices, small businesses, and medium-sized businesses that have not yet settled on a particular broadband solution.

IAD Management

Since the customer premise IAD is the gateway to various networks, it must support multiple operating environments and varying customer requirements. At the low end of the spectrum is the minimum function IAD that supports voice and data to a centrally controlled network. This is in stark contrast to the next-gen IADs that distribute intelligence and services from the edge of the network with multiple transport and physical access methods. The new products also address provisioning, installation, commissioning and ongoing operational support, which are often overlooked costs of IAD installation and maintenance. Of course, fewer truck rolls and the more remote operational support from the *Network Operations Center (NOC)* tend to result in lower costs and more savings.

The key for the future success of IAD technology is how easy these devices will be to manage and provision. Vendors are spending a lot of time and money trying to develop features for an IAD that allow for remote management.

In the future the customer will be able to self-install IADs that have plenty of customer-selectable services, such as security, encryption, back-up and e-mail services and other specialized data services that only needs a PC to activate through the IAD locally via a web interface.

The ability to enable service providers or their customers to provision IADs through simple interfaces is a popular notion that many vendors are beginning to offer. For example, VINA through its acquisition of Woodwind Communications Systems, offers software-driven IADs referred to as soft IADs for next-generation services such as VoDSL.

Soft IADs

IADs are getting smarter and more software-centric, resulting in new benefits for both the customer and the service provider. Software-based solutions like soft IADs help by adding remote capabilities to improve the initial CPE installation, provisioning and ongoing configuration changes as users upgrade or modify their service subscriptions.

The solution to many of these issues is in the IAD software architecture. While the "box" is still necessary, it's the software that delivers the real leverage and value.

For example, soft IADs can be configured to interoperate with more than one vendor's equipment, allowing the service provider to select best-of-breed core and edge products. When used with conventional DSLAMs and Class 5 switches, soft IADs function merely as aggregation devices with no intelligence, since their signaling is derived from the Class 5 switch. Installation packages for soft IADs working in a conventional ATM-VoDSL environment tend to include drop-down menus where the installer can select the brand of DSLAMs and voice gateway used.

In recent years there has been an industry movement to take the intelligence out of the core network (switches and SS7) and move it closer to the customer. In keeping with this idea, soft IADs keep increasing in "intelligence" and are starting to interoperate with softswitches via protocols such as MGCP, Megaco/H.248 or SIP. Already soft IADs can offer such CLASS features as call waiting, caller ID, and call waiting with caller ID.

These more advanced soft IADs — many working with next generation IP networks — support remote provisioning. The implications of this are that an upgrade or change no longer involves physically swapping out a card, so a truck roll to the customer premise is unnecessary. It also means that modifications can be made seamlessly and transparently to the network without changing the box itself, so the customer's business no longer need be disturbed.

Hence, the soft IAD model delivers flexibility, migration and growth opportunities for both the customer and the service provider.

Choosing an IAD

IADs represent one of the most important elements in the service provider selection process. Look for a service provider that offers an IAD that can support today's services and tomorrow's expanded services. Service providers should always seek the best IADs to protect the network from upheaval that may be caused by the natural evolution of technology and changing customer requirements.

The scope and size of IADs are directly related to their physical attributes, but their intelligence determines their ultimate value. Also note that an IAD's support for provisioning, installation, commissioning and ongoing operations is an often overlooked consideration during service provider evaluation.

The intelligent soft IAD can satisfy the most customer and business criteria. Intelligent soft IAD benefits include:

- The ability to offer value-added service to customers.
- Minimization of initial costs, ongoing support and replacement costs.
- Scalability and flexibility as customers' needs or infrastructures change and evolve.

The Future

The evolution of IADs has been going on for a few years and will continue to be driven by the business community's needs to simplify communications.

There are more IAD features to come. Value-added services are driving the innovations, so the IAD vendors are compelled to provide the ability to integrate various other services.

Many vendors are targeting the lower-end IAD market and have internal development efforts underway to build such IADs for the residential and SOHO markets. Somewhat like the DVD player, the device will sit inside the home and will likely be replaced every few years as more functions and features become available.

Wrap-up

The VoDSL service provider will provide an IAD for installation at the customer premise. That IAD will furnish, at a minimum, a data interface (e.g. Ethernet) and a number of voice interfaces (e.g. RJ-11 jacks providing standard analog interfaces for POTS). This device represents the network demarcation point: on one side it terminates the DSL circuit and on the other it delivers services. The IAD is responsible for converting between the analog voice signals on the user side and voice signals on the network side.

The IAD must also operate a signaling protocol towards the voice gateway to indicate when voice calls are present. Traffic streams carrying voice are multiplexed together with signaling to control the voice streams and other ATM, IP or Frame Relay data and carried on the DSL link to the network.

The installed IAD should be compatible with circuit-switched, packet-switched and packet voice technologies. For example, it might be an IAD that can interoperate with both conventional GR-303 interface compliant gateways and packet voice platforms. GR-303 compliant gateways convert packet voice (whether transported via IP packets or AAL2 ATM cell encapsulations) to circuit-switched voice (like a T-1) suitable for handoff to the legacy TDM network. But, currently, a number of leading VoDSL service providers are installing IADs that interoperate with both GR-303 gateways and packet voice platforms. And in order to deploy VoDSL in a next-generation, softswitch architecture, IADs must be able to also support VoIP, MGCP, Megaco/H.248, SIP and other, new telephony protocols and features. If the IAD offered by the service provider doesn't provide support for these platforms, at some time in the future the service provider will be forced to do a complete change out of the CPE, which is costly for the service provider and disruptive to the customer.

With an IAD running at the helm, legacy networks can be integrated into evolving network infrastructures, enabling service providers to provide budget-constrained customers the power of wide-area communications and allowing them to be competitive in today's ever-changing business environment. In particular, these new services enable SMEs, which often lack the resources required to install and manage multiple communications devices, to compete effectively with their larger counterparts in the global market place.

Residential Access Devices

There are other access devices besides the "pure" IAD that are sometimes more suitable for the residential environment. We will discuss in this section some of the various devices that might be found in the marketplace.

A *residential gateway* supports high-speed Internet access, as little as one voice line, and has mandatory lifeline support. Also, the residential gateway needs to support some form of home networking, whether it's Ethernet, Powerline, 802.11b wireless or something else.

The market for residential gateways, which is defined as devices that connect the service provider's WAN to a network on the customer's premise (these devices could actually be called "residential IADs") is rapidly expanding. Within five years I believe every high end residence will have a residential gateway device in operation.

To exploit the residential gateway opportunity, a few newcomers are joining existing vendors in the marketplace. One such company is 2Wire Inc. (San Jose, CA — 408-428-9500 — www.2wire.com). It offers a sophisticated gateway line that enables a household to share a single broadband connection for high-speed Internet access throughout the premise so multiple PCs and peripherals such as Internet radios, video recording devices, security devices and so forth can simultaneously connect to the outside world.

2Wire's HomePortal 1500 is a bit pricy at $599 per unit but it does incorporate a DSL modem supporting 8 Mbps downstream and 1.5 Mbps upstream, a network hub, router, a business-grade firewall and value-added software. The HomePortal 100 is more reasonably priced at $349 but doesn't have a built-in DSL modem although it is engineered to work with a service provider's DSL access device (whether a "modem" or an IAD).

2Wire touts its HomePortal as enabling the residential user to gain access to streaming audio and video, broadband gaming and value-added services like automatic backup, content filters and unified messaging.

Gaming Consoles as IADs

One device that many homes already have in place is the gaming console, such as Sony Computer Entertainment Inc.'s (Foster City CA — 650-655-8000, www.scea.sony.com) PlayStation and Microsoft's XBox. The gaming console would need very few modifications to be a host for residential broadband access to a variety of services. As the family entertainment platform, the Playstation, Xbox or like device could be upgraded via software, have DSL access capabilities added, and thus become a residential gateway for scheduling, entertainment and e-mail access.

Network Interface Devices and More

Some vendors and industry experts feel that it might be difficult for residential customers to plug a variety of devices such as telephones, computers, faxes and TVs located in various rooms in their homes into an in-home device. Many think that it makes more sense to put the access equipment outside the house and have all devices in the house access it by simply plugging into existing outlets. That eliminates the wiring issue, plus keeping the DSL equipment just outside the premises tends to result in fewer interference problems between the DSL connection, splitters and security alarms.

In fact, just a few years ago, there was quite a bit of buzz about evolving the *network interface device (NID)* on the side of the home into a residential gateway with multiple interfaces on both the home and network sides. Although any efforts along those lines seem to have fizzled out because of a dearth of broadband connections to the home and the expense of such units, the idea is once again on the drawing boards of many forward-thinking vendors.

A third potential location for a residential gateway is an in-home wiring closet where such a device would connect devices such as modems, faxes and telephones to the in-home wiring. In fact, Efficient Networks Inc. (Dallas, TX — 972-852-1000, www.efficient.com) is working with another vendor to include cross-connect boxes in new homes' wiring closets so a DSL modem or IAD device can sit there and have access to all in-home wiring.

Receiving Devices

The telephone handset was the standard office communications device for the entire 20th century, and it's still in use. No matter how "antiquated" handsets are, they must "become friends" with your IAD. The same can be said for the more modern fax machine and other office communications devices.

Analog Instruments

Traditionally, a standard analog telephone handset connects to a telephone line through an RJ-11 jack in the customer premise, which runs to the local Class 5 switch and out into the PSTN. In order to use a VoDSL network in conjunction with legacy analog handsets, the IAD must have some kind of interfacing functionality in order to map the telephone's signaling and voice traffic into the VoDSL access network and again at the voice gateway to map the signaling and voice traffic back into the service providers network.

Faxes

The receipt and transmission of faxes via VoDSL technology present a unique challenge to VoDSL providers. The IAD provisioning VoDSL must use vocoders in conjunction with fax relay to handle non-voice signals such as analog modem and fax tones.

Because vocoders are designed for the coding of voice based on various unique characteristics of human voice, they achieve varying results for other audio signals such as music, fax and modem tones. In fact, some speech codecs are unable to pass fax tones at all. In these situations the gateway devices must discriminate between voice and fax in order to offer special handling of fax transmissions.

Packet networks address this problem by implementing Fax Relay, according to the ITU T.38 standard. With fax relay, instead of running a speech coder on the fax signal, the DSPs in the gateway nearest the sending fax machine emulate a receiving fax machine and terminate the fax message into a digital format. The gateway then packetizes the digitized fax message and sends it to the gateway nearest the receiving fax machine. The receiving gateway then emulates a sending fax machine and sends the fax data to the receiving fax machine.

Fax Relay Support

Indeed, one of the key features of a VoDSL service is support for fax relay. Fax relay provides reliable real-time fax service between two analog fax machines over a packet network. Equipment at both ends of the packet network "spoofs" the analog fax machines such that they operate as if directly connected over a PSTN connection. The VoDSL equipment performing fax relay functions must handle the effects of network delay, jitter (variable delay), and lost packets while preventing the fax machines from timing out. Standards protocols such as T.38 and AAL2 exist for interoperability between equipment vendors. Proprietary techniques are often used to improve the interoperability between different fax machines that are subjected to long delay and other packet-network effects. Fax relay is supported by the following standards:

Fax Modem Pumps: V.17, V.29, V.27ter, V.21

Fax Relay Protocols: T.38 (TCP/IP), AAL2 (ATM)

Fax Machine Spoofing Protocols: Proprietary

Fax relay systems demodulate (receive) the fax transmission at the near-side gateway, then transmit the fax across the packet network according to the fax relay standard, and finally modulate (transmit) the fax from the far-side gateway to the receiving fax machine.

Digital and IP Instruments

IADs with VPN capabilities can use PBX extension services to provide voice capabilities to teleworkers and remote call center employees, allowing them to use corporate digital phones for call redirect, voice mail, conferencing, speed dialing, and other CLASS features from remote locations.

Many VoDSL service providers with the proper IAD at the customer premises can also offer the capability to turn digital telephones (which are common on many SME desktops) into feature and application rich IP handsets. This allows businesses to make the jump to VoDSL without having to install new handsets. There's even technology that can be brought in-house so a business can connect digital phones to an IP PBX.

An IAD or other device, which contains the necessary technology, would sit where a PBX or telephone key system would typically reside in a network, connecting to a building's telephone wiring rack. If it's a device other than an IAD (or a low-end IAD without routing capabilities), an Ethernet port on the device would link to a router, from which the VoDSL service would be piped in from a service provider.

Aside from providing basic PBX functions, such as multiple lines, call hold, transfer and conferencing, such a technology box-set could provide IP phone-like applications to digital handsets, controlled by an IP softswitch in the carrier network. Such applications include "click-to-dial," which lets an end user place calls by clicking on contacts from, say, a Microsoft Outlook or ACT address book. Text messages can also be displayed on handsets which sport a LCD screen along with other IP phone functions that go beyond traditional digital handset features, e.g. a list of previous calls received and calls placed, and the ability to call back missed calls with one button.

Video Conferencing Devices

Primitive video conferencing, in the form of a simple "Videophone" that could work between two points, was first demonstrated in a purportedly "consumer" milieu at the 1964-1965 New York World's Fair. However, what we would now consider *bona fide* video conferencing was only really made possible when the telecommunications networks could handle the tremendous amount of video and audio information that must be transmitted back and forth between a number of systems simultaneously.

Since most acceptable "business quality" video conferencing sessions require a bandwidth of about 384 Kbps (the equivalent of three ISDN BRI lines), it was not until the appearance of DSL that such bandwidth could be made cheap enough and could be delivered over preexisting copper twisted pair phone wires to the home and business.

Videophones

Now that VoDSL networks are being widely deployed, the progression from conversational VoDSL to video-enabled VoDSL is relatively easy. It's already available in Microsoft Corp.'s NetMeeting and Polycom's ViaVideo.

When AT&T demonstrated a working telephone with two-way video at the 1964-1965 New York World's Fair, thousands of visitors experienced a science-fiction vision becoming real. At last you could not only talk to people over the phone, but you could see them at the same time. At the time, AT&T optimistically (over-optimistically, as it turned out) predicted that this new "Picturephone" system would be in widespread use within a few years. According to AT&T, it would start with businesses, since they could better afford this new technology, and then to residences as the technology became more affordable and people began to see the "benefits" of this innovative technology. It is difficult for me to think of another technology that got the tremendous exposure and "buzz" that the Picturephone did. Everyone was talking about it. There were jokes about pajamas, dropped towels, and bad hair days, and so on.

But, neither the idea nor the technology caught on with the buying public of the 1960s. Not for want of trying, however — AT&T and others introduced, and re-introduced Picturephone-like services for years. As recently as 1997, AT&T tried to re-launch the Picturephone concept with its Picturephone Meeting Service, but once again, the public wasn't ready quite ready (or couldn't afford it given the expense of the existing technology).

Nevertheless, the idea of video over a telephone instrument stayed on the minds of many within the telecommunications industry. Analysts throughout the 1960s, 70s, 80s and 90s would periodically predict that "within a couple of years" every home and business would have a picturephone-like device as their central communications fixture.

Well, finally, it's becoming a reality. There are several reasons for the upswing in public acceptance of the Picturephone, now known generically as the videophone.

First, there was the September 11th tragedy which not only pushed people to look for ways to meet without traveling but also caused businesses to consciously begin developing a more distributed work environment for their employees.

Second, the hardware for this technology has matured and it's cheap.

Third, the bandwidth is readily available at a reasonable cost, thanks to DSL.

Fourth, the vendors have gotten the message and have begun offering technologically superior, reasonably priced instruments for a variety of markets. For example:

British Telecom has begun offering the mm215 videophone from Motion Media Technology, Inc. (Severn Bridge, Aust Bristol BS35 4BL, England — 44 (0) 1454 635400, www.motion-media.com), which is designed specifically for BT's customer base.

InnoMedia (San Jose, CA — 408-432-5400, www.innomedia.com) offers an alternative to the costly high-end video conferencing systems that only large corporations can afford, with its all-in-one device, which includes a snap-on Webcam, a 4-inch TFT color LCD screen, and a slim handset. The camera has a fairly high frame rate (15 frames per second) and a built-in speakerphone. It has video-out connectivity for a television, VCR, or projector, and video input for hooking up another video camera, VCR, DVD player, or digital camera.

InnoMedia advises that its videophone works best in a QoS local area network environment with a fixed IP address or through a high-speed connection, such as DSL. But, once the videophone is properly set up and configured, it works like a regular phone. The user just punches in the IP address (not a regular phone number) for the video device being called and you're ready to play George Jetson. Yet, if you're shy or in your pajamas (either alone or with somebody else's spouse) or just don't want to be seen, there's a privacy feature that can block the video. In that case, the device can be used just like a regular telephone. The device can also easily be moved from room to room for use in many different settings and situations.

Polycom also carries several videophone type products. These products have the look and feel of a traditional phone with a built-in video display.

Of course, to use the video capabilities of any videophone, the recipient must have a videophone as well. However, many videophones will still transmit voice when the recipient is not video-enabled.

In Summary

VoDSL offers economical quality voice and data services to the small business and residential market. A VoDSL service provider that offers a variety of bundled service options can provide its business customers the same business-class voice service normally affordable only to large enterprise customers. Thus, there is no doubt that bundled services will be an integral part of most VoDSL service providers' portfolio. Such services will include multiple voice line support, Internet access, and video communication. VoDSL services provide an excellent foundation for these applications, as they extend the in-place telecommunications network to the customer's premise.

Key to the migration from circuit-switched to packet-switched is the ability to use the customer's legacy CPE. The IAD is the enabler. It provides transparent support of all *Custom Local Area Signaling Services (CLASS)* features such as call waiting, caller ID, and three-way calling. Dynamic allocation of bandwidth with voice prioritization ensures toll quality voice service. Compatibility with existing telephone equipment, including fax machines, analog telephones, modems, and PBX and key systems ensures ease of use.

CHAPTER SEVEN

Considerations When Choosing a Service Provider

Voice and data communications are a business's lifeline. Choosing a single provider for these services requires serious forethought and performance of stringent due diligence. Yet, as you will learn, putting all of your communications eggs in one basket can sometimes be a smart decision. In this vein, VoDSL service is quickly gaining acceptance, especially in the SME market due to the sizable benefits it offers via bundled services.

Service Provider Types

The author's glib use of the single term "service provider" (or "network operator") ignores the fact that it is actually a generic term that can apply to more than one kind of organization. Who your "service provider" happens to be depends upon whether you live in a large metropolitan market space, what the prevailing state and federal legislation is, how far your business is situated from a central office, among other factors.

At the local level, one finds three types of service providers offering connectivity services for voice and data. These include *Incumbent Local Exchange Carriers (ILECs)*, which at one time had a monopoly on local services; the state-certified *Competitive Local Exchange Carriers (CLECs)*, which compete either by building their own facilities or by leasing "unbundled" facilities for resale from the ILEC; and the *Internet Service Providers (ISPs)*, which also provide access to the Internet.

The ISP could be a subsidiary of the ILEC, but state regulation means that these ISPs must be customers of the ILEC, just like the rest of the population. This arrangement maintains both the conceptual and physical split between basic transport and enhanced services. Besides providing pure Internet access, most ISPs now offer web and application hosting for businesses looking to take the worry out of operating, managing and supporting costly in-house file servers. Enhanced services such as unified messaging and VoIP are being targeted at the residential and small business customer. VoIP is also being used to offer savings on international calls, and once Quality of Service (QoS) is assured we can expect to see more businesses adopt this technology. In fact, CLECs became ISPs by offering data services (an easy transformation) and some ISPs became CLECs by offering Voice over IP (VoIP) (a not-so-easy transformation).

Network unbundling, the process of breaking the network into separate functional elements, opens the local access area to competition. CLECs select the unbundled components they need to provide their own service and pay the ILEC for their use (ILECs complain that the payment is just a small, token one). If the CLEC considers the unbundled price to be too expensive, the service provider will furnish its own private resources or turn to an *Alternative Access Vendor (AAV)*, which can offer private line service between an entity and facilities at another location. Many service providers who enthusiastically entered in the CLEC market in the late 1990s are now bankrupt. ILECs tended to drag their heels when it came to local loop unbundling, some CLECs had to wait too long until top priority central office sites became available, while others fell victim to the costs of co-locating their equipment with those of the ILEC. It costs a new competitor much more than it costs an incumbent to offer DSL and/or VoDSL services, and by the end of 1999 it became difficult, if not impossible, for new competitors to borrow money.

Yet, next-generation carriers are still in existence, and new breeds of CLEC have emerged, such as the *Building CLEC or BLEC* that wire multi-tenant apartments or commercial buildings with DSL or fiber-optic cable to provide *Shared Tenant Service (STS)*.

"Deconstructing" the Service Provider and its Capabilities

There are many questions that must be answered before any final commitment can be made. Can the VoDSL service provider effectively deliver voice traffic with the quality that customers, particularly business customers, demand? Can the service provider's equipment deliver the same call quality and feature sets found with PSTN equipment?

Once an analog voice signal is converted into some kind of alternate format (Channelized Voice over DSL, Ethernet, ATM, IP or Frame Relay); a service provider needs to ensure that there is sufficient bandwidth to transmit voice in a timely and stable manner. This is difficult on IP or Frame Relay networks, where packets travel across the network haphazardly and can arrive late and out of order. We've seen how carriers have addressed the problem by layering their DSL services on top of underlying *Asynchronous Transfer Mode (ATM)* networks. The ATM protocols set up "rules" for sending data over a DSL link that includes *Quality of Service (QoS)* features which ensure a clear path from point to point, thus allowing a higher quality of service that's on par with the PSTN.

As we've seen in previous chapters, to deliver VoDSL services, the service provider must provide an *Integrated Access Device (IAD)*, which is installed at the customer's premises. IADs consolidate multiple information feeds (voice and data) onto a single line. Many IADs currently being provided by service providers were originally designed to carry data traffic only — what about the IAD offered by your service provider? There are new multiservice IADs now making their way to the market, and customers should, if possible, go with a VoDSL service provider that offers and supports such state-of-the-art equipment.

Also, as we'll see, the customer should find a service provider that offers a VoDSL solution which integrates seamlessly with the overall DSL and broadband network infrastructure, one offering characteristics such as the following:

- Support for multiple high-quality voice channels on a single DSL line.
- Ability to offer commonly used traditional voice services: call waiting, caller ID, three-way (or N-way) conference calls, etc.
- Provisions for automatic detection of fax calls including line speed and protocol adjustment.
- The ability to use a conventional modem over the DSL circuit.
- Use of industry standard interfaces so as to enable to use of existing customer premise equipment: telephones, fax machines, PBXs, key systems, etc.
- Interface with GR-303 telephone switch infrastructure if necessary.
- Multiservice ability to utilize all standards-based transport technologies: IP, ATM, Frame Relay, etc.
- Guaranteed toll quality voice.
- A price competitive with existing PSTN services.

To deliver a seamlessly integrated VoDSL solution, the service provider must invest in or partner with vendors and network operators that provide quality equipment and services enabling the deployment of both voice and data services. This starts with the IAD at the customer's premise and the DSLAM at the local central office or other location where the unbundled copper loops are terminated (of course, a very small business or SOHO could use CVoDSL, which doesn't require as much equipment and removes a few headaches, at the cost of restricting the number of available voice channels from 24 to four or perhaps eight).

Transportation Mode

One item to be considered is the transport mode. Although the transport protocol controversy was discussed in detail in Chapter 5, let's quickly review what's at stake and how it affects the choice of a service provider.

As this book is being written there's been a great debate going on within the telecommunications industry — the IP versus ATM debate.

ATM

Designed to operate with ATM and SONET transport technologies still prevalent in the network core, ATM-based DSLAMs work with ATM's fixed-size, 53-byte cells that are indistinguishable from each other with respect to source, destination and type of service. ATM-based DSLAMs cannot distinguish whether a cell is carrying a voice packet or just an e-mail packet, and many ATM-based VoDSL systems must provision separate end-to-end *Permanent Virtual Circuits (PVCs)* for each kind of traffic, and for each subscriber, to maintain a specified level of quality of service.

This might lead to considerable provisioning and scaling problems as service providers sign up more and more subscribers for multiple, concurrent services such as VoDSL.

IP

Whether at the central office, remote terminal or basement-phone room in a multi-tenant building, to avoid upgrade-related disruptions in service, you might want to look for a VoDSL service provider that has standardized on DSLAMs with IP service intelligence. IP-optimized DSLAMs offer not only cost-effective access solutions for VoDSL in today's legacy ATM circuit-switched networking environment, but they offer the only DSLAM solution that can migrate seamlessly to the softswitch SIP / MGCP / Megaco-H.248 environment. IP-optimized DSLAMs work with whole Internet packets and can sort voice. Unlike wholly ATM-based DSLAMs, they can scrutinize the packets' datagrams, which enables them to tell the difference between various "grades" of traffic (voice, video, data, etc.) and identify where the traffic is coming from, where it's going, and how important it is. Because IP-optimized DSLAMs can make these kinds of distinctions, DSL service providers who standardize on them can aggregate traffic from different subscribers on shared paths through the network.

No one knows whether carriers will transport voice over pure ATM (AAL1 or AAL2), IP over ATM (AAL5), IP over MPLS or Ethernet for their access and/or core packet network infrastructure. All of the possibilities offer advantages and disadvantages, and next generation softswitches must support whatever technology is needed. It all comes down to when or how deeply IP will push into what has been ATM's domain, and the vast array of IADs available reflects this uncertainty — some are based on TDM/ATM while others support IP-based services via DSL (and ironically, over ATM AAL5). Some hedge their bets by supporting every protocol possible. Also, today's service providers provisioning voice and broadband services over DSL currently can choose between three kinds of DSLAMs: ATM-based, IP-based, and those that can handle both. Ultimately, both IADs and DSLAMs will probably be developed that can handle any transport protocol, though the trend will probably be toward an IP-centric environment.

If and when the day comes that QoS is readily available for IP at carrier scale, the new generation of edge switches deployed will support a move away from ATM at the core and at the premises.

It is thought by nearly all experts that eventually (one hopes sooner rather than later) "convergence" between the IP and traditional telephony networks will finally be complete. Just when and how this will occur are questions that are still difficult to answer. The process has been slow and the world will be using a hybrid network for the foreseeable future. Will service providers eventually package all VoIP and other IP services over ATM or will they offer a pure IP play, or send IP over something like Ethernet? No matter whether the "winner" is IP or ATM, it's obvious that the biggest factor in the evolution of the core network — and the access network where VoDSL dwells — is the overall migration to packetized communications (the exception to this is channelized VoDSL). And as next-gen networks are built with softswitch functionality, SIP, MGCP and Megaco/H.248 protocols will all have to be accommodated in the network — and when this happens the IAD will be ready and waiting to host them.

Still, perhaps in the end it really doesn't matter if applications ride on ATM or IP, or if voice is packetized at the point of network access, or somewhere else in the public network. Technology agnosticism may prove to be the order of the day, particularly among "less informed" organizations. After all, from the customer's standpoint, they don't really care about the underlying technologies, as long as they can get the services they need, the quality is good and the price is right. Tom Nolle, president of the network consulting firm CIMI Corporation, has been quoted as saying that network managers don't care "if little silver Tinker Bells carry their data packets from one place to another."

Network Infrastructure

One would think that, in an effort to create the most efficient type of end-to-end network, the access network (VoDSL in the local loop) should eventually use the same transport protocol as the core network. While an elegant all-ATM or all-IP architecture is appealing in an abstract, almost aesthetic sense, physical realizations of such architectures turn out to be somewhat less than utopian. Tradeoffs for efficiency and performance differ in the two networks for "purist" implementations of any given transport protocol or technique.

The core network consists of mostly fiber optic cabling, and there are various ways how traffic can be multiplexed onto such fiber. Some experts assert that a 100% Internet-like backbone is "just around the corner" (it isn't). Others claim that ATM is the wave of the future. The core / backbone has undergone a progressive evolution:

1. The Circuit-Switched Time Division Multiplexed (TDM) Backbone was the "network model of the 1970s," whereby all traffic is assigned to high quality, fixed trunk bandwidth that is calculated based on the peak traffic rates of each input circuit.

2. The Hybrid Backbone was the "network model of the 1990s" (as well as the network model of the present day) and is vaguely similar to the circuit-switched TDM backbone. It uses both ATM and IP, as well as SONET, *Wave Division Multiplexing (WDM)* and *Dense Wave Division Multiplexing (DWDM)*, with the traditional voice and modern data networks still basically separate. All bursty data traffic tends to be groomed onto a packet-switched, *Statistically Multiplexed (StatMux)* network such as Frame Relay or the Internet, providing better utilization of network backbone resources. Fixed rate traffic remains on a circuit-switched TDM network. SONET-based TDM *Digital Cross-Connects (DXC's)* assign fiber bandwidth to the packet-switched and circuit-switched traffic on a circuit switched basis, as does WDM and DWDM.

3. The Packet-Switched StatMuxed Backbone is the "pure Internet" next generation network model that is said to be the the network model for the future communications infrastructure (particularly by IP zealots). Both fixed rate and variable rate traffic is packetized and StatMuxed onto high speed trunks prior to insertion into large scale, long-haul fiber networks. SONET-based DXC's and WDM / DWDM may appear in the network as needed.

4. The Asynchronous Transfer Mode (ATM) Backbone is also claimed to be the future of communications (by ATM enthusiasts, naturally). All traffic is divided into fixed size cells and StatMuxed onto high speed trunks prior to its transfer to fiber. Fixed rate traffic is assigned *Constant Bit Rate (CBR)* virtual circuits capable of providing TDM-like QoS. Bursty data traffic is assigned Variable Bit Rate, Available Bit Rate, or Unspecified Bit Rate virtual circuits and is StatMuxed onto trunk bandwidth not reserved for CBR traffic. Once again, optical systems such as SONET-based DXC's and WDM / DWDM are used extensively.

The future of the core network appears to be something of a mystery, so it should not affect your choice of a service provider, who has more reasons to worry more about the core network than you do (in all likelihood, it will remain a hybrid structure with a slow growth toward an all-packet, but multiservice, network). Instead, potential VoDSL customers, the business customer in particular, should take a close look closer to home — at the service provider's own network infrastructure. This is important, for in some geographical areas a customer will not have a choice of VoDSL providers and will have to work with whatever is available. The service provider has made a bet over what the future of the world's communications infrastructure will be, and you are along for the ride. For instance, many ATM-based VoDSL solutions were based on the *Broadband Loop Replacement Service (BLES)* that made use of the standard GR-303 digital line interfaces, which, as we learned in a previous chapter, is the specification for all functional aspects of a *Digital Loop Carrier (DLC)* system that aggregates analog phone lines and outputs them as T-1 circuits. The centralized GR-303 voice gateway was developed to enhance the capabilities of the DSL-enabled local loop by converting packet voice, which is transported via either IP packets or AAL2 ATM cell encapsulations over broadband (or perhaps even fixed wireless), to circuit-switched voice in the GR-303 format suitable for handoff to the Class 5 switch and the rest of the legacy TDM network. The voice gateway thus performs the functions that are needed to interface, using the correct format, with the existing voice network. This was necessary since all local exchanges used standards that were in place for many years (i.e. GR-303 for ANSI countries like the U.S. that use T-1s and V.1 and V5.2 for ETSI countries in Europe and other locales that use E-1s). Even today, these interfaces might be the preferred way of communicating, since calls can be easily concentrated over them with existing equipment. Also, these interfaces are characterized by the fact that the traditional telephone service remains part of the local exchange, guaranteeing the service transparency of established telephone services.

Next Gen VoDSL Solutions

A service provider who isn't using ATM-based BLES VoDSL is betting instead that the Class 5 switches will become totally unnecessary and that the world will be using an IP or at least multiservice infrastructure based on softswitches, which are actually various switching and media transcoding components distributed throughout the network.

The advantages of using a VoDSL service that ties into an all-packet (IP) next-generation converged communication network are in its softswitch architecture / platform, which bypasses the Class 5 switch. The complexities associated with a central office switch are well known. They are difficult to configure, and adding a new enhanced service is a nontrivial exercise. On top of that, (enhanced) services must be co-located in the central office with the switch, which is a very expensive solution.

Advantages of IP-based next generation networks include the following:

- Migrating of functionality away from proprietary central office switches to commercial off-the-shelf hardware and software.
- Utilizing open standards at the hardware level (such as CompactPCI computer backplanes for durable, fault tolerant packaging) as well as at the software level (such as TCP / IP, XML / VoiceXML, SOAP, etc.).
- Managing telephony services from the customer premises via a web browser, which is easier to do than programming a PBX, and quicker than waiting for the phone company to make moves, adds or changes to Centrex.
- Offering applications rarely or never encountered in the landline PSTN world. Enhanced services is perhaps the most important feature softswitches and media servers enable. It's easier and cheaper to design, build, and maintain these services in a next-gen network than in the traditional PSTN. These would be services such as call logs for inbound, outbound and missed calls; the ability to click to return calls or click to e-mail a response; and "personalized call treatments" such as defining "VIP calls" that can reach the called party wherever he or she is (while sending non-VIP calls to voice mail). End-users could also group incoming calls into categories, allowing some calls to reach them during weekends, travel, vacations or during any arbitrarily defined time interval. Voice portals, such as the VoiceCaster from Channel Access (Los Gatos, CA — 408-378-5500, www.channelaccess.com), voice browsers and wireless web access will provide users with access to web content through voice commands. These services will evolve and will be integrated with personal computers, PIMs and mobile phones.
- Delivering telephony applications over xDSL with an applications-enabled softswitch can be done without bringing a conventional PBX or key system into the scenario at the customer premises. These applications will work with a customer's existing business phones by plugging them into the IAD (or using IP adapters and/or IP phones).
- Distributing the gateway and signaling functions throughout the network, which drives down the cost of providing VoDSL service by letting carriers avoid the toll charges, interconnect fees and backhaul charges that are inescapable with the centralized, TDM circuit-switched-based approach.
- There is also an added bonus — because they share resources that may be added as needed, next-generation distributed VoDSL networks are highly scalable. Carriers can use them to deploy VoDSL services at the enterprise level, regionally, nationally or globally.

These key advantages are translated directly to lower cost, a larger available talent pool, systems that are easier to provision and deploy, and a quicker time-to-market for those developing new applications for the next-gen network.

Thus, the installation of broadband voice that connects to an applications-enabled softswitch is easier to deploy for the service provider and its customers, and will deliver capabilities beyond legacy telephone systems.

The cost advantages of a service provider that ties into softswitches and trunk gateways are compelling, but carriers who opt for these solutions now or who plan to migrate to them in the future need to augment them with the proper kind of IADs and DSLAMs so as to ensure a high end-to-end QoS for voice calls.

In addition to the challenges in architecting networks with end-to-end QoS, service providers must ensure that the rollout of such networks cause no disruption to their existing voice service revenue, which currently represent about 80% of their overall revenue source.

With more than $650 billion of worldwide revenue generated by traditional voice and fax services and a more than $250 billion installed base of traditional equipment infrastructure in the U.S., carriers will have to spend even more money to deploy next-generation packet switches that seamlessly interconnect and competitively function as TDM-based PSTN switches as well as support voice over ATM (VoATM) and VoIP.

The packet switching industry is arguably still in its early stages and the IP versus ATM debate will continue as the industry matures. Both long-distance carriers and competitive local exchange carrier need to choose a solution that will protect them through the evolution — which means they must choose a solution that seamlessly interconnects to their TDM-based PSTN network today and also supports both ATM and IP voice services. In addition, equipment vendors need to provide solutions that support a carrier's choice.

Once again however, one sees a general trend toward moving voice services to packet networks both to reduce costs and to provide more value-added services in an increasingly competitive environment.

Many service providers are drawn to packet-based equipment primarily because it promises to be both far less expensive than legacy equipment and a great deal more efficient in transporting packet traffic. In a study by General Bandwidth (Austin, Texas — 512-681-5400, www.genband.com), port costs for packet-based equipment were found to be declining faster than traditional equipment costs.

General Bandwidth says that in one study, packet equipment was found to be 70% less expensive than traditional voice equipment, and data access lines were 60% to 80% cheaper than voice lines. Packet network maintenance was 50% less expensive,

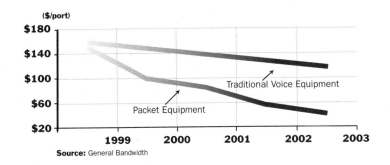

Fig. 7.1. Rapidly declining port costs for packet equipment.

while provisioning was 72% lower in cost.

One concept that perhaps more than anything else foretells the future of networks is the "multi-application platform" as embodied in devices such as the G6 from General Bandwidth. Although appearing at first glance to be an ATM switch, the G6 uses a high-powered (38.4 Gbps) switching fabric to act as both a voice gateway and a sort of super-aggregation point, enabling service providers to mass deploy *Voice over Broadband (VoB)* services to business and residential customers, regardless of access technology (DSL, cable, wireless) or voice protocol (ATM or IP). It can aggregate and groom voice and data traffic originating from VoB networks, *Passive Optical Networks (PONs)* and next generation *Digital Loop Carriers (DLCs)*. The G6 can also interoperate seamlessly with third-party softswitches via an integrated media gateway controller.

Technical Aspects of the Provider's VoDSL

Aside from examining general architectural issues, the potential VoDSL customer must dig deeper into the service provider's network infrastructure before making a commitment to subscribe to its VoDSL service. There are many items you want to consider before going with a specific VoDSL service provider. What follows is an outline of what to look for.

Interoperability

Interoperability should be the *modus operandi* for VoDSL service providers. You obviously want a service provider which has DSLAMs and/or softswitches that allow customer IADs to take advantage of all the services offered by the network. Vendors, such as Marconi, Cisco and Paradyne are offering sophisticated multimode xDSL line cards that allow for the delivery of voice and data services simultaneously over a single copper loop. These line cards allow a download speed of around 8 Mbps and supports a line reach of up to 18,000 or more feet from the CO.

Conversely, if you have an advanced IAD, it will be able to size up the network architecture and services and configure itself accordingly. For example, the Model 4753 G.SHDSL IAD from Netopia (Alameda, CA — 510-814-5100, www.netopia.com) connects to devices such as ISDN telephones and PBXs and supports the G.SHDSL standard (ITU G.991.2) for symmetrical DSL transmission as well as the Emulated Loop Control Protocol (ELCP) and voice gateway-specific voice signaling protocols. This makes the IAD compatible with the DSLAMs and voice gateways from a broad range of vendors including Alcatel and the now bankrupt Jetstream.

Other IADs are tailored for a specific platform; for example, Cisco's 6705 IAD works with the sophisticated Cisco 6732 Full Access Device, which offers not only xDSL service but TDM and ATM protocol interface support, GR-303 and TR-008 traffic grooming, SONET transport, and *Digital Cross-Connects (DXCs)*. The 6700 Series is based on a new multi-fabric switching architecture and delivers an end-to-end network access system which simplifies service provider provisioning and management by decreasing the number of network elements.

Scalability

The service provider you select must have a scalable infrastructure. "Scalable" can mean several things, but look for a service provider that can handle the provisioning of an influx of new customers, and also sudden bandwidth demands. Failure to take these items into account can lead to voice and data traffic congestion and degradation in service, particularly voice quality.

In general, ATM-based loop replacement VoDSL systems are not quite as scalable as all-packet networks that eliminate the need for the Class 5 switch (but Class 5 switches and voice gateways are still needed to provide connectivity between older parts of the PSTN and packet networks). Thus, a principal reason a service provider would migrate to a packet-based infrastructure is because of its promised scalability. To justify its deployment the service provider must assure customers that it has enough equipment to scale its VoDSL network to large volumes of traffic. This requires high-density voice gateways driven by hardware power per voice channel, channel density and voice features supporting true toll-quality voice to support very high volumes of traffic without degradation of voice quality.

Flexibility

The service provider should have in its equipment arsenal products with the flexibility to add new services and to react to a continuing standards evolution. Look for a service provider that has built its VoDSL service atop a VoDSL gateway platform that's based on solutions which support features beyond the typical analog *Pulse Code Modulation (PCM)* voice, i.e. features such as low-bit-rate codecs and fax relay. VoDSL customers need a service provider that can add services without the disruption and expense of replacing equipment (all of which could impact the customer). Find a service provider that offers quality VoDSL based on a platform that integrates voice software with programmable DSP-based access communications processors optimized for voice applications.

Compression

The type of service that can be offered over a local loop is highly dependent on the length and quality of the loop; both affect the amount of bandwidth that can be provisioned to the customer. With symmetrical DSL technology the ratio of downstream to upstream bandwidth is the same, but with asymmetrical DSL the downstream to upstream bandwidth is approximately 10:1, with the upstream bandwidth being the limiting factor for VoDSL services.

Today's ATM-based VoDSL deployments typically utilize either PCM or *Adaptive Differential PCM (ADPCM)* that is a reduced bit rate variant of PCM encoding, consuming 64 Kbps or 32 Kbps of symmetrical bandwidth, respectively. Actually, although the original G.721 ADPCM coder converted a 64 Kpbs A-law or mu-law pulse PCM channel at 8000 samples per second to-and-from a strictly 32 Kbps channel, the newer G.726 ADPCM is more flexible and uses waveform encoding to convert a 64 Kbps PCM channel to a 40, 32, 24 or 16 Kbps ADPCM channel.

These codecs provide voice channels with quality similar to that of PSTN phone calls, but multiple voice channels occupying lavish bandwidths can quickly consume the total available upstream bandwidth. For some customers (such as residential or SOHO) the reduction or elimination of bandwidth available for data transfers while voice channels are active may be acceptable. But for many businesses, a significant reduction in data bandwidth due to voice channel activity isn't acceptable.

Voice can be further compressed through the utilization of *Low-Bit-Rate (LBR)* codecs such as G.729A/B and G.723.1a. LBR codecs reduce the amount of bandwidth needed per voice channel, i.e. from 32 Kbps with standard ADPCM to as low as 8 Kbps with G.729A/B. Voice services utilizing G.729A/B would require approximately 24 Kbps for three voice channels as opposed to 96 Kpbs with the default setting for ADPCM. Even with six G.729A/B voice channels, there would still be significant amount of bandwidth left for data service. Companies such as Texas Instruments/Telogy are devising new DSPs and LBR codec software to allow channels of narrow bandwidth to carry high quality voice.

With such a reduction in the amount of bandwidth required per voice channel, LBR codecs provide significant advantages such as:

• Services can be offered on long loops, which can be advantageous to a customer that's situated on a long local loop.

• A higher voice-channel density over the local loop, which is good for customers in a densely populated metropolitan area.

• Higher bandwidth availability for data services over the local loop, which is a godsend for customers that transmit a lot of data — graphic designers, architects, and so forth.

• Lower speed DSL connections can offer equivalent voice-channel density, enabling residential and SOHO customers to obtain some of the benefits of VoDSL using their existing ADSL "Lite" service.

Reliability and Quality

Obviously, the service provider should have reliable equipment, no matter what it costs. The failure of a small, inexpensive piece of network equipment can halt service as easily as a big, expensive one can. But with competition increasing, costs are critical. Thus, to offer competitive prices to subscribers, service providers must find and purchase high-value network equipment that can be affordably maintained. Moreover, during the purchasing process, the service provider must be cognizant that engineering, installation and maintenance labor costs can quickly overshadow the original equipment costs.

While taking the aforementioned to heart, it's advisable to find a service provider that has invested in modular, customizable equipment that can grow and change at the pace of the network. The most competitive service providers evaluate all types of network equipment carefully and select reliable, flexible components that can be easily installed and maintained. Once the right equipment selections have been made, they can be replicated over and over throughout the expanding network, keeping service levels high and costs low. When it comes to selecting network equipment, the supporting equipment is as critical as major network elements — even the smallest advantages add up to better service for the customer.

Dependability is foremost, which will be reflected in the service level agreement (SLA) you sign with the service provider. Telephony has always been a lifeline service, but data services are now just as critical. Subscribers will quickly switch carriers if the service turns out to be unreliable.

While customers may be attentive to their SLAs with service providers when choosing a service provider, they should also worry about the contracts the service provider has with its equipment vendors. Service providers should have broad, continuous service coverage for their own network equipment upon installation, have a 24x7 technical hot line for rapid response to networking problems, and should be able to achieve easy migration to new platforms with minimal cost and impact to the service provider's operation

Another important issue is the service provider's standards for VoDSL voice transmission quality. It should be the same or as close as possible to those of POTS, i.e. packet-based voice should be truly transparent. Advertisements for packet-based voice implementations are always formulated using the terminology associated with high-quality PSTN voice services — "carrier class" or "toll quality." These terms are universal and set the expectation for quality for service providers and their customers.

Software is a critical ingredient of high-quality VoDSL systems, with the most important software features that must be implemented for carrier class systems being as follows: echo cancellation, voice compression, packet play-out, tone processing,

NEBS Certification

One of the best indicators of reliability is the NEBS Level 3 certification. The *Network Equipment-Building System (NEBS)* or "NEBS Criteria" were originally formulated by Bell Labs in the 1970s, further developed by Bellcore and were made public documents in 1985. Central office equipment manufacturers were the prime target audience back then, but today NEBS compliance has also become the benchmark of excellence for "nex-gen" service providers, such as CLECs, ISPs, and ASPs.

Of course, any company that can bend sheet metal around a backplane can say that they've got a "NEBS compliant" PC, but only top-notch companies actually have a sample finished system "NEBS certified." Performing actual NEBS certification on a product costs over $100,000 and is fascinating to watch. For example, at Underwriters Laboratories Inc.'s testing facility in Research Triangle Park, NC, there's a servo-hydraulic "shake table" that can simulate earthquakes measuring 7.0 on the Richter scale on samples weighing up to 5,000 pounds. NEBS equipment will withstand earthquakes, fires and other disasters.

voice activity detection (with silence suppression and comfort noise generation), fax relay support, packetization, and network management.

Echo cancellation, packet play-out, tone processing, and voice activity detection are key in meeting quality expectations. But to provide true carrier class toll-quality voice, these features require a robust and in-service-hardened implementation. If a service provider is offering a VoDSL solution that's missing any of these features, the results will be voice that sounds less than POTS quality.

Echo Cancellation

A hardened line echo canceller that can properly cancel echo is one of the hallmarks of high-quality VoDSL service. Some minor echo effects can be detected even in a POTS network, but the delay is less than 50 milliseconds and the echo is masked by the normal side tone, which is generated by the telephone instrument itself, so it doesn't interfere with the phone service. In a packet-based network, however, the delay is typically greater than 50 milliseconds, so echo related problems will immediately appear. For modem calls, we want to turn off echo cancellers, but if a packet delay of more than 100 seconds occurs, then the echo canceller thinks that the call is finished and will re-activate itself. Fax machines can turn off echo cancellation with a special tone in the 2010-2240 Hz frequency range (it's one of the annoying sounds they make when you call them up by accident).

The ITU sought to address echo cancellation with its original standard for echo cancellers, Recommendation G.165. However, after several years of further intensive research, the ITU established an important series of new and more comprehensive tests, which led to the more stringent set of requirements that is found in Recommendation G.168, which appeared in 1997 and was later revised in April 2000. The fundamental requirements for echo cancellers conforming to this Recommendation are listed in ITU G.168 Section 3.2, as follows:

1. Rapid convergence.
2. Low returned echo level during single talk (one speaker).
3. Assured double-talk detection.
4. Low divergence and clipping during double talk (duplex speech).
5. Proper operation during fax transmission.

The ITU standards do provide a series of objective performance tests, but testing to today's standards is just the beginning, for tests alone don't describe how to implement an echo canceller nor do they address the subjective nature of evaluating an echo canceller. You've got to actually listen to a voice channel hosted by the service provider to determine the overall quality of the system.

Ask your VoDSL service provider if it has implemented echo cancellation techniques with the following attributes:
- Ability to remove echo well, including removing echo at the start of a call as well as preventing echo during a call.
- Handles double-talk (both sides talk simultaneously) well, including not clipping the voice at the beginning or end of a spurt of double-talk.
- Handles background noise well, including high background noise and variable background noise.
- Exceeds the ITU recommendations, i.e. G.165 and G.168 and provides the ability to support future ITU echo cancellation recommendations, e.g. G.168-2002.
- Be field proven ("field hardened").
- Supports up to a 128-millisecond tail (a specification for carrier class quality) including support for multiple reflections over the entire 128 ms tail.
- Ability to dynamically track echo path changes. This is needed not only to support redundancy, but also for conferencing, call transfers, and permanent off-hook connections.
- Works properly in the presence of a four-wire connection and low hybrid attenuation.
- Has built-in configurability and instrumentation.

Packet Play-out

Packet play-out addresses the effects of network impairments on the voice, such as loss and delay of packets, including variable delay that distorts the timing sequence of the original voice. Components of human speech such as phonemes have a duration of 32-64 millisecond intervals, so any cell loss creating gaps at this level will become audible (cell loss of a duration of 4 to 16 milliseconds is not considered objectionable nor even noticeable to most listeners).

Ask the VoDSL service provider if its packet play-out algorithms can:
- Compensate for packet loss, delay and jitter?
- Adapt for lowest delay?
- Reside in DSP for system scalability?
- Be configurable and upgradable?
- Provide comprehensive network management statistics?

Tone Processing

Dual Tone Multi-Frequency (DTMF) is the technical term describing push button or

Touchtone dialing (Touchtone was a registered trademark of AT&T until its breakup in 1984). Tone processing is essential for call setup and termination as well as handling Interactive Voice Response (IVR) caller functions such as accessing voicemail, placing credit card calls, answering surveys and polls, etc. A service provider's VoDSL architecture should support the following key elements:
- Reliable tone detection, i.e. no false detects and no failure to detect.
- Early detection to minimize delay and to prevent in-band tone leakage. Leakage can lead to false tone generation at the remote end.
- Different detection requirements based on network application and system architecture, e.g. dial digits, fax and modem detection, call progress tones, etc.
- Support for bi-directional tone detection and generation, if the CPE doesn't perform these functions.

Fax and Modem Support

When scrutinizing a service provider, one should immediately determine how it handles analog fax and modem calls (most analog modems are used for transferring data but they can also transmit and receive faxes).

Many businesses are not aware that the process of sending analog fax and modem calls over packet-based VoDSL is a technological challenge for service providers, all of whom have resorted to various solutions.

Before reading the full analysis that follows, what you need to know "up-front" is the following:

If you're going to be installing an IP-centric VoDSL system and will be connecting to a next-gen packet network (not a Class 5 switch), then look for a service provider (and overall system) that uses the T.38 standard for supporting real-time fax. The service provider should also use the new *Modem-over-IP* (V.MoIP) or "modem relay" specification that will be finalized as an ITU standard just as this book goes to press. Older systems will use what's called *modem pass-through*, an imperfect system which sets up a bandwidth-wasting 64 Kbps channel (latency is a problem in the modem pass-through systems and this degrades throughput rates). There are actually two forms of V.MoIP modulation support for V.MoIP compliant gateways, Universal and V8. The most sophisticated and best performing V.MoIP solution is Universal Modem Relay, while V8 is basically a less complicated feature-reduced subset of Universal.

Whatever the system, keep in mind when formulating service level agreements that the following parameters are required to make a successful modem (both via fax and the basic analog modem) connection possible on a clean packet network: less than 50 millisecond one-way delay, no packet loss, and minimal jitter.

Fax Support

Fax is data, so it naturally belongs on packet-switched data networks — not on circuit-switched voice networks. However, most existing fax equipment has been engineered to transmit over such ordinary circuit-switched networks. Thus, when sending a conventional fax, your fax machine goes off-hook, dials, and the phone network completes a circuit connection over phone lines to the receiving fax device. You pay for the entire circuit connection.

The whole procedure is described in the T.30 standard fax handshake protocol, which establishes and manages communication between two fax machines. There are five phases to a fax call over the PSTN: call set up, pre-message procedure (selecting the communication mode), message transmission (including phasing and synchronization), post message procedure (end-of-message and confirmation) and call release (disconnection). Nailed-up fax connections are wasteful, inefficient and limiting.

To send a Fax over Packet (FoP), particularly Fax over IP (FoIP), using a conventional fax machine, demands that a "smarter" technology be built into the VoDSL IAD and the service provider's media gateway. This is because the gateway must be smart enough to figure out what type of call is about to pass into or out of the packet network and suitably process the call packet stream. Fax over Packet technology must therefore enable the interworking of standard fax machines with packet networks as well as the interoperability of certain classes of packet-enabled fax machines and PCs. It does this by digitizing the fax image from an analog signal and carrying it as digital data over a packet network, such as the Internet (IP), ATM and even Frame Relay.

There are two ways of sending fax over packets: *store-and-forward* and *real-time fax*. The older "store-and-forward" technique became an ITU protocol standard called T.37. (It's the protocol most commonly used by IP fax servers.) It's not as easy or transparent a process as using a regular fax machine, since the fax image is sent as an e-mail attachment (which means you can broadcast faxes to many people easily).

At first glance, T.37 appears quite logical and attractive, since it merely involves demodulating the analog fax signal as close to the source as possible and forwarding the "real" data bytes as a graphic image file in an e-mail. (The graphic data is attached to a Multipurpose Internet Mail Extension (MIME) file and passed through a predetermined Simple Mail Transfer Protocol (STMP)-based e-mail server.) The e-mail is then sent whenever the system finds it convenient to do so (which means there's never a "busy" signal).

This sounds less complicated than any kind of real-time fax process we could think of, which would ultimately involve sending "bytes carrying a digitized modulated analog signal representing a data byte." Yet, T.37 does have some limitations. For example, in a T.37 environment, the service provider or network operator knows only that the network "consumed" the fax, not whether it was actually received by the intended party. T.37 does have "full-mode extensions" that include mechanisms for ensuring call completion through negotiation of capabilities between the transmit and receive devices, but this adds complexity to the system. Also, the network operator would have to intervene if, for example, the destination number was incorrect, the machine was out of paper, was turned off, etc. So, if you typically need a "fax confirmation," a service provider using this type of technology would not work for you.

If your main concern is using conventional fax machines, look for a service provider offering real-time fax capabilities via Demod/Remod, or so-called "spoofing"; both supported by the T.38 protocol, which can work in conjunction with SIP or H.323 packet call control protocols. Either technique appears to users as if they are making an ordinary fax call. They just don't know about the IP "trunk" or other type of connection that sits in between the two fax devices. They dial the number and the receiving fax device on the other end hears the T.30 ITU international fax standard signals it would normally hear.

Demod/Remod: Doing real-time faxing via Demod/Remod (also called fax relay) involves de-modulating the fax modem's carrier wave signal, parsing it into binary 1s and 0s and loading them into IP packets that travel through an IP telephony gateway. The receiving fax-over-IP gateway reverses the process, re-modulating the modem carrier wave and feeding into the fax device, thus yielding the original fax. This process is very delay sensitive, however, and it's generally used only on networks with little or no IP latency, such as those with underlying ATM transport and voice gateways (just as an IAD can use ATM AAL5 to carry VoIP packets reliably in the local loop), since delays of even just a second can cause the session between the two fax devices to time out.

Spoofing: On the other hand, spoofing techniques can be used for fax transmissions over IP networks characterized by high levels of packet latency and "jitter," or variable levels of delay of packet receipt, such as badly designed LANs and networks that tend to be overloaded (e.g. networks commonly used for conferencing). The original form of spoofing compensates for both long delays (up to five seconds) and the effect of jitter by using the T.30 fax protocol which pads the signal with occasional packets resembling "I'm still connected" flags to keep the session active even though IP network delay may be causing some packets to take a long time in getting to the far-end fax. The receiving fax device is thus fooled into interpreting the incoming transmission as coming from a conventional, real-time, circuit-switched network. Two separate T.30 sessions, one for each fax with spoofing between, are used. There also are some proprietary spoofing techniques.

Typically, a receiving fax machine will assume that a transmission has been aborted if it has not received information after a period of time, because of its nailed-up telephony circuit origin. But with spoofing, just before the receiving fax machine times out, spoofing algorithms intervene by feeding it "fill lines." These fill lines produce a white streak on the fax, which on its own makes little or no difference to the readability of the fax. But it does let fax machines reliably fax over IP in real-time even though they may encounter up to three seconds one-way delay.

But there is a downside to spoofing — elongation. Elongation means that a two-minute fax might take two minutes ten seconds because of all the flags being transmitted as well as system delay, the number of pages, and transitions between compression formats. A system delay of 400 milliseconds can become a delay of about two seconds for the first page and one second for each additional page. If you send a lot of faxes, spoofing results in more bandwidth used, fewer free ports available at any one time, i.e. network congestion. Also, if you send a lot of international faxes, your costs can increase dramatically.

While both T.37 and T.38 were approved by the ITU in June, 1998, the nice thing about the T.38 real-time fax protocol standard is that the two endpoints aren't aware that a media gateway even exists. The sender simply loads the paper in the fax machine (that's plugged to the IAD) and it comes out at the destination. Success or failure is immediately reported, just as it would be when operating over the PSTN.

T.38 is the "official" fax transmission protocol selected for use with the H.323 protocol suite. Thus, if you have an IP-centric VoDSL system and you want to send faxes using packets end-to-end (without resorting to a Class 5 switch or the PSTN) T.38 should be part of the system.

To summarize, the VoDSL equipment performing fax relay functions must handle the effects of network delay, jitter (variable delay), and lost packets and at the same time prevent the fax machines from timing out. Standards protocols such as T.38, ATM AAL5 and ATM AAL2 exist for stability and interoperability between equipment vendors; although proprietary techniques are sometimes used to improve the interoperability between different fax machines when it comes to using a spoofing technique.

No matter what your fax needs may be, the VoDSL service provider's fax relay techniques should consist of the following minimum functions:
- Fax modem pumps: V.17, V.29, V.27, V.21, etc. (and now the preferred V.MoIP).
- Fax relay protocols: T.38 (IP) used with AAL2, AAL5, or other transport.
- Fax machine spoofing protocols: Can be proprietary.

Analog Modems

Just as fax equipment and "T" fax protocols are really designed to run over the PSTN, so are analog modems and the ITU's "V" series of modem standards (V.90, V34, etc.). Analog modems are designed to operate in networks with impairments that degrade the *Signal to Noise Ratio (SNR)*. If a naive customer connects an analog modem to a simple voice port on a non-ATM based IAD or residential gateway (such as an IP / Ethernet connection) and makes a call over a packet network (not the PSTN), then the call is treated as if it were a standard VoIP call, with the modulated modem signals treated as if they were voice signals. A simple concept, but by making a call in this way, the modem will experience forms of impairment that it would never encounter when working over the PSTN. In particular, two of these impairments are packet loss and jitter. Packet loss occurs when a packet does not reach its destination in time to be played out because of network congestion or other environmental factors. Jitter refers to the variability in packet arrival times.

An analog modem interprets packet loss on an IP network to be the equivalent of a "signal dropout" on the PSTN, which would mean that the call has been disconnected. Thus, when packet loss occurs, these modems will either retrain at a lower speed or simply disconnect. Random packet loss of as little as 0.1% or the bursty loss of several packets will cause connection failures, reduced throughput, and abnormal call disconnects. Packet jitter varies the round trip delay of the network and will cause the same kind of problems as does packet loss. Additionally, the average round trip delay is greater on a VoIP network than the PSTN, which results in lower data throughput rates. The reduction occurs because the error correcting protocol limits the amount of data a modem can send without receiving an acknowledgment from the remote analog modem. The added round trip delay can diminish effective data throughput by as much as 60%. This means that a normally 50 Kbps modem connection can be lowered to 20 Kbps.

Moreover, all of the G.7xx series codecs, with the exception of G.711, introduce too much distortion for analog modems to connect over an IP network.

Modem pass-through: Until recently, the most popular way to deal with analog modems on an IP network was *modem pass-through* in which a special VoIP channel is configured that uses a high bandwidth (64 Kbps) G.711 codec to prevent distortion of the modem signal. To support this feature, a gateway monitors the data stream to distinguish between voice and data calls. The voice gateway is reconfigured to modem pass-through mode once the determination has been made that the call is an analog modem. The analog modem (and the destination equipment) will think that they're communicating over a regular PSTN DS-0 circuit.

For modem pass-through to work, the gateway must deactivate the *Voice Activity Detector (VAD)* used for silence suppression and comfort noise to prevent distortion of the data stream. As for echo cancellation, low-speed analog modems (V.22bis and below) and fax machines use echo cancellation, and so echo cancellation for these connections must remain active. High-speed analog modems, however, disable the PSTN echo cancellers by adding phase reversals to their answer tones. The PSTN network detects these phase shifts and will disable any echo cancellers. In a voice over packet environment, the same information needs to be used to disable echo cancellation.

In a typical real-time application, data is produced and transmitted at deterministically-spaced time intervals. On the PSTN, end-to-end delay is relatively constant, while packet networks cause a variable amount of delay that changes the time interval spacing of the packets carrying the data — packet jitter — that can wreak havoc with analog modems. To eliminate the effects of packet jitter, a receiving host can employ a "jitter buffer" (a delay buffer) to delay the play-out of packets in order to reconstruct the original intended timing of the packet arrivals. Increasing the size of the jitter buffer increases the magnitude of the packet jitter that can be dealt with, and a jitter buffer can even be made adaptive so that it can detect a rise in packet jitter and make itself larger. But since analog modems will malfunction if the end-to-end delay suddenly changes in mid-call, what's best to use for modem pass-through is the largest acceptable, non-adaptive jitter buffer. When the jitter exceeds the jitter buffer size, then packets arriving later than can be handled by the jitter buffer must be discarded as their play-out time has already passed.

Dealing with packet loss in the modem pass-through environment is an almost impossible task. One attempt at a solution involves sending redundant packets in G.711 mode, a technique wherein each packet sent by the gateway contains not only the voice payload of the current packet interval, but also the previous packet interval, and possibly intervals from even more adjoining packets. If a packet disappears somewhere in the network, the redundant information contained in the succeeding packet is used to fill the gap. Compared to when a single previous payload is used, however, a redundant payload nearly doubles the channel bandwidth, and two redundant payloads triple the bandwidth. Ironically, such an increase in bandwidth usage can also increase packet loss, since packet loss usually occurs because of network congestion, which itself can be triggered by a shortage of spare bandwidth. Besides, typical packet loss occurs during quick bursts, and redundancy will offer little or no improvement.

V.MoIP or Modem Relay: Since the procedure of packetizing digitized modem signals for transmission over packet networks does not achieve the same level of quality that users currently enjoy on the PSTN, the ITU standard that appeared in 2002 to fix this is *Modem over IP (V.MoIP) or "modem relay."* In this model, local gateways perform the demodulation / remodulation of modem signals. Modem relay creates two distinct connections: one between the receiving modem and the ingress gateway and another between the egress gateway and the receiving modem. The data traveling over the IP network is user application content, not digital modulated modem signals.

In other words, V.MoIP / modem relay converts analog modem signals to data right at the customer premises (presumably the IAD). The transport across the service provider's network would be the demodulated data payload transported via packets. Then, with V.MoIP, it would be converted back to the PCM representation of analog at, say, the CLEC's point of interconnection with the backbone. This is all done transparently.

V.MoIP / modem relay is more reliable than modem pass-through, since by terminating the modem connection at the edge, gateway V.MoIP / modem relay solutions avoid subjecting the modulated modem signals to IP impairments in the packet network. For example, packet loss and jitter do not affect the V.MoIP / modem relay connections, so call connectivity and abnormal disconnect rates are enhanced, and the shorter round trip delays of V.MoIP / modem relay improve data throughput even more.

Furthermore, since V.MoIP / modem relay solutions send only content data across the IP network, this reduces modem traffic by between 50% to over 80% over what can be achieved with modem pass-through. This is because modem pass-through must continuously transmit / receive the modulated modem data even if no content is conveyed between the modems. Bandwidth usage is also drastically reduced because a modem pass-through solution demands a minimum of 64 Kbps of full duplex data over the IP network for the entire duration of the call regardless of the amount of actual user data traffic being generated. For low speed modems the differences between the two technologies is even more pronounced. A V22bis connection would use over 25 times more IP bandwidth if modem pass-through were used instead of modem relay.

The V.MoIP standard has defined two different levels of modulation support for V.MoIP compliant gateways, Universal and V8. V8 is, in essence, a subset of Universal. A Universal V.MoIP / modem relay gateway is a gateway that can terminate the following modulations locally: V.92, V.90, V.34, V32bis, V32, V22bis, V22, V23, and V.21. On the other hand, a V8 V.MoIP / modem relay gateway can only terminate V8 initiated modulations. A universal gateway can support the lower speed modulations using either V.MoIP / modem relay or modem pass-through but a V8 gateway can only support these modulations using modem pass-through.

The Universal solution thus results in better connectivity, higher throughput, and more efficient use of IP bandwidth. The V8 solution is simpler than the Universal solution while still providing modem relay support for high-speed modulations.

Mindspeed (Newport Beach, CA — 949-579-3000, www.mindspeed.com), a Conexant Business, is ready with V.MoIP products. A leading provider of access, edge and core silicon and systems software for switches, hubs, routers and other communications equipment, Mindspeed has developed enterprise voice access processors that can be incorporated into systems that will serve the converging voice and data networks. Enterprise solutions incorporating these products can support real-time fax-over-IP (T.38) and Modem over IP (V.MoIP) while supporting a wide range of VoIP and VoATM functions.

Transparency of Service

For a business scrutinizing a service provider's abilities in terms of fax and modems calls, the basic theme to look for is *transparency of service*. End users aren't supposed to know that they're sending data through a VoIP segment in the network. Let's take the example of a VoIP trunk segment between two central offices, a model that could be used by service providers to save on toll charges. We've seen that, if you happen to be sending a fax or connecting to an ISP across such a segment, many systems have to send the traffic just as if you were using normal, G.711 compliant, long-distance trunks carrying PCM bytes, a wasteful, troublesome process.

The job of T.38 and V.MoIP is to render the interposing IP network transparent to the PSTN endpoints. The endpoint fax or modem is not "aware" that the corresponding gateway is anything other than transparent. To achieve this, the media gateway, whether it be a local access gateway or a distant trunking gateway, must respond to voice and data calls differently. First, it must "sniff" the call's media stream to determine the call type, then respond accordingly. At a minimum, it must ensure that low-bit-rate vocoders are not used in conjunction with modems, and that silence detection and comfort-noise generation are shut off, since they won't work with analog modems. But without the benefit of such fax- or modem-protocol spoofing, the gateway will revert to G.711 full duplex, which automatically allots at least 128 Kpbs of channel capacity for the payload data (T.38 needs only a maximum of 14.4 Kbps bandwidth for faxes).

The bottom line is that you should find a service provider that has media gateways offering toll-quality voice via voice-specific call-stream processing. The gateways should also provide transparent toll-quality fax transport via fax-protocol-specific processing (T.38) and offer toll-quality modem transport through transparent data-modem-specific call-stream processing.

Network Management

Fundamental to any viable communications system is the service provider's ability to discover, isolate and remedy problems as quickly as possible to minimize or eliminate the degree to which the customers are impacted. So ask the VoDSL service provider about its network management capabilities. Its network management system should be able to do the following:
- Configure on a per-channel basis.
- Provide the ability for the customer to subscribe online to voice service and the web-based activation of extra channels.
- Provide per channel statistics and status reporting. This is important as it relates to the maintenance of Call Detail Records (CDRs) and billing (billing for VoDSL is much more complex than billing data or Internet services — the system must handle interpretation and labeling of CDRs, third party billing for resellers/ISPs, do invoice processing, and ideally should provide a web interface for retail customer information).
- Provide for automatic configuration in the switch/gateway.
- Perform per channel real-time trace and diagnostics.
- Provide Bellcore (Telcordia) test line support for diagnostics.
- Provide redundancy support.

Voice solutions based on the integration of software and DSP-based access communications processors enable service providers to offer true toll quality for their VoDSL service. Voice compression is perhaps the key component, so look for a service provider whose migration strategy includes the use of LBR codec technology.

Bureaucratic and Other Issues

Purely technological considerations, however, are just one set of challenges. The service provider should have made things easy on itself by making sure its whole system incorporates or has taken care of the following items:
- To help maintain service and keep the customers happy a robust help desk and trouble ticketing procedure should be in place. This enhances the service provider's market credibility or "brand."
- Interconnect agreements should be ironed out with the incumbent carrier (licensing, service level agreements, options for planned expansion of interconnections, etc.).
- Regulatory obligations should be fulfilled (adherence to physical plant standards, ability to allow government wiretapping tapping, etc.).
- Support for Telephone Number Portability. If a customer doesn't move to a new

premise, then changing one's PSTN voice carrier (from MCI to Sprint, let's say) and keeping the same telephone number is no longer a problem — it's a procedure that merely involves changing a number in a common database on the SS7 network. If you physically move to a new location, however, you may fall outside of the serving area for DSL. Even if your new location is served by the same central office as before, the CO staff will still have to take down the DSL connection and "rebuild" the service at another location on the frame, which can take several days to carry out.

• Whoever the service provider may be (perhaps a CLEC), like any phone company, it must adhere to strict regulations and tariffs as defined by regulators. The service provider must also be capable of updating the common databases as customers change so everyone stays informed. All of this can be a hassle and explains why many ISPs have stopped trying to recast themselves as CLECs.

Some Service Providers Use Outsourcing

You may discover that your service provider is one in name only, having abrogated many of its responsibilities via outsourcing. Since owning a voice gateway involves significant voice expertise and resources, it's not too surprising that some service providers have followed an outsourcing trend in the market that involves the DSL operator becoming a "partner" of a *Voice ASP (Voice Application Service Provider)* or becoming a "voice wholesaler" in its own right, with the Voice ASP being responsible for the voice gateway and the termination and origination of the circuits at the switch.

For you, the customer, this could be a good thing, since smaller service providers working on their own might not ever be able to marshal the kind of resources needed to serve customers properly. Still, in such a case you must inspect the capabilities of the outsourcer as you would the service provider itself. For example, when dealing with a Voice ASP, a key issue is one with which the reader is now familiar — whether the interface to the switch is a nonpacketized one, such as a GR-303 or V5.2, or whether it is fully packetized such as ATM or IP, and perhaps doesn't need a voice gateway at all.

Customers should also determine the stability and nature of the relationship between the service provider and the outsourcer. A service provider can delegate varying degrees of responsibility and functionality to an outsourcer. This is because VoDSL providers must weigh a wide range of factors when deciding whether or not to pass along their headaches to outsourcers.

There are various business models where differing proportions of functions are divvied up between the service provider and the wholesale Voice ASP. Somewhere between the extremes of "doing everything yourself" and "outsourcing nearly every-

Functions Divided Between VoDSL Provider and Voice ASP When VoDSL Provider is Acting as...

	Distributor	Reseller	Partner	Voice Wholesaler
DSL Provider	• Transport only	• Transport, Brand & Billing	• Transport, Brand Pricing & Billing	• Transport, Brand Pricing, Billing & Product Management • Technical & Voice Expertise • All voice service features
Voice ASP	• All voice service features, Brand, Billing & Pricing	• Termination & Origination • All voice service features • Pricing	• Termination & Origination • All voice service features	• Termination & Origination

Fig. 7.2. Outsourcing Tradeoffs.

thing," companies such as KPN in Europe believe that a mix between wholesale/partner model appears to create a "win/win" situation for both voice and DSL partners. So know what your service provider is handling in-house and what it is outsourcing, as this may very well be the deciding factor in your decision-making process.

In Quest of the Full Digital Loop (FDL)

Prior to the appearance of G.SHDSL, the most popular kinds of DSL in the U.S. have been ADSL and ADSL Lite (G.Lite), which currently deliver voice as regular analog POTS or "Voice Under DSL" as we have so amusingly referred to it. When such ADSL networks were being built, service providers had to install, as a minimum, five new equipment elements:
- A DSLAM to multiplex the channels.
- A possible POTS splitter to separate any analog voice and data signals (in ADSL).
- Power panels to power and protect the network elements.
- Racks or cabinets.
- A voice gateway that connects to the Class 5 switch.

Some ADSL line cards for DSLAMs are now built with chipsets that merge both ADSL and POTS services onto the same chip set, which eliminates the need for a splitter by collapsing that functionality onto the chipset. Eventually, however, ADSL and

ADSL Lite should be able to operate without an overlay, with all voice being carried in-band as data — VoDSL. This type of Voice over ADSL (minus POTS) architecture has been called a *Full Digital Loop (FDL)* by Alcatel and others. It's a concept that is thought to be well-suited to mass deployment for residential and SOHO customers as it offers a clean, high bandwidth data access for everybody. The FDL makes it possible to eliminate expensive, cumbersome POTS splitters, achieve a major breakthrough in ADSL density at the central office, and enable the service provider to achieve major savings in the cost of ownership. Enterprises both large and small are now able to buy integrated voice, data and Internet services from one provider and receive one bill.

The splitter at the central office is expensive and is unlikely to become significantly cheaper since it is constructed using many non-semiconductor components (e.g. coils and capacitors with high voltage isolation requirements) that do not follow Moore's Law. Moreover, because of the splitter, each subscriber requires two pairs to be connected to the DSLAM (one pair to the customer's premise, one pair connected to the voice switch). The resulting cable volume is becoming a bottleneck for DSLAM densities. In some countries, the requirement to be able to bypass the splitters for line testing further complicates matters.

Note: Moore's Law is named for Intel's co-founder, Gordon Moore, who observed that the pace of microchip technology change is such that semiconductor circuit densities continue to double every 12 to 18 months. As technologist Raymond Kurzweil has further noted, Moore's law is quite expansive and substrate independent, and has continued for at least the last 100 years, through five "paradigms" of computation: mechanical, electromechanical, vacuum tube, transistor, and integrated circuits. Furthermore, data showed that the "doubling time" for certain measures of transistor complexity has actually decreased from three years down to one year over the last century of doublings. So instead of slowing down, technological progress is accelerating.

An FDL architecture (in which POTS is nowhere to be found) eliminates the need for two patches per subscriber to be wired at the main distribution frame, allowing more subscribers to be connected to the main distribution frame, thereby facilitating the widescale deployment of ADSL.

With FDL, all voice is VoDSL and is carried digitally within the DSL channel, so FDL customer premise equipment will typically offer both voice and data services carried over the DSL link. For example, in the residential case, the CPE might offer both an Ethernet and two analog short-haul telephone interfaces. When delivery of the voice and data services are combined with termination of the DSL circuit, the voice gateway can be highly integrated. Residential users might benefit significantly from the provision of interfaces other than Ethernet, such as Home Phoneline Networking Alliance (HPNA) and wireless LANs (e.g. 802.11b). On the other hand, business users will benefit from more voice ports and the delivery of voice services over digital interfaces (e.g. primary rate ISDN) integrated with the delivery of data services over Ethernet.

Such a totally non-analog DSL network will now have to provide emergency power feeding to the customer premise equipment. Early studies how shown that by limiting the link bandwidth in emergencies and by reducing and/or eliminating power to other functions (e.g. the Ethernet), it is possible to operate the gateway solely from power supplied by the DSLAM.

Another challenge in hammering ADSL into a full digital loop is working around the way ADSL modem circuitry in IADs deal with line noise that causes bit errors. Interference that is short in duration but of a large magnitude is known as *impulse noise*. Impulse noise can be caused by lightning or by a motor turning on and creating a power surge. This results in a short, but very strong disturbance of the line. Impulse noise can last from microseconds to several tens of milliseconds and results in a burst of bit errors that is hard for decoders to correct.

The main sections of a typical ADSL modem are the Digital Interface (e.g. ATM), the Framer / FEC plus Encoder / Decoder, the DMT Modulator, and the Analog Front End (AFE). The Framer multiplexes serial data into frames, generates *Forward Error Correction (FEC)*, and "interleaves" data. The ADSL modem splits incoming data into two streams or logical "paths" — a *Fast Path* and an *Interleaved Path*.

Interleaved Path: Data streams can be allocated to one or other of the paths (in some cases both paths can be used for different data streams). Data passing through the interleaved path has the additional "interleaving function" performed on it, which boosts the ability of FEC function to fix errors. This works as follows:

Prior to interleaving, the ADSL modem circuitry sets up the data stream for normal forward error control by organizing the aggregate data stream created by multiplexing downstream channels, duplex channels, and maintenance channels together into blocks, and attaches an error correction code to each block. The receiver then corrects errors that occur during transmission up to the limits implied by the code and the block length. To interleave the channel, the transmitting unit additionally buffers the data blocks, then creates "superblocks" of data by interleaving data within subblocks, re-ordering the individual bytes of the subblocks in a different sequence so that adjacent bytes become scattered about and separated in time.

In other words, the interleaving process rearranges data so that those samples that were located contiguously in time are now spaced far apart. De-interleaving is performed at the receiving end to put the bytes back in the original order. The benefit of interleaving is that if data symbols are corrupted by impulse noise on the line during transmission, the corrupted bytes are not in continuous blocks after de-interleaving,

What Does it Take to Provision VoDSL?

Although already discussed in detail in the book, here's a synopsis of what it takes to provision the most popular form of VoDSL from both the service provider's and the customer's point of view.

The service provider provides the customer with an Integrated Access Device (IAD) with "voice" ports that can be used for voice service. At the central office or other point of presence (POP), the service provider's DSLAM typically demultiplexes the voice cells and runs them through a voice gateway, where they are put back into pure PCM byte-interleaved format, presented to a Class 5 circuit switch, and sent over the PSTN. The data cells are carried as packet or cell traffic to their ultimate destinations, which could be the Internet or a corporate Intranet, through the ISP.

An IAD provides support for between 2 and 24 analog or T-1 digital voice ports over a single DSL line. These ports can connect to any analog telephone, fax machine, analog modem, or key telephone system, and some IADs offer support for digital PBX systems. The IAD can also support data connections via Ethernet, USB, or ATM25, and some IADs support customer-located routers or frame relay devices via V.35 connections.

The service provider's voice gateway, which resides in the service provider's network, receives traffic from the IAD in packet format, usually by way of a DSLAM, then it reconstructs the calls to be received by the operator's Class 5 voice switch through standard TDM trunks using a GR-303 interface. Some gateways can support T-1 trunking, which enables calls to be delivered directly to an operator's Class 3 or 4 long distance switches, effectively bypassing reciprocal charges for Class 5 switch termination. For operators outside of the North American market, the voice gateways have V5.2 interfaces.

Customer support is often neglected when it comes to provisioning DSL. To help, Turnstone Systems Inc. (Santa Clara, CA – 408-907-1400, www.turnstone.com) is developing ways to automate the physical copper cabling plant with new interfaces to back-end support systems from multiple OSS vendors. Turnstone's Copper CrossConnect box sits between a DSLAM and patch panel in an access provider's network and enables remote troubleshooting. When used with Turnstone's CrossWorks software suite, the copper plant can be controlled through software from any location, such as a centralized network operations center, reducing on-site technician visits and streamlining operational procedures.

which reduces the number of corrupted bytes in each Reed-Solomon FEC codeword, and helps reduce the probability of encountering completely uncorrectable codewords.

Hence, the use of interleaving combined with coding spreads out these errors in time to improve decoding performance. Interleaving extends the ability of the forward error correction function so as to allow the receiver to correct any combination of errors within a specific span of bits. The typical ADSL modem interleaves 20 milliseconds of data, and can correct error bursts as long as 500 milliseconds. Tests indicate that this level of correction is suitable for MPEG2 and other digital video compression schemes.

The Interleaved Path, with its combination of interleaving and error correction coding, protects against bit error bursts, but introduces an additional delay (latency), typically between 10 and 60 milliseconds, depending on the interleaving depth. This means that the Interleaved Path should only be used with data traffic which is sensitive to errors but not to delay.

Fast Path: ADSL's other type of channel, the "Fast Path" doesn't use interleaving; so it's not very good at error control, but it suffers from less delay. So, when doing voice over ADSL one must make a choice between having good end-to-end delay or else having good error control. It's impossible to achieve low delay and low packet loss over ADSL. One is stuck with a trade-off.

A better solution is to choose a service provider that allows something other than ADSL to be used for VoDSL. Although ADSL and ADSL Lite got a head start in terms of number of DSL deployments, the future appears to belong to G.SHDSL, which works with pretty much all the VoDSLs, even channelized VoDSL (CVoDSL). G.SHDSL has a G.hs (handshake) protocol that's used during pre-activation to negotiate the most efficient service type (T-1, E-1, ISDN, ATM, or IP framing) during start-up (training) to avoid unnecessary overhead and latency on the G.SHDSL link. This is welcome news to owners of slower links, since some services such as VoIP and VoATM do not work as well at very low bandwidths. It also helps to enable transport of native TDM if toll-quality voice is to be transported over the DSL link. For networks that use IP in the core, G.hs can negotiate IP framing all the way to the edge, eliminating the overhead associated with ATM cell headers when the primary traffic type is IP.

In Summary

Although VoDSL is a serious revenue opportunity for all telecommunications providers, no one solution for deploying VoDSL resolves all issues or is ideal for all situations. For example, a service provider with an established ATM network infrastructure and a plethora of ATM engineers available to maintain and support the network would most likely select the VoATM solution to deploy its VoDSL services. Yet, a service provider that is new to the industry would probably go with an entirely IP-based network (perhaps VoIP over Ethernet) and even an established service provider might opt to migrate to IP or to have coexisting IP and ATM networks. Whereas, a service provider, such as an ILEC with both an IP and TDM infrastructure that is adding data services to its network (or an ISP installing voice over campus-based or in-building copper) might consider a channelized voice and data solution as its most expedient option.

In most, if not all of these cases, the service provider will consider its options in view of the existing architecture and skillset available. Look, therefore, for a service provider that can take advantage of the existing PSTN facilities (narrowband and broadband network infrastructure) and deliver high quality voice connections; in other words, voice quality that's comparable to POTS and offers customer premise equipment that fits your specific needs. At the same time, keep in mind that the VoDSL solution should also allow for the addition of future services.

So, at this point in time, you might do best to keep your focus on finding the best short-term solution and simply look for features — after all, no technology is permanent — they're all ultimately transitory technologies, bridge technologies to the future.

CHAPTER EIGHT

VoDSL - A Compelling Business Case

Once upon a time, the idea of simply having a broadband xDSL connection for data access to the Internet was novel and tantalizing. Unfortunately, the upfront equipment costs and phone-line upgrades required for DSL also make it difficult for service providers to compete in the Internet access market space as these costs overshadow the $40 or so a month that the typical DSL subscriber is willing to pay, thus rendering any potential return on investment nugatory.

Service providers realized that they needed to devise and deliver new value-added services to supplement DSL's vanilla high-speed Internet access so as to generate more subscription-fee revenues. Soon broadband virtual private networks (VPNs), remote network management, wireless LAN support, and voice-over-DSL (VoDSL) services appeared. Of all the possible DSL value-added services, VoDSL service is the likeliest candidate for universal adoption and presents a compelling business case, with its cost-saving capability of supporting up to 24 toll-quality phone lines and high bandwidth Internet access over a single pair of copper wires.

By its nature, VoDSL is best suited for small businesses. Neither *Incumbent Local Exchange Carriers (ILECs)* nor the few surviving *Competitive Local Exchange Carriers (CLECs)* really expect to supply VoDSL to larger enterprises because of its four-to-24 line range. Most large enterprises will already have architected their own dedicated, private network, perhaps based on higher capacity T-3 lines.

A 2000 Cahners In-Stat Group research report found that the target customer group for VoDSL service would be the millions of businesses with 20 to 49 employees, that competitive providers claim have been historically underserved by the incumbent telcos.

When a business hears that the new-fangled VoDSL "will save you money," visions of cheaper per-minute long distance charges may immediately spring to mind. This is not necessarily the case, since VoDSL is an access technology, and may only result in cheaper long-distance phone calls if a "bucket of minutes" is offered as part of a service bundle or if the telephone system is used to plug into a "toll-bypass," all-packet, "next-generation" network.

Voice over DSL has been called "the killer application" in as much as it has the potential to greatly reduce the costs associated with delivering a total integrated communication solution over a subscriber's local loop. Indeed, some of the savings and benefits brought about by VoDSL can be quite subtle. For example, traditional three or more phone lines are normally needed for a telco to furnish a customer with multiline *Centrex* service (PBX-like call control outsourced to the central office). With VoDSL, however, the telco can offer such a service over a single copper local loop.

Note: In the U.S., Centrex is a term used to describe a central office exchange service offered by service providers whereby up-to-date phone facilities at the service provider's central office are offered to business users, thus alleviating the need to purchase their own facilities and the upgrade and upkeep thereof. Centrex works by the service provider partitioning a part of its own centralized capabilities among Centrex customers. In most cases, Centrex replaces the in-house PBX system, while providing the customer with as much if not more control over the same services than they had with their PBX. Typical Centrex services might include direct inward dialing (DID), sharing of the same system among multiple branch locations, self-managed line allocation and cost-accounting monitoring.

Any cost savings derived from VoDSL over equivalent non-integrated communication services is just the tip of the iceberg, however. VoDSL interworking with a next generation architecture incorporating softswitches and IP-based media servers also makes possible customized, novel applications.

VoDSL Call Centers

For many small businesses, VoDSL's ability to transform one line into 16 or 24 for a reasonable fee will be a welcome relief. This will be most evident in small businesses, many of which well find the establishment of a call center advantageous.

Call centers are now referred to as contact centers, or even "multimedia contact centers" since they now incorporate voice and non-voice media technology i.e. chat, intelligently directed e-mail, web-call back and collaboration / co-browsing (also called "symmetrical browsing" where a web-browsing customer shares a page with a customer service representative, who can show information, describe details by voice or chat and even push information from other Internet sources).

In reality, however, most contact centers today are still "call" centers in that they support voice as the primary contact channel. VoDSL offers an easy and relatively inexpensive way of adding sufficient lines to a business so that a small call center can be established. The fact that the VoDSL system also supports a high bandwidth data link to the Internet is advantageous, since the link can be used to support the multimedia contact center paradigm.

VoDSL will do much to bolster the development of the "casual call center," or "informal call center," terms coined by telecom industry researcher Blair Pleasant during her tenure at the PELORUS Group. (Raritan, NJ — 908-707-1121, www.pelorus-group.com). As it was originally defined in the author's *Computer Telephony Encyclopedia* (CMP Books, 2000):

> So-called "knowledge workers" are humming away (thinking, calculating, writing, programming, doodling, etc.) when their phones ring. They answer. Callers are looking for information or help. These workers (hopefully) do their best to oblige. Certainly it's not as regimented as the formal call center, where well-armed agents wait ready and willing for calls and caffeine-addled supervisors drool over real-time call-processing performance pie charts. Still, the ubiquitous "informal" or "casual" call center does represent a vast arena for computer telephony technology to flourish in. Indeed, it's one that potentially dwarfs the traditional call-center market. How far have we gone in exploiting this killer opportunity area? Well. . . .we've only just begun.

The more phone lines that become available, the more an entire enterprise becomes a call center. Ultimately, the call center can spill out of the physical confines of a business, thanks to the networking innovation of the Virtual Private Network (VPN), which we will examine toward the end of this chapter.

"In-Building" DSL and Building Local Exchange Carriers (BLECs)

VoDSL's ability to extract greater benefit from ordinary copper telephone wires by "multiplying phone lines" and integrating packetized voice telephony services with high bandwidth data can perhaps best be appreciated when it is implemented in situations where there is both a high density of users with sophisticated needs and a limited number of copper wires leading to those users. The most extreme example of this is a call center operation, which we examined in the previous section.

Yet another example of such a scenario is "in-building" DSL, which enables high-bandwidth access for commercial and residential buildings in geographical regions where these buildings' tenants cannot be reached with existing local-loop xDSL services. If you own a small or medium sized business that's not situated in a metropolitan area, chances are that it occupies a suite of offices among those of many other companies, all housed in a building that is just out of reach of the service provider's DSLAM (or is served by a telco's remote terminal), thus ruling out the availability of xDSL service.

But don't worry. As we shall see, there's a special kind of service provider that can bring your particular "pigeon hole" both xDSL and VoDSL service.

Alphabet City: MTUs, MDUs and MHUs

The in-building DSL market space can be segmented into *Multi-Tenant Unit (MTU)*, *Multi-Dwelling Unit (MDU)*, and *Multi-Hotel Unit (MHU)* sub-markets. MTUs consist mainly of commercial properties with business customers as tenants, whereas MDUs consist of apartment and condominium complexes with residential customers as tenants. The MHU segment is the hospitality or hotel marketplace where guests (particularly laptop-equipped business travelers) need high-bandwidth access to Internet and corporate VPNs while staying in hotels or resorts far from a DSLAM or a switch.

But even if an MTU (or MDU or MHU) is within range of a DSL-equipped central office so that local loop xDSL service can be deployed to each tenant independently, it actually makes more sense to go with MTU broadband service delivery, which essentially creates an in-building local loop via the breakout of a high bandwidth line (T-1 to T-3, or OC-3) by a basement mini-DSLAM into smaller in-building xDSL lines to each tenant's office. This is because there are many benefits to be gained by concentrating subscriber (tenant) traffic locally in each MTU.

You might be tempted to treat your MTU-based business as a standalone entity, since such a traditional scenario doesn't require DSL concentration equipment within the building and allows for a dedicated xDSL loop to the central office (provided, of course, that you're in range of the central office). However, if every business in your building has the same idea and also signs on for VoDSL service, then individual xDSL-enabled local loops must be deployed to each subscriber, which means that VoDSL customer premise equipment (an IAD, router, etc.) must be deployed to each subscriber too, which increases truck roll and provisioning expenses per subscriber. All of these new xDSL lines must lead off to an equivalent number of xDSL ports at the central office, where co-location space is typically more expensive and/or difficult to negotiate, and which is reflected in xDSL and VoDSL recurring monthly local loop charges.

The Building Local Exchange Carrier (BLEC)

A voice and data service provider who specializes in coverage for the in-building market space is termed a *Building Local Exchange Carrier (BLEC)* or *Building Service Provider (BSP)*. The BLEC concept predates VoDSL, xDSL or even digital networks. In New York City in the early 1980s, with more of an interest in retaining and winning tenants than a desire to enter the telecommunications market, Rockefeller Center's landlords began querying their existing tenant base to find out what sorts of amenities might increase the property's value. In a business where location is proverbially everything, one wouldn't have guessed that one of the most high profile buildings in midtown Manhattan needed anything radical in the way of value-added services. Interestingly, the tenants wanted less, not more of the customer premise telecom infrastructure they were supporting. Specifically, they wanted to be free of the PBX metal and associated wiring tangles that were taking up space on high-rent square footage, and eating up time and talent resources that could be better spent elsewhere. Thus was born Rockefeller Group Telecommunications Services (RGTS), one of the first BLECs.

Today, RGTS is a thriving business in its own right, expanding from its roots in hosted PBXs to offer everything from local and long distance voice to Internet access to VPNs and e-mail outsourcing. The company now operates not only in the buildings owned by its parent, but also in 36 other sites around the New York City area and is expanding across America. At a time when a competitive carrier market did not even exist, Rockefeller and a handful of others in the "shared tenant services" space recognized a shortcoming in the level of service being offered by the phone company to occupants of multi-tenant office buildings, and came up with a customer service-centric, fully outsourced model to fill the gap. By assuming the role of the internal telecom staff for the companies that they served (so that the companies could focus on their own core competencies), RGTS essentially became a kind of Voice Application Service Provider, or Voice ASP.

Internet access has become just as much a commercial utility as phone service. But for small and medium-sized businesses in multi-tenant unit (MTU) offices, getting anything faster than a dial-up line has been a generally painful experience, and the idea of sending voice over DSL seems downright exotic. Also, the number of possible channels for voice and multimedia is limited by the building's wiring. DSL and VoDSL seemed a promising alternative to costly and slow-to-deploy T-1 lines.

Without them, options for these customers to move beyond basic telephone connectivity and toward enhanced services and hosted applications would be essentially non-existent. Despite radical post-deregulation shifts and the rise of the Internet, the fact is that telecom carriers, incumbent and competitive alike, still don't want to touch anything past the communications pipes that terminate at the customer premise, and they are not eager to offer different services to widely varying premises.

It's not surprising, then, to see the service provider community jumping at the opportunity to capture the building market space, feeding the same latent demand that prompted the Rockefeller Group to start offering PBX service alongside heat and air conditioning. Everyone, ILECs included, more or less agrees that there are thousands upon thousands of small to medium-sized businesses located in MTUs across the country who remain underserved when it comes to broadband voice and data access.

What many of these providers share, despite significant differences in their business models, is a level of service that goes well beyond mere connectivity. This distinguishes them from the majority of LECs and ISPs in two key ways: first, by offering truly differentiated services on top of their basic broadband access connections, in an ASP or ASP aggregator-like fashion; and second, by backing up those services with "high-touch" and often on-premise customer support networks. It doesn't hurt, of course, that by establishing a physical presence in the buildings themselves, these providers — call them BLECs, BSPs, or what-not — can also beat the older and slower carriers to the punch in signing up new tenants.

Most of the initial BLEC players were (and are) facilities-based local carriers, who own and operate not only the equipment in the buildings, but also heavy-duty POPs and often private IP networks, sometimes even DSLAMs and Class 5 telephone switches.

Under most early in-building xDSL service scenarios the BLEC treats each MTU-based subscriber as a standalone entity — each tenant gets a dedicated DSL loop within the building and a deployment of customer premise equipment, but the BLEC recognizes that MTU-based subscribers are co-located, and so it concentrates the xDSL loops in-building to provide greater deployment flexibility. Early implementations simply resorted to the traditional non-VoDSL ADSL technique of concentrating only the data and letting the voice travel as a traditional analog service back to the central office (VoDSL could be added on later). Thus, no central-office loop termination is required (at least not for data) and an economical over-subscription of WAN uplink(s) is an option. The WAN uplinks themselves can be traditional, high bandwidth pipes (DS-1 / T-1, DS-3 / T-3, etc.), as well as new fixed wireless WAN uplink options to save costs or quicken the installation time. In the future, optical *Fiber To The Curb or FTTP* will also be an option at the core of metropolitan areas. Although, in the U.S., the percentage of office space located inside the core central city area or central business district dropped from 66% in 1979 to 47% in 1999 and it is estimated that only 10-15% of these buildings are in reasonable reach of a fiber splice point.

DSL Concentrators — such as Copper Mountain's CopperEdge 150 and 200 — offer a choice of WAN interfaces ranging from T-1 to DS-3 (45 Mbps) to OC-3c (155 Mbps), as well as Ethernet (10/100 Mbps).

It is possible, in practice, to configure in-building systems that yield extremely high in-building DSL speeds or many VoDSL channels (depending on mini-DSLAM performance), and yet have the whole model potentially less costly overall, since there will mostly likely be lower recurring monthly WAN charges.

This means that the BLECs' possible profit margins can be reasonably high. BLECs typically don't have to share revenues with an ILEC or another local loop provider and the *Integrated Access Device (IAD)* can be purchased as subscribers are signed up, making deployment "pay as you go" and thus reducing some of the up-front capital outlay necessary to start generating revenue on service delivery. By installing the mini-DSLAMs or other concentrator equipment in the basement of the building, BLECs avoid the astronomic co-location charges incurred with CO-based DSL deployments. Also, by eliminating any interaction with the telco having to do with qualifying each individual subscriber line, there are dramatic savings in administrative costs. Finally, there are also significant savings in customer acquisition costs — to sell the service a BLEC representative can just walk in the front door on a "cold call" or else put up in-building posters or use mail drops (it is estimated that MTU customer acquisition cost can be reduced by 90% over traditional mass-market promotional strategies).

But, on a relative scale, BLECs actually *do* have some high up-front capital expenditures associated with their infrastructure. Indeed, they find themselves facing some of the same logistical and economic issues as a regional service provider, only on a smaller scale: The in-building concentrator and backhaul line with sufficient capacity must be purchased up-front, before there are subscribers, and there will be at least one truck roll per customer. This is why most BLECs are targeting only higher-end buildings in dense metropolitan areas, where they can be reasonably sure of a favorable return on investment.

However, in the U.S., there are also over 200,000 MTUs with fewer than six tenants; even small and mid-sized business customers in a sparsely-occupied building want an affordable bundle of "big company" services. If it isn't available, they'll look for office space that does offer these amenities. For example, a January 2001 survey of 454 telecommunications decision makers of business tenants in U.S. MTU buildings found that, "Four-out-of-ten businesses will consider moving elsewhere at lease renewal if their important telecommunications needs aren't met at their current location." Add to the large portion of MTUs still underserved (particularly smaller buildings and office parks, and those in less populated cities), the vast number of residential communities and apartment complexes or multi-dwelling units and you can see a growing business case for BLECs.

It's important to bear in mind, when looking at the BLEC market space, that owning one's own network is not a requirement for BLECs, as long as they can provide the necessary infrastructure within the building. Even those BLECs who do own part of the central office infrastructure know that the major competitive factors going forward will be the application services and personalized support that they can offer their customers, not just platforms and the size of communications "pipes." As it becomes increasingly difficult for competitive telecom carriers to secure investors (or stay in business at all, for that matter), having a lightweight, strategically planned network infrastructure is key for all levels of provider.

BLEC-specific Equipment and the Features They Offer Customers

Equipment vendors have also perked up to the MTU opportunity, and have begun to manufacture products tailored to the needs of in-building service providers. Like the range of BLEC business models, the products aimed at this space, while relatively new, already run the gamut. There are differences in architecture, core technology, and scale. Some common aims or requirements for this diverse set of products, however, have also emerged. In-building providers, by and large, look for equipment that is physically compact, collapses multiple voice and data access functions into one unit, and supports differentiated, and typically IP-based services over a common network infrastructure. A fast-growing group of vendors, from startups to some of the biggest names in networking, are now tackling these problems at hardware and software levels in unique and interesting ways.

The problems DSL and VoDSL have encountered in local loop deployments, while clearly not insurmountable, have left many potential customers frustrated. What BLECs and equipment vendors alike have realized, in the meantime, is that MTUs can be an optimal environment for DSL technology, and are relatively free of the stumbling blocks faced in CO-based networks. By locating DSL termination equipment in the basement of a multi-tenant building, the service provider can distribute broadband service using the building's existing copper telephone wiring. This wiring has, for the most part, already been installed in the building's risers to connect tenants to the PSTN. In this scenario, the local loop connection back to the central office relies on standard T-1 or T-3 lines, which are easy enough for the provider to order and provision, and can be oversubscribed to achieve economies of scale. Ethernet, optical technologies and broadband wireless access technologies can also be used to connect to the regional access network and thence to the central office, the PSTN and the Internet. Since the distance between a customer's premise (the tenant's office) and the point of aggregation (the building basement) is much shorter than the network extending from the CO to the customer, tasks such as loop qualification are not nearly as onerous here as they are in more conventional xDSL networks.

At the same time, it is generally impractical to use the same xDSL equipment you would locate in a CO to distribute service within a multi-tenant construction. It would not make sense, for example, to bring a 200-port DSLAM into a low-rise office complex with a dozen small business tenants. The equipment footprint must also be within scale: Even more than in the co-location facilities that house both ILECs and CLECs — space in equipment rooms of multi-tenant buildings is at a premium, and providers want to do anything possible to avoid the big metal chassis that typically occupy central office sites. The makers of central office DSL equipment were among the first to recognize these differences in requirements, and to begin developing products to specifications defined by the MTU environment. Particularly strong here were the pioneer vendors, such as Copper Mountain (Palo Alto, CA — 858-410-7305, www.coppermountain.com) that early on identified MTUs as a large potential niche for their products.

Copper Mountain's early DSLAM systems, the CO-scale CopperEdge 200 and the somewhat smaller CopperEdge 150, quickly found a home in various MTUs, MDUs, and hospitality environments. One specimen of the 200 model, for example, which can support up to 192 ports per chassis, can currently be found operating in the basement of the Empire State Building. Copper Mountain offers an even more compact, eight-to-twelve-port pizza box-sized unit, known as the OnPrem 2400. This vendor takes more of an IP-centric approach to DSL concentration than traditional DSLAM vendors, by including routing and OSI Layer 3 intelligence in its platform (in a typical DSL/ATM solution, there is no option other than a Layer 2 model from the IAD's subscriber port right through to the service provider's POP). As a result, Copper Mountain's products lend themselves well to convergence applications, and the vendor has a number of partnerships in place with Voice-over-DSL and Voice-over-IP gateway vendors, as well as with softswitch developers.

Most importantly, Copper Mountain realized early on that an MTU opportunity existed, and did its homework on the business case for multi-tenant DSL services. Copper Mountain has also established a professional services and support organization around its products, to help educate the market and assist service providers in planning their deployments.

Companies such as Copper Mountain started out building equipment for central offices; their product evolution, as a result, has been a progressive scale downward, from large, multi-slot chassis to self-contained, 1U high, rack- and wall-mountable units. Larger networking equipment companies such as Lucent and Cisco have gone a similar route within their own xDSL product lines. Other vendors, however, who sell particularly into small business and residential MTUs, have moved in the opposite direction — starting with low-cost, entry-level systems and scaling up performance and port density.

Notable startups in this subcategory are Tut Systems (Pleasanton, CA — 925-460-3900, www.tutsys.com), Avail Networks (Ann Arbor, MI — 734-761-5005, www.avail.com), and ARESCOM (Fremont, CA — 510-445-3638, www.arescom.com).

One new, advanced delivery option centers on platforms built with a new category of product called an *Integrated Concentration Device (ICD)*, such as the ingenious, pioneering Frontera platform developed by Avail Networks. Under this model, an ICD is deployed in the MTU which concentrates both xDSL and voice CPE in-building. This is a new alternative for MTU-based VoDSL, which is now available only when an ICD is part of the architecture and is in operation. Like our previous scenario, it recognizes that MTU-based subscribers are co-located. This model has all of the benefits of the scenario mentioned earlier, but it also has several new advantages made possible by the ICD's unique architecture, such as the following:

Enhanced QoS controls: A next-generation ICD employs hardware-based traffic management for QoS control, meaning very low, predictable latency and packet delay variation (jitter) characteristics (this guarantees proper prioritization of packets initiated at the ICD, and also preserves the QoS of any tenant-based VoDSL IAD traffic flows).

Multi-subscriber VoDSL capability: An ICD's subscriber-side voice ports allow multiple tenants' phone equipment / lines to plug directly into the ICD for centralized VoDSL IAD functionality.

Built-in "E911" voice support: An *Uninterruptible Power Supply (UPS)*-backed ICD allows the connection of multiple tenant IAD E911 POTS ports into the ICD's integrated voice ports for VoDSL support even if building power is lost.

Flexibility: An ICD allows the use of discrete VoDSL IADs at some tenant premises, while allowing other tenants to share the VoDSL IAD support integrated into the ICD.

Cost savings: Providers can avoid the cost and provisioning expense of placing new VoDSL CPE at the subscriber premises by reusing existing DSL CPE (e.g., SOHO

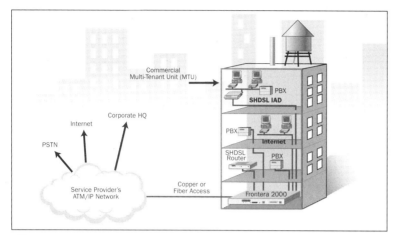

Fig. 8.1. Avail Networks' system is based on a new product called an Integrated Concentration Device (ICD) which sits in a building's basement and concentrates in-building the tenants' xDSL data and voice traffic. In this scenario, the multi-subscriber service delivery capabilities of Frontera systems are leveraged in such a way that a single Frontera platform can serve multiple subscribers (in this case, separate SMEs in a commercial MTU) via "in-building" DSL. This approach allows each subscriber to receive a complete range of voice and data services while leveraging equipment and access bandwidth costs across multiple subscribers. Avail's system delivers broadband data and voice services to smaller clustered business tenants. It also bundles multiple services for these tenant communities over a single copper, fiber, or wireless access facility — resulting in the need for fewer copper wire pairs at the CO while allowing converged access over NxT-1 / IMA, DS-3, OC-3, and Wireless Local Loop uplink(s).

routers) when adding VoDSL on top of data-only DSL services. This savings can be passed on to the customer.

Provisioning simplicity: An ICD allows for the provisioning of multiple subscribers' voice and data services in one platform, adding new capabilities over time via remote management.

Avail's approach to incorporating an ICD is straightforward. Avail's aim is to pack all of the core functionality of a central office DSLAM, including ATM switching and IP edge routing, into one compact and inexpensive box. The concentrator unit can be paired with an ATM-based IAD, from Avail or a third party, or can terminate lineside analog voice and broadband data connections directly, on the same box that holds the WAN interfaces. In the latter scenario, an Avail Frontera ICD could be the only product a service provider needs to locate in a building: the box simply sits in the basement, aggregating traffic from either xDSL modems / IADs, Ethernet hubs, or POTS lines, and backhauling this multiservice traffic to the central office via a single high bandwidth T-1 or T-3 uplink.

Avail's Frontera in-building concentrators support interfaces for both the widely deployed industry-standard SDSL and the international standard G.SHDSL (reconfiguring the ports for one type of DSL or the other is done via a simple software configuration change). This eliminates the need for hardware replacement for those providers deploying SDSL CPE today, but who are migrating toward the G.SHDSL standard. Avail's entry level system, the Frontera 2100 platform, is expressly designed to enable multiservice access deployments in multi-tenant subscriber environments, and offers eight software-selectable SDSL / G.SHDSL subscriber ports and dual T-1 / E-1 uplinks. The system is driven by the OpenVoB D1 Implementation Guide, an industry document that provides a set of implementation guidelines that ensure a common level of interoperability between gateway and IAD products based on the ATM Forum's Loop Emulation Service (LES) specification. By adhering to these guidelines, any IAD or voice gateway is, in theory, able to operate with any other gateway or IAD on the network.

Avail's Frontera platforms are flexible. Although this discussion has focused on xDSL and VoDSL, Avail's Frontera platforms can support multiple packet voice

At the Edge of the Edge: Multi-Tenant CPE

Unlike the gear a BLEC would locate in the basement of a multi-tenant building, there is no very clear or unique set of requirements for the equipment needed inside tenants' offices. Largely, this endpoint device is determined by the type of network a provider has put in place and the applications being offered thereon. In DSL deployments, for example, the endpoints are much the same in an MTU as they would be in a CO rollout – modems or routers for data-only, and integrated access devices (IADs) for converged voice and data. Nonetheless, a few CPE vendors have found a particular niche for their products in the MTU or BLEC market space, and are working closely with partners to help address their carrier customers' needs.

Veteran conferencing vendor Polycom (Milpitas, CA - 408-526-9000, www.polycom.com), which first got into the DSL router and IAD market through an acquisition, says it has enjoyed success among BLECs in general. Polycom's products in this space range from standard DSL endpoints to boxes with higher port densities that integrate router and LAN switching functions, as well. This appeals to BLECs, who want to offer their customers as self-contained and flexible a solution as possible. Another key issue is self-provisioning – a problem that all DSL vendors are working hard to crack, and one which Polycom reports it is continuing to refine with some of its BLEC customers. On the company's roadmap is VDSL, which is attracting a good deal of attention from the BLEC community, and is slowly starting to emerge in platforms from a variety of equipment vendors.

Chapter Eight

options — including the ATM Circuit Emulation Service (CES) for TDM-to-packet voice interworking, native VoDSL, and VoIP transport — along with DSL aggregation and enterprise interfaces such as Ethernet and Frame Relay. It is a true multiservice delivery platform.

When Tut Systems announced its IntelliPOP solution for MTUs in 1999, the company was already shipping products to multi-dwelling units and hotels, as well as a larger CO-style platform that was deployed by the Rockefeller Group, among others. IntelliPOP pulls together a number of Tut's individual platforms into a soup-to-nuts package, including multiplexers, CPE, management software, and even a physical transport system for the building's risers. IntelliPOP is more focused on the commercial office MTU market than earlier industry products, which were targeted mostly for use in residential buildings.

For starters, the system has more IP capabilities: The multiplexing ("mux") component is a router, switch, and concentrator all wrapped up in one; and while ATM is still used for transport to and from the CPE, the platform can do packet-level filtering and classification. Also, using its own proprietary methodology, the system can transport data and voice over copper or fiber in the building's risers. Current versions (Tut 5212 /5224 IntelliPOP) support an Ethernet-style, Cat 5 cable infrastructure, as well as the newly emerging Very High Speed DSL (VHDSL or VDSL), which supports up to 15 Mbps symmetric or 26 Mbps downstream over existing copper wires, enabling the delivery of advanced data, video and voice services (making it an early industry example of voice-over-VDSL). Until now we haven't mentioned VDSL much, since the standards and products relating to it are just starting to appear. Once it becomes established, users will revel in being able to run applications that luxuriate in super-high bandwidths. We'll meet up with VDSL again in the final chapter on the future of VoDSL.

ARESCOM is among the more recent of DSL vendors to enter the MTU market, having broadened its product line significantly from customer-premise DSL modems and routers. Starting at the top, a major addition to ARESCOM's DSL broadband access solutions is the ARESCOM MTU/MDU CDS 6000 DSL Service Platform, a sophisticated MTU / MDU-customized DSL broadband service platform, which will work with a complete family of easily-expandable and scalable multi-access and multi-service controller and concentrator modules. The CDS 6000 functions as what ARESCOM calls a *DSLAR (DSLAM + Router)*. The CDS 6000 is placed at the base (or in the basement) of the MTU / MDU, and from there it supports advance routing features, subscriber bandwidth management functions, as well as a wide range of WAN interface options, including OC-3/STM1, T-1/E-1, 10/100 Base T, and V.35.

The CDS 6000 Platform operates at both Layer 2 and Layer 3, and a service provider can choose between different flavors of DSL, including SDSL, G.SHDSL, ADSL, and VDSL. ARESCOM stresses open architecture in its product: the CDS platform can interface to third-party billing, management, and CPE systems, as long as they are standards-compliant. Some of the value-added networking services the ARESCOM CDS 6000 MTU / MDU Platform is capable of delivering include the following: VPN, teleworking, web hosting, e-commerce, toll-quality VoDSL, Internet-ready Hotel Service, Video-on-Demand, digital TV service, videoconferencing, a security service, and much more.

Video-on-Demand, of course, is something that providers in the MDU and hospitality verticals would especially like to get their hands on, and ARESCOM thinks it may even surpass voice as a "killer app" for MTU broadband. The company has at least one partnership in place with a vendor of video servers, from which it could receive and distribute demultiplexed and decoded cable or satellite video feeds.

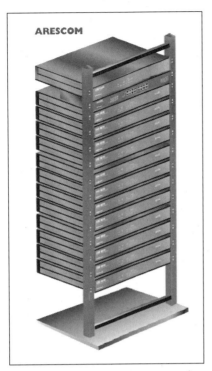

Fig. 8.2. ARESCOM's CDS 6000 platform for servicing multi-tenant units.

Building tenants are guaranteed a full range of network services at the fastest broadband speeds for the lowest prices. The CDS 6000 also offers comprehensive network management software to enable easy management of remote CPE equipment at the MDU site, easy plug-n-play installation, automatic configuration, and transparent integration into corporate voice and data network.

The Evaluation Process

As a BLEC customer, things should be rosy. Expect reduced, all-in-one bundled pricing, a single point of contact and simplified billing. Nonetheless, the customer shouldn't skimp on the due diligence process.

Bandwidth

What a prospective customer should make sure of in any in-building architecture is that the "high bandwidth" backhaul link to the central office really does have high bandwidth. A typical business-class video conferencing application requires about 400 Kbps full duplex, while an MPEG2 video stream will take about 6 Mbps — multiply that by the number of business tenants in a building and one immediately realizes that the total bandwidth requirement for the backhaul link becomes quite large. (Moreover, the available bandwidth generally needs to be symmetric because customers may be hosting web servers or running heavy-duty communications applications.)

Also, since multiple subscribers will be sharing this backhaul bandwidth, whichever WAN backhaul is selected, quality of service (QoS) and bandwidth management at both the building and the metro POP are critical for success. If the service provider charges the tenants for an offering of 2.3 Mbps (for G.SHDSL) or 10 Mbps (Ethernet) but has a backhaul connection that's a T-1 (1.5 Mbps), the tenants will never experience line speeds greater than 1.5 Mbps. Furthermore, as additional tenants access the network, the bandwidth that each tenant is allocated is significantly lower than 1.5 Mbps (VoDSL or VoIP over Ethernet voice quality would be pretty bad, too).

Calling 911

Practically all BLEC VoDSL systems are going to be working in a crowded business or residental environment where there is a higher density of people than one would encounter in a residential environment. Users may be high up in buildings and may not have easy access to and from the premises. For these reasons you must make sure that the system you're going to be using on a daily basis can handle 911 emergency calls. The system should be able to make such calls even if there is a power outage and the CPE (such as the IAD) finds itself without a supply of electricity.

Prior to the appearance of VoDSL, DSL was initially marketed as a high speed alternate to dialup modems for Internet access, so early ATM-based VoDSL IADs have voice ports added to DSL data access technology almost as an afterthought. Adding voice to xDSL is typically handled through ATM multiplexing, so existing IADs — particularly those following the "local loop-replacement" paradigm — are designed to be unobtrusive and to act as transparent as possible for voice circuits. As we've seen in prior chapters, the kind of VoDSL facilities that rely on voice gateway equipment provides a GR-303 compatible interface to a Class 5 PSTN switch and relies primarily on transparent pass-through for voice feature sets and functions. In order to provide dynamic call setup and teardown, proprietary methods for relaying call control functions such as off-hook, ring, and hook flash, are often supported between the gateway and the IAD. DTMF "touch tone" based features and Caller ID information are often simply passed "in-band" within the voice channel.

By simplifying the architecture in this way, the basic transparent channel to chan-

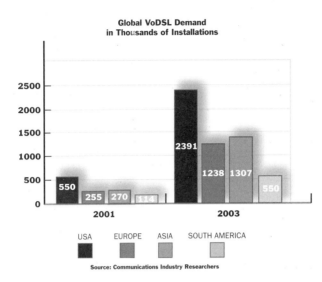

Fig. 8.3. VoDSL demand is expected to grow rapidly once economic and other issues are taken care of.

nel methodology does not provide any flexibility and poses a lifeline resilience problem should power go out or the DSL link goes down. Even though the voice trunks are transparently extended over the DSL and replicated locally by the IAD, they no longer have CO line power and can't be used. This could leave the carrier exposed to potential lawsuits should a customer be prevented from making a 911 call which would otherwise have completed based on their previous analog trunk lines.

The are several options:

1. Maintain a separate analog phone line for a conventional analog phone that derives it's power from the central office in the PSTN's customary way and not from the power mains at the customer premises.

2. If the system is based upon ADSL, the analog "voice under DSL" service is still available.

3. Use batteries or a whole Uninterrupted Power Supply (UPS) power distribution subsystem for the phone system.

4. Some IADs can be configured so that a power failure will cause all VoDSL calls to switch to a POTS line (be it a separate line or the "voice under DSL" line) and make all the system's VoDSL phones become extensions to the analog line.

5. Some systems have been demonstrated with DSLAMs that can provide enough electricity to power highly efficient IADs over a line. In particular, one should use G.SHDSL, the modems and range extenders of which can be powered from the central office over the copper local loop link.

6. The analog trunk(s) could be put to use. They do not need to remain idle in wait for an emergency call, power failure, or DSL service outage. They can instead be utilized to route any outbound call, which the CLEC does not wish to handle over the DSL connection.

Another characteristic of transparent pass-through IADs, is that all calls must be handled through their associated gateway. There is no flexibility to route calls selectively, attach to multiple gateways or PSTN switches, or enable a carrier to route calls based on revenue criteria such as whether or not a given call can be completed within their network versus a call which must be handed off to another carrier for completion.

This is why a good BLEC-type organization in its function as an *Integrated Communications Provider (ICP)*, will have moved (or will be moving) away from older technology and follow the "convergence" trend — building a flexible, packet-centric system (whether it's ATM, IP over ATM, IP over Ethernet, IP over MPLS, or just everything over Ethernet) that manages to support a reliable Quality of Service on individual packet streams within a given communication link to the outside world. If they don't already, many such systems will eventually implement *Multi-Protocol Label Switching (MPLS)* on their platforms as well as *DiffServ* and other emerging packet QoS standards. These moves reflect not only a view of the types of service offerings that will take root in the multi-tenant market, but also illustrate how the "building area network" is slowly becoming an intelligent edge to the larger IP network.

Can a BLEC "Outgrow" DSL?

For the largest BLECs in top-of-the line commercial office space, DSL, for all its advantages, still cannot always satisfy customers' and property owners' demands. Yes, one can configure multiple concentrators in the building, but the fact remains that all DSLs (except for VDSL, which has yet to enjoy wide acceptance) "top out" at just a few megabits per second, so it's hard for DSL to compete with alternatives like Ethernet, which can run up to 10 or even 100 times as fast. Until recently, the biggest drawback of using Ethernet to reach customers in multi-tenant buildings is that while almost every building has wiring for phone service, few come equipped with the Cat 5 cabling required to carry 10/100 Ethernet (let alone Gigabit Ethernet). This means that a BLEC looking to offer its customers direct Ethernet-to-WAN connectivity typically needs to gain access to the building's internal riser system, and lay the cables itself — an expensive and potentially hard-to-negotiate proposition.

Still, Ethernet has proven appealing to major BLECs in new buildings, where deals can be worked out with property owners in advance of construction, or in buildings where enough tenants have agreed to a high enough level of service to justify the cost of new infrastructure. Ethernet's native support for IP services, high-speed transport capabilities, and low-cost components help make it a desirable choice where available. An Ethernet switch also offers amazingly scalable service, at least more so than a mini-DSLAM, since most DSLAM architectures are really at heart circuit-switched access concentrators rather than genuine packet switches. When vendors attempt to enhance the flexibility and intelligence of their ATM-based mini-DSLAMs by aggregating "same-class" traffic from multiple subscribers onto a single *Permanent Virtual Circuit (PVC)* over the WAN, the packet classification engine in the mini-DSLAM is normally done by software instead of by hardware, as is the case with Ethernet Layer 3 switches. This can reduce performance, since latency is introduced into the system as each packet is converted to ATM cells, and *vice versa*. But even providers whose current business model centers on Ethernet admit that it is simply not realistic to assume that the same technology can work, and be profitable, in every type of multi-tenant building.

DSL and Ethernet Make a Great Team

In previous chapters we examined Nortel Network's Business Communications Manager (BCM) which uses Ethernet, not ATM, to transfer IP packets containing voice over xDSL. Industry manufacturers have recently speculated that perhaps a technology could be found for BLECs that can successfully combine the best aspects of both DSL and Ethernet. And, as it happens, Net to Net Technologies, Inc. (Portsmouth, NH — 877-638-2638, www.nettonet.com), a leading provider of Ethernet-based IP DSL broadband solutions, is one of the companies that has stepped up to the plate to bring speculation to fact.

Net to Net is interesting in that, while nearly all of the world's DSLAMs are based on ATM technology, Net to Net provides DSLAMs, CPE, and backhaul solutions that are all Ethernet-based and are optimized for IP transport. This allows carriers and service providers to provide broadband access solutions without the need for ATM, which in turn drastically reduces the capital and operational expenditures associated with broadband deployment.

One example of Net to Net's technological prowess occurred in June 2002, when the company demonstrated its new IP Video-over-DSL solution at the SUPERCOMM Expo in Atlanta, Georgia. Net to Net collaborated in its demonstration with Minerva Networks (Santa Clara, CA — 408-567-9400, www.minervanetworks.com), which supplied video head-end hardware, software and third-party set-top boxes, while Net to Net provided the Ethernet-based DSL access equipment.

The Net to Net Video-over-DSL solution consists of the company's IP DSLAMs and ADSL access modules, which support *IGMP Multicast Snooping* and provide bandwidths of up to 10 Mbps downstream out to 10,000-12,000 feet from the central office. When used in conjunction with video head-ends like Minerva Networks' and third-party set-top boxes, the solution allows telcos to deliver bundled voice, video, and data all over a single pair of copper.

Note: IGMP (Internet Group Management Protocol) is a protocol that runs between hosts and their immediately neighboring multicast routers; the mechanisms of the protocol allow a host to inform its local router that it wishes to receive transmissions addressed to a specific multicast group. And multicast snooping is the capability of some Ethernet switches to respond to multicast join messages by presenting multicast traffic to ports were the "join" message originates. This feature minimizes gratuitous multicast traffic from wasting port bandwidth.

Also, in late 1999, 3Com (Santa Clara, CA — 408-326-5000, www.3com.com), first demonstrated a Visitor and Community Network (VCN) system, which is built around a technology the company refers to as Ethernet-over-VDSL. Using VDSL technology, the VCN system can carry switched data at up to 10 Mbps (the same rate as standard Ethernet), symmetrically, across existing copper wires, and it leaves a separate channel open for voice. Tenants have the advantage of plug-and-play access from their Ethernet hubs to a VCN Access Point, which in turn connects to the VCN Access Concentrator, located in the basement or at another central point in the building. The Concentrator converges up to 24 Access Points, which can each be located as far as 4,000 feet away from the box. (Since VDSL is the underlying transmission carrying the Ethernet packets, Ethernet's shorter distance constraint of 100 meters, or about 330 feet, do not apply. In theory, VDSL can be used to deliver IP/Ethernet connectivity transparently over telephone-grade wire at up to 10 times the distance of traditional 10/100 Base-T on Category 5 cabling.) From there out, the service provider has a number of options, depending on the type of network and the kinds of applications it is looking to deploy.

In 3Com's model deployment architecture, the Access Concentrator would hook up to a set of standard Ethernet switches, which the company also supplies, with its SuperStack product line. These switches then act as gateways to a partitioned local area network, housing application, media, and content servers to which tenants can gain access through the VCN system. One of these servers, in fact, forms a key component of the system itself, for which 3Com has partnered with a company called Solution, Inc. (Halifax, NS, Canada - 902-420-0077, www.solutioninc.com).

The SolutionIP server plays several key roles in provisioning and operating a Visitor and Community Network, including IP address and NAT administration, interfacing to billing and property management systems, and pushing a web-based portal to users on the network, from which they can order and connect to services. The service portal concept has become popular among BLECs, both for office and residential or hospitality environments, though its function in each differs according to the type of tenants using the system. Other servers connected to this "basement LAN" might offer things like streaming media, video-on-demand, enterprise applications, and, of course, voice-over-IP. 3Com's own NBX telephony servers, for example, are purely Ethernet connected devices, to which tenants can interface with IP phones hanging off the VCN Access Point.

The most important point about the architecture 3Com was proposing with VCN is that one can create a system that's effectively all IP and all Ethernet, while still conforming to today's infrastructure constraints by leveraging VDSL. Using largely off-the-shelf and relatively inexpensive systems, a service provider could install VCN in a building and start offering a highly diversified range of voice and data applications to tenants with little or no physical network overhaul. As Ethernet starts to move into the local loop as well, the business model could get even more interesting, requiring minimal access equipment in the building itself, and co-locating application servers in a data center or an ASP's POP. The abundance of cheap, metro area bandwidth in such a scenario lends itself well to creative opportunities for distributing services in and around city centers.

Ethernet may predominate in the local loop (over xDSL) giving you a whole range of new applications that can be deployed from the metro area, right into the building. The bottom line is that service providers need a strong infrastructure to deliver services today that will also be future-proof. A one-megabit network won't cut it.

Another technology in this area is EtherLoop (short for Ethernet Local Loop), which is currently semi-proprietary technology from Elastic Networks (Alpharetta, GA — 678-366-9221, http://elastic.com). EtherLoop uses the advanced signal modulation techniques of DSL and combines them with the half-duplex "burst" packet nature of Ethernet to achieve high bandwidth transmissions over ordinary copper phone lines. EtherLoop modems will only generate bursts of hi-frequency signals when there is some packets to send. The rest of the time, they will transmit only a low-frequency (ISDN-speed) management signal. EtherLoop can measure the ambient noise between packets. This means that interference can be avoided on a packet-by-packet basis by shifting frequencies as necessary. Since EtherLoop is half-duplex, it is capable of generating the same bandwidth rate in either the upstream or downstream direction, but not simultaneously.

The original EtherLoop could achieve a data transfer rate of 4 Mbps up to a range of 6,000 feet or 400 Kbps at 18,000 feet. In June 2001, Elastic Networks announced the new EtherLoop2 technology that can expand the capabilities of EtherLoop exponentially by delivering services at rates up to 100 Mbps and at distances up to 21,000 feet, exceeding conventional xDSL rates by more than 90%.

Etherloop technology enables a "copper bridge" to advanced optical Ethernet networks. Etherloop2 remains an IP-native Ethernet solution that can adapt to existing copper lines, avoids interference with other signals and has flexible bandwidth symmetry. EtherLoop's adaptability eliminates the need for loop conditioning and continuously maximizes performance in an unpredictable and constantly changing loop environment. EtherLoop has been successfully deployed by service providers in the

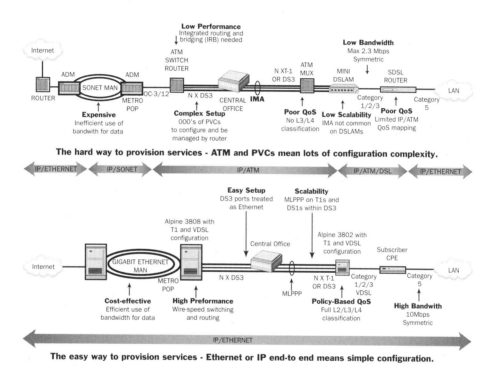

Fig. 8.4. Extreme Networks makes a case for the ease of provisioning and maintaining large-scale Ethernet networks that are delivered over super-high bandwidth VDSL.

carrier, MDU/MTU, hospitality and international markets.

The user connection consists of a PC or EtherLoop-compliant device that's connected to an EtherLoop modem that, in turn, connects to the phone line. This is the typical local loop that leads to the central office where an EtherLoop multiplexer receives the signal and forwards packets to an Ethernet switch that, in turn, delivers the packets to one of several destinations: the Internet, a private intranet, or an ATM / Frame Relay network. Home users are assigned static or dynamic IP addresses.

EtherLoop not entirely proprietary. Elastic Networks will license the technology to manufacturers who can design and create their own EtherLoop solutions.

Etherloop2 is but one of various other competing techniques designed to liberate Ethernet from the distance limits that restrict its use to local networks in buildings, thereby enabling it to both serve huge buildings and to replace the BLEC's T-1 or T-3 backhaul line to the CO. For example, Extreme Networks Inc. (Santa Clara, CA — 408-579-2800, www.extremenetworks.com) has developed an "Ethernet DSLAM," the Alpine, that provides a riser-independent transport solution that leverages data-grade Category 5 LAN cable and fiber where it exists, but also provides the extended reach of VDSL over the voice-grade copper cabling that can be found in all buildings. This eliminates any need to rewire the building while still offering a minimum 10 Mbps symmetric bandwidth to each subscriber — over four times the subscriber bandwidth available with SDSL.

Extreme Networks' Ethernet DSLAM and it's higher performance Ethernet core switches will also let metro-area service providers carry Ethernet traffic over very-high-speed digital subscriber line (VDSL) circuits set up in the local loop, bypassing more expensive and slower legacy network protocols used for local access. Aside from operating on VDSL circuits, they will carry Ethernet traffic at speeds as fast as 45 Mbps on T-1 or T-3 access circuits, according to Extreme.

One feature pioneered by Extreme Networks is the *virtual Metropolitan Access Network, (vMAN)*, as Extreme calls it. Similar to a *virtual LAN (VLAN)* but designed for the service provider, a vMAN does a double encapsulation of Ethernet traffic, letting users run Layer 2 protocols inside an IP-based VLAN. This lets a customer link several buildings serviced by the same service provider into its own Layer 2 MAN. Another of their innovations is called *IP-TDM (IP Time Division Multiplexing)* which is a packet-scheduling algorithm that lets the Extreme Alpine switch guarantee latency for small packet sizes.

The 800-pound gorilla in networking, Cisco Systems, also is developing and deploying extended-reach capabilities for its Ethernet switches and intends to dominate the upcoming "long-haul Ethernet" market (since Cisco estimates that it has a 60% share of the Ethernet switch market, it should make a fortune off of upgrades). The first generation of such Cisco products will carry 10/100 Mbps Ethernet traffic on a regional or metro-area ring, with individual customers connecting to the ring via Gigabit Ethernet links. Another competitor in long-haul Ethernet will be the Shasta division at Nortel Networks.

So don't be too surprised if your BLEC is using a mysterious-sounding Ethernet / DSL hybrid.

The above discussion should make clear to the reader than no one VoDSL implementation or strict BLEC architecture will be either universal or of great longevity.

Only platforms based on flexible, true multi-service architectures will succeed as the foundations for toll-quality voice, broadcast-quality video, and other QoS-intensive, highly profitable services in the MTU.

If you can't beat 'em. . . .

Of course, now that you've finished reading about BLECs, you may no longer wish to be a customer of a BLEC — instead, *you might actually want to become a BLEC!* You could become the founder of a service provider to supply broadband services to your building! Imagine being able to determine the fate of all the voice and data communications in your little section of the city skyline.

It's an enticing thought. After all, the potential MTU (Commercial) market size is huge, the U.S. Department of Energy estimating that in 1995 there were 150,000 buildings in the U.S. that house business customers, accommodating more than 1.5 million separate tenants. The breakdown is as follows:

- Buildings with 6-10 establishments/businesses: 91,000
- Buildings with 11-20 establishments/businesses: 28,000
- Buildings with 20+ establishments/businesses: 31,000

IAD and DSLAM manufacturer Occam Networks (Santa Barbara, CA — 805-692-2900, www.occamnetworks.com) lists four basic types of MTU environments, categorized by square footage and the number of tenants. Each type has a specific broadband infrastructure requirement:

1. Skyscrapers or buildings with an area of more than 500,000 square feet, typically housing 20 or more tenants. This environment requires access concentrators with maximum flexibility and expandability from a few up to hundreds of ports. This category also encompasses buildings with a large number of potential "subscribers" such as hotels.

2. Mid-sized buildings range in area from 75,000 to 500,000 square feet and house 11 to 20 tenants. These types of buildings require smaller form factor equipment, reducing both the cost and the amount of space required for the equipment.

3. Small buildings with an area between 25,000 and 75,000 square feet typically have six to 10 business tenants. These buildings require the most cost optimized equipment.

4. Campus environments such as office parks and universities are very similar to MTUs, with a similar subscriber environment and able to use identical or very comparable infrastructure for service delivery.

As for the MDU (residential) market size, there are more than 860,193 residential buildings and apartment complexes in the United States, with an estimated 8.6 million tenants to be served in these buildings.

According to a 1999 study by the Yankee Group:
- Buildings with 4-9 tenants: 622,850
- Buildings with 10-19 tenants: 107,170
- Buildings with 20+ tenants: 130,173

And in 1998, the American Hotel and Motel Association estimated that there are over 51,000 motel and hotel (MHU) properties in the U.S., with a potential 3.9 million guestrooms needing high-speed access.

There is also an unexpected market appearing regarding cruise ship lines. Imagine having several thousand captive customers on board an ocean liner each needing satellite, microwave, or harbor line broadband connections.

Any way you look at it, in-building voice and data DSL is a winning situation for everybody: Tenants, BLECs, building owners, telcos, etc. Tenants get high bandwidth voice/data access, service providers make money off of premium services, and building owners and property managers stand to benefit from higher occupancy rates than their counterparts who can't offer the level of services enabled by an MTU multiservice delivery infrastructure.

Virtual Private Networks (VPNs)

VoDSL makes it possible for your employees to be productive while not actually being physically present in the office. Local Area Networks (LANs) and voice services can be extended far from the enterprise, and may be carried over a DSL connection to a teleworker site. This is achieved with the latest craze in networking, the VPN.

As it's name implies, a Virtual Private Network (VPN) is a *private* voice/data network, run over *public* network facilities. VPNs can connect two geographically separate LANs so that they can behave as a single network, thanks to use of a secure, encrypted "tunnel" (using a protocol such as Microsoft's IPSec) that the VPN sets up between the LANs and through a public network such as the Internet.

Currently the two most popular uses for VPNs are company-company and employee-company connections. The first allows partnering companies to participate in the same private network so they can share and exchange market and transactional information. The second creates a secure link over the Internet between a corporate LAN and a remote user's PC. Let's take a closer look at how an employer-company VPN would work.

The user first connects over a dialup or xDSL broadband connection to an ISP using the Point-to-Point Protocol (PPP) and the ISP assigns the user's PC (the "VPN client") a

TCP/IP address, a Domain Name Server (DNS) address, and a default gateway. The user is then connected to the Internet and the VPN client software is activated, which then proceeds to wrap the PPP packets normally used in dial-up connections with additional tunneling protocol headers, enabling the VPN packets to travel securely over the Internet. For example, in the case of Windows 2000, the Layer 2 Tunneling Protocol (L2TP) uses Microsoft's IP Security (IPSec) to encrypt data for transmission through the tunnel. L2TP wraps a PPP packet with an L2TP header and User Datagram Protocol (UDP) header, making it safe to transmit the packet over the Internet. IPSec encapsulation then encrypts all of the packet except for the IP header, which is used to provide the necessary addressing information between the VPN client and VPN server.

The data typically flows from the client to an ISP's router, through a firewall, across the ISP's network, perhaps traversing some additional ISPs' networks, to the company's router, to a firewall or proxy server, and finally to the destination corporate server running L2TP and IPSec.

Technically, when the VPN is active, the user's previous, direct Internet access is disabled and all subsequent communication goes through the corporate LAN. Therefore, if the user wants to browse the web while the VPN is active, the user is actually reaching out to the web and browsing through the firewalls / proxies from inside the corporation, not directly from the remote PC to the web.

Since the VPN makes it appear ("virtually") as if the PC is on the corporate LAN, VPNs can thus serve as one of the key enablers for teleworking, putting employees out of harm's way, scattering them around the world, if you like. In theory, a VPN lets you build out as your business grows.

According to the International Teleworkers Association and Council, there were 28 million teleworkers operating at home in 2001. And, according to market research firm In-Stat, more than half of the U.S. workforce at any one time perform their jobs at what could be considered "remote sites." The bulk of these workers are in small remote branch offices or are teleworkers (there are nearly three million remote offices in the U.S. business market today). Employee benefits include flexible work schedules and the elimination of commuting, while employers are said to benefit from increased employee satisfaction, productivity and savings in real estate. By 2005, the research firm In-Stat expects more than 60% of the workforce to be classified as remote, of which 35 million will be teleworkers.

Growth in this area is expected to take place at varying rates due to an interesting split between small and medium enterprises (SMEs) and larger companies. SMEs generally lagged behind in accepting telework during the 1990s, while in larger companies teleworking is much more widespread for economic reasons. According to

Alcatel (Paris, France — +33 (0)1-40-76-10-10, www.alcatel.com), the bulk of the market for teleworking consists of mid-size and large enterprises, where employee numbers range from 200 to more than 10,000. However, the trend shows a more even spread of teleworking across the different company sizes than in previous years, as the effects of governmental initiatives take effect and smaller enterprises find ways to support teleworking.

Therefore, at the same time small and medium-sized businesses discover VoDSL, they will also hopefully use it to discover the advantages adopting a teleworking environment.

Of course, the usefulness of your teleworkers depends to some degree on the capabilities and features of your VPN.

Software-based VPNs are generally more flexible than hardware-based VPNs, and are used in situations where remote sites encounter a mixture of VPN and non-VPN traffic, such as web surfing. Software systems have less performance than hardware-based systems and tend to be more difficult to manage. Still, it is rather tempting to just download some client software, run it on some PC, and have a VPN up and running.

Firewall-based VPNs, on the other hand, can leverage the firewall's existing security mechanisms, including restriction of access to the internal network. They also do address translation and can execute powerful authentication procedures. Most commercial firewalls "strip out" services that may pose security problems, thus hardening the host operating system kernel. Firewall-based VPNs are inexpensive and are great for intranet-only implementations.

Hosted VPN Services such as OpenReach (Woburn, MA — 888-783-0383, www.openreach.com) allow you to dump most of your VPN headaches on a friendly service provider. In the case of OpenReach, you can purchase one of their pre-configured systems, or you can simply download their software and use it on your PCs. OpenReach technology is based on a packet-filtering firewall.

Also, the Hosted Business VPN Service (HBVPN) from eTunnels (Seattle, WA — 206-239-9800, www.etunnels.com) does deals with service-oriented channel partners to resell a variety of low-cost, end-user-managed Internet-based VPN services to business without too much fuss.

End-users of the eTunnels system can securely extend corporate LAN or WAN resources to mobile and/or telecommuting workers, build and easily manage branch-to-branch VPNs between satellite offices, and quickly set up secure business partner and/or customer Extranets.

With eTunnels, there's no hardware to buy, configure, or maintain. The eTunnels system replaces hardware VPN implementations with a hosted, pay-as-you-go application- and user-oriented system that makes it easy for VPN users to securely connect to VPN resources such as file shares, printers and applications.

Setup and administration for the channel partner is kept simple and straightforward via bundled professional services and service templates, customizable GUI client and server software and pre-built web-based administration tools. Billing and customer technical support is transparently handled by eTunnels.

The eTunnels Business VPN Client is a Windows 95/98 and Windows NT/2000 GUI software client with an integrated mini-browser that's installed on the end user's workstation. This client software manages the secure connection between the end-user clients and servers using IPSec tunneling.

The eTunnels Business VPN Server is a Windows NT/2000 or Linux 2.x software service or daemon that's installed on corporate application, file and print servers. This software manages the secure connection between the server's resources and each end-user workstation, using IPSec.

The eTunnels Network Server co-ordinates the setup and teardown of the IPSec tunnels between eTunnels network elements (clients, servers and gateways) by handling Internet Key generation and exchange through a secure SSL conversation with the network elements.

The eTunnels Web-based Company Management Tool is a secure website through which the end-user can configure and view the company-specific aspects of the eTunnels VPN-On-Demand service.

Finally, eTunnels' "VPN-On-Demand," is a new approach to creating security relationships that makes dynamic provisioning of VPNs faster than ever. Called a "zero configuration" technology, eTunnels' VPN-On-Demand is easy to deploy and manage, and won't put a strain on IT resources. The VPN-On-Demand platform makes it easy to rapidly build and manage multiple, overlapping, and secure networks that can scale to any number of people. The platform delivers versatile solutions for a variety of secure networking challenges including branch offices, mobile workers, partner networks, and wireless local area networks.

VPN-On-Demand offers dramatically increased efficiencies over traditional VPN technology. Features include remote configuration capabilities, central administration across any number or type of VPN element, granular manageability down to individual users or applications, independence from network / topology / devices, transparency to intervening address translation / firewalling / proxy services, end-to-end security, and support for multiple, concurrent and overlapping VPNs. The system offers independent manageability among multiple IT authorities within a partner extranet.

Voice over VPNs

One new major aspect of VPNs is the addition of voice. In the VPN world, until recently, the addition of voice was literally that, running voice over a data network that just happens to virtually extend beyond the physical brick and mortar business. Voice on packet networks and VPNs usually takes the form of what is called a *PBX Extension*. A PBX Extension gives people outside the company the ability to call workers in the company (or call out of the company) as if they're still attached to the PBX phone network in the office. To set up a PBX extension, one can use a company's real PBX communicating through an IAD over a packet network, or a company could simply use the PBX-like functionality found in softswitches and the next-generation network to provide dial tone to anyone on a VPN.

An ATM / VoDSL-based PBX extension using an ATM network infrastructure can be installed by a service provider with only minor changes to the legacy network, lowering costs and dramatically improving the ROI. No new phones, networking gear, or PC software are required at the customer end of the network, and the service provider investment is limited to a voice gateway in the central office and a voice-enabled IAD.

As is usually the case with ATM-based xDSL, ATM AAL2 *Virtual Circuits (VCs)* will carry the voice between the gateway and the IAD, creating secure voice VPNs to the corporate PBX, while data will travel via ATM AAL5 (unless both voice and data has been converted to IP). For example, Copper Mountain's DSL products can forward IP packets securely at Layer 2, combining PBX extension and a corporate VPN for data is a breeze, creating a secure, converged voice and data solution for remote workers, allowing them direct access to their organization's corporate PBX, e-mail, and the company intranet, just as if they were sitting in the corporate office.

Also, the IAD/gateways are able to determine and route voice calls based on their destination. For example, a residential call made from a phone plugged into a certain IAD voice port can be identified as a call that should be routed directly to the PSTN, while a phone call made from a second phone plugged into a different port on the IAD is recognized as a business call and so is routed to the corporate PBX, allowing the teleworker to take advantage of his company's corporate dialing plans, four-digit dialing, etc.

A PBX extension over VoDSL using the existing ATM infrastructure offers many benefits, but other examples of PBX extensions are based on IP over DSL (or some other form of broadband) which bypass the central office Class 5 switch and use the softswitch technology of the next-generation network. The on-net voice bearer traffic between packet voice endpoints is forwarded by a host of distributed routing elements, including the IAD, an IP-aware DSLAM, and routers located in the core of the network. The downside to such solutions is that the service provider (and ultimately the rest of the world) must adopt a "convergence philosophy" and make a major upgrade in the telecom infrastructure.

Here's an example of a PBX extension that uses a real, legacy PBX (it doesn't even need a VPN to work, though you'll want to use one with this product to achieve a high level of security).

In early 2002, Nortel Networks introduced two new voice-over-IP products built specifically for enterprise teleworkers and call center environments. The Remote Office 9110 and 9115 devices enhance and extend the features of Nortel's Meridian PBX and Succession IP-PBX systems, and completes the line that includes Nortel's 9150 Branch Office product.

The products work with Nortel Meridian digital phones. You just take the 9110 (it's a circuit card) and snap it onto the bottom of nearly any of the Meridian 2000 phones. The 9115 is a more elaborate PDA-sized adapter that can be affixed to both Meridian 2000 and 3900 series phones. You then plug the phone into any broadband connection that will transport IP — xDSL, cable or satellite. The phones let teleworkers enjoy the same functionality they have in the office: audio conferencing, Caller ID, call transfer and dialing plans.

To prepare your company's network for this enhancement, you need Nortel's Reach Line Card. The cards can support a mix of Remote Office 9110 and 9115 Office and 9150 Branch Office users. One card supports 16 to 20 teleworkers, or up to eight branch offices.

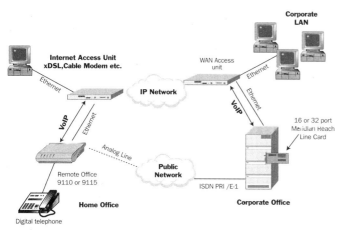

Fig. 8.5. Nortel's 9110 and 9115 work with ordinary Nortel Median phones and PBXs to provide teleworkers VoIP and Ethernet over a corporate VPN. In case of a power outage, the system can revert to using analog POTS voice channels.

The phones can connect to the system via IP / broadband only, analog only, or both. The system does not depend on a high-speed connection, so teleworkers with even a primitive old phone can connect to the system. Workers waiting for their xDSL connection can continue using an analog connection until broadband becomes available.

Interestingly, one can use both broadband and analog connections. When using both, the analog connection serves as a backup to the IP connection, providing a quality of service (QoS) failover capability. To be precise, the user can set a "quality threshold" between one and 10. If call quality falls below the threshold for a preset period of time, the system automatically dials the analog line, confirms a good connection, then moves over the call. Concurrently, it continues to monitor the IP network, and if that connection is restored it will move the call back to IP.

Some other VoIP vendors (such as MCK Communications) offer similar VoIP products that allow dual connections over analog and IP broadband, but the ability to transition live calls from IP to the PSTN to maintain the audible quality of a phone call is still rather novel. This means that a teleworker can make a "serious" phone call with a customer or a supervisor with complete aplomb.

Nortel hopes that the QoS feature will be extremely attractive to call centers that are planning to "virtualize" their operations but are afraid that the sound quality on a packet network won't be acceptable to customers. A travel agency or a help desk could marshall remote agents to handle its call center queues, even if they're on the other side of the world. Of course, in terms of security, the connection should be run over a secure, encrypted VPN. That should be no problem, since Nortel's Remote Office can operate with Nortel's Contivity family of security products, which can provide a high level of security for a customer's network. Contivity supports firewall services, authentication of endpoints over IP, and end-to-end encryption of VoIP packets via IPSec as they travel across the public Internet.

The Nortel Remote Office system even takes into account the importance of being able to dial 911 and getting through. When a teleworker dials 911 the call is automatically routed over the analog line so emergency services can locate the office. An optional button lets teleworkers route local calls over the analog line, rather than routing them to the network and back again. And a call-cloning feature lets a worker's phone simultaneously ring in corporate and home offices, obviating the need to forward calls or give out multiple contact numbers.

As stated previously, until recently, companies added voice to their VPNs themselves, since the VPN simply acted as an enlarged version of their network data network. Eventually, however, the VPN industry "discovered voice" and voice has come to the world of outsourced, managed VPNs. Perhaps the first product to appear in this area was the Clarent VPN Telephony Suite from Clarent (Redwood City, CA — 650-306-7511, www.clarent.com) which was introduced in 1998. The VPN Telephony Suite can run on a Celeron-based PC with 64 MB of SDRAM that's equipped with telephony connections such as a digital T-1 / E-1 or analog POTS lines (RJ-11 jacks). The system gives service providers a novel way to add value and flexibility to the corporate communications services they offer by enabling voice and fax to travel over

VPNs. The product is essentially a voice and fax over IP gateway for connecting telephone voice calls and fax over VPN and public IP networks.

Clarent's product doesn't generate the VPN itself, but acts as a front-end that handles multiple gateways and does VoIP processing by efficiently structuring packets and getting them out to the right gateways.

Each call is routed to a gateway within the VPN if one is designated. If the VPN doesn't cover certain cities or countries, corporate traffic can still be routed over public IP voice networks. Calls can access the public network from within the VPN, but public calls cannot access the VPN from outside. This allows low-cost voice and fax communications anywhere in the world, least-cost routing, and the security of a closed, private network.

In terms of voice quality, Clarent's technology improves packet loss recovery, reduces latency and delivers the optimum sound quality within the lowest bandwidth even while other traffic consisting of fax and data also travels along the single, secure network. For VPN service providers, adding voice and fax to the VPN leads to a closer bond with the customer, allowing the provider to move beyond simple bandwidth leasing to become a complete one-stop shop for communications service on a global basis. Also, service providers can now offer more competitive rates to customers because the cost of administration, billing, and account maintenance of multiple services are now consolidated into a single system.

The Clarent Command Center provides billing and management for networks of multiple VPN domains, providing hundreds of thousands of simultaneous telephone calls through a single point of administration. Different inbound and outbound charges can be set for each VPN domain. Free Calls or Call Blocking options can be set for individual corporate customers. In terms of call detail recording, the system records both inbound and outbound call information including customer account, time and duration of call, destination and call termination code.

More recently, in 2002, NetCentrex (San Jose, 408-521-7400, www.netcentrex.com) announced availability of an end-to-end VoIP VPN solution. As part of the NetCentrex IPCentrex family of managed services, the VoIP VPN solution is a software-based voice and video VPN application based on NetCentrex' flagship CCS 3.0 application softswitch. It includes new provisioning and system management tools and graphical user interfaces to improve the ease of deployment and operation by service providers around the globe.

NetCentrex's new end-to-end VoIP VPN solution is meant to provide an abundant return on investment (ROI) for service providers and their enterprise customers. It allows service providers to increase their revenues by offering voice on top of data VPN services, reduce customer churn by becoming more of a "one-stop shop", and dramatically lower operating expenses compared to TDM voice technologies since far fewer switching centers are now needed. The VoIP VPN solution then becomes a platform for other managed services such as IP telephony, IPCentrex, video-conferencing, and virtual contact centers, increasing minutes of use on the service provider's network. Enterprises can take advantage of free on-network VoIP VPN calling between sites and reduced cost and complexity compared to managing a "home-grown" or premise-based voice VPN system.

Centrex DSL is not VoDSL

Some remote workers who try using packet-based next-gen phone systems complain that some of these "futuristic" IP telephony environments just don't have the kind of features that they're grown accustomed to back at the office. Now, for remote workers, SBC Advanced Solutions, Inc. (Plano, TX — 972-578-5603, www.sbc.com) has introduced *Centrex DSL*, which gives users the same features as an office phone and a DSL data connection over one line. The data connection can run to either the Internet or back into a company's network.

Before the reader becomes confused and orders the service, consider that SBC Centrex DSL is not VoDSL. It essentially consists of a shotgun marriage of Digital Subscriber Line broadband network access coupled with traditional analog Centrex voice services. SBC Centrex DSL provides remote workers a high-speed connection to the public Internet or their corporate LAN, along with the voice calling feature set of their desktop telephone simultaneously over the standard copper wire that delivers telephone service. The service is available in five different DSL speeds.

SBC Centrex DSL will be offered through CLECs and service providers. It became available to customers in Connecticut in February 2002, then became available to customers in Illinois, Indiana, Michigan, Ohio and Wisconsin in June 2002, and customers in Arkansas, California, Kansas, Missouri, Nevada, Oklahoma and Texas in July.

During trials, prospective CLECs were asked to maintain these parameters in their infrastructure in order to be able to add DSL transport service to their existing or new Centrex service:

1. Centrex should be working only on analog lines (not digital lines such as ISDN or T-1).

2. Lines should have local loops that are 12,000 feet or shorter.

3. Lines must reach the customer directly from a Central Office (CO), not a Remote Terminal (RT).

Chapter Eight

A VPN in Every Business and Home?

The original forms of DSL were meant to deliver high-bandwidth Internet access at a low entry cost to both service providers and end-users. Once such services were established, however, it made possible other, higher-revenue services. First voice, then VPNs, PBX extensions and teleworking. Service providers continue to structure a complete value chain around these services by offering even more related services — such as enterprise voice services functionality and enhanced security — which demand little or no additional investment on the part of the service provider. For example, all that is needed to deliver simple VPN teleworking services is to leverage the existing DSLAM (DSL access multiplexer) installed base.

One expects that in the near future, businesses both large and small will be offered various "remote worker service packages" by service providers who wish to maintain their competitive advantage, reduce churn, and harvest profits from high-value services.

ASPs

During the late 1990s, everybody thought that no one would own software anymore. Instead, every time you needed to run a spreadsheet program or play a game, something called an *Application Service Provider (ASP)* would send the software over the Internet to you. Upon finishing with the software, it would literally disappear, leaving only the data file you were working on.

Technically speaking, an ASP deploys, hosts and manages access to a packaged application to multiple parties from a centrally managed facility. The applications are delivered over networks on a subscription basis. This delivery model speeds implementation, minimizes the expenses and risks incurred across the application life cycle, and overcomes the chronic shortage of qualified technical personnel available in-house.

The ASP model of software delivery was felt to be not just the future of Mankind, but the greatest thing since proverbial sliced bread. While Windows XP has a tiny bit of ASP ambiance to it, the whole ASP industry has not lived up to the high expectations of its proponents.

Once reason the author thinks that ASPs didn't explode in popularity was because most of the world's applications consist of bloatware written by inept programmers, and in order to deliver bloatware in a reasonable amount of time over a public network, you *definitely* need broadband, specifically xDSL or cable modem. It took too long for xDSL and other forms of broadband to become available to the public, so many ASPs "died on the vine." Now that ADSL and G.SHDSL have grown in popularity, however, perhaps we will all bear witness to the Second Coming of the ASP

In the short-term however, as deployments of xDSL services and cable modems continue to increase throughout the world, the only "service provider" to become wildly popular will probably be a denizen of the very unbusiness-like entertainment realm — namely, *Video-on-Demand (VOD)*. VOD over Internet Protocol (IP) networks will grow to a total of more than 17 million users, generating over $1.9 billion (U.S.) in subscription and pay-per-view revenue during 2006, according to In-Stat/MDR (www.instat.com). In a report entitled, "Consumer Oriented Video-On-Demand Via IP Networks" (#IN020022MB), the high-tech market research firm reports that, as consumer-oriented VOD services over IP become more pervasive, revenue generated by family-oriented VOD services will eventually surpass those of adult content sites, which currently dominate the VOD-over-IP market.

"Consumers who have broadband Internet connections represent a strong growth market for VOD services," says Gerry Kaufhold, a Principal Analyst with In-Stat/MDR. "Adult content Internet VOD services are way ahead of cable TV VOD deployments in terms of total subscribers and revenues. However, four new family-oriented VOD-over-IP service efforts are underway: CinemaNow, Intertainer, MovieLink, and Movies.com." CinemaNow has deployments in North America, Taiwan, and Singapore, with more to come. Intertainer is rolling out their service in 35 major U.S. cities that have strong broadband deployments in place and the other two services are expected to launch before 2003.

"Several million movie streams per month are currently being served up for free, but as the major movie studios enter the fray with premium movie titles, pay-per-view and subscription services will gain traction, helping Hollywood figure out what the market is for 'on demand' content, and help engineers and software programmers to develop efficient delivery systems and workable Digital Rights Management solutions," says Kaufhold. Slated to generate approximately $460 million worldwide in 2002, the adult content segment of the market (representing over 98% of revenues) will serve as a barometer for the future success of the market as a whole. By the end of 2004, the number of subscribers and pay-per-view participants, regularly using family oriented "on demand" IP services, will out number the users of adult content services, and, by 2006, family oriented "on demand" services will overtake the adult content sites in terms of annual revenues.

In-Stat/MDR has also found that by 2006, about 40% of worldwide consumers who have high-speed Internet connections to their residences will be using "on demand" services for which they pay monthly fees, bringing $1.9 billion to Hollywood. The

Chapter Eight

North American market has the lion's share of consumer broadband connections deployed, and, by 2006, will represent over 7.6 million VOD users, generating over $820 million in revenues. Still, Asia, especially South Korea, Taiwan, Singapore, and others, will represent about 37% of worldwide VOD-over-IP subscribers by 2006, producing over $700 million for movie studios. Europe will provide about 15% of worldwide VOD-over-IP revenues in 2006, and the rest of the world will bring in about 4.7%.

Also along these lines, Minerva Networks, which we discussed previously, and TelStrat (Appletrees, Hollybush Lane, Burghfield Common Berks, UK: +44-870-135-0162, www.telstrat.com) have a partnership to provide a full range of IP television services to local service providers. Video services, including live television, near-video-on-demand (nVOD), video-on-demand (VOD), and personal video recording (PVR), will be provided over DSL via Minerva's IP Television headend through TelStrat's Inteleflex Multi-Service Intelligent Access Platform (MIAP)

TelStrat's super-versatile MIAP delivers and manages narrowband and broadband voice, data and video services over any network architecture, including fiber, copper, and wireless. The video services utilize the platform's SONET OC-3 / OC-3c Uplink and Restorable Ring and Video Delivery over ADSL functionality. MIAP can handle both electrical and optical lines, and has software driven network intelligence capable of handling a multitude of services from Voice to Data to Video over a choice of protocols. Inteleflex allows the aggregation and integration of service offerings over a concentration of equipment and access facilities, which reduces network deployment costs and complexity. The impressive Inteleflex provides all the functionality of a Packet Loop Carrier (PLC), a Digital Subscriber Loop Access Multiplexer (DSLAM), a Digital Access Cross-connect System (DACS), a Dial Up Data Router, a T-1 Concentrator, a Voice over Data (VoDSL, VoATM, VoIP) Gateway, and a SONET Restorable Transport and Uplink and Video Delivery Platform, all in one multi-service Intelligent Access Platform.

Minerva Networks, on the other hand, brings to the table a series of complete IP Television headend solutions, enabling the delivery of broadcast-quality television and video services over IP networks. As the central point of subscriber management, media aggregation and video distribution, an IP Television Headend consists of Minerva iTVManager video services management software, Minerva VNP analog encoders and Minerva MediaGateway direct video broadcast (DVB) to IP demultiplexers. Through Minerva's IP Television Headend combined with TelStrat's Inteleflex MIAP, network operators can provide a broad range of television services including live television, VOD, nVOD, interactive advertising and PVR. Minerva also offers a suite of professional services to facilitate the deployment of end-to-end video networking solutions.

So, essentially, the partnership brings Minerva's powerful IP Television solutions to TelStrat's networking technologies, providing the IP-based video distribution systems that can be used by xDSL service providers to compete in the video services market. The Minerva/TelStrat solution is thus an integrated platform that offers these providers the ability to bundle voice and data with compelling revenue generating video services and counter competitive pressure from regional cable and satellite providers.

So at least you'll know what to do with your xDSL connection during your lunch break.

In Summary

As we've seen, VoDSL is ideal to help small and medium-sized businesses. A company that wants to establish a call center and also have high-bandwidth Internet access can now take an existing phone line and magically turn it into 16 or 24 VoDSL lines, with broadband access to the Internet to boot. And once you can move data at broadband speeds, many applications become possible: You can extend your LAN out to your employees via a VPN, then add to that an enhanced teleworker package by extending your PBX, thus creating a private phone network over your VPN. ASPs may even get their second wind thanks to VoDSL. Once a business has experienced the benefits of (and ponders the tantalizing application possibilities of) sending voice and data over broadband, there's no turning back!

CHAPTER NINE

Why VoDSL Will Succeed

With the millions of residential, SOHO, and SME customers who are situated on a local loop capable of supporting DSL, the market potential for VoDSL with its bundled services cannot be ignored. VoDSL service providers have taken note. They realize that there is a huge potential market for VoDSL because of the millions of SOHOs, SMEs, satellite/branch offices and high-end residential users who have heretofore been cut-off from high-value voice and data services such as PBX call processing, VPNs, audio and video conferencing and so forth.

For the customer, the benefits include reduced cost of doing business, more and better telephony and communications applications, and the ability to share the same resources for voice and data. While the SME is the primary candidate for VoDSL services, the SOHO, satellite/branch office users are also clamoring for affordable multiple voice and high-speed data connections. Then there's also the growing residential market that needs more than one or two telephone lines.

For example, VoDSL lets SME, SOHO and residential customers get as many as 24 voice lines and high-speed Internet access over a single copper twisted pair at a price far lower than the monthly cost of the same service via POTS, IDSN or T-1/E-1 service. The average monthly rate is expected to be in the range of $200 to $500 for the typical eight to 24 line service. That's far less expensive than equivalent POTS, plus the customer gets high-speed Internet access.

On the service providers end, the providers that offer a flexible varied approach to provisioning VoDSL benefit from not only VoDSL's reliability, ease of deployment and affordability but also increased revenue gained through a new-found ability to offer attractive bundled services to their customers. By adding VoDSL technology to their DSL infrastructure, service providers can realize a significant boost in revenues. For example, the average POTS line generates around $70 per month in revenue when you factor in both local and long distance services, but when that same line is conditioned so it can be used to support multiple services (multiple voice lines, data via VPN, Internet access, unified messaging, Centrex, and so forth), the service provider can increase that revenue stream from $70 to a few hundred dollars per month. While, this may sound like bad news for the customer, it's not. The customer, who originally paid $70 for "plain vanilla" POTS, now has "enterprise class" communications services for a pittance — at least when compared to the price tag all of these services would carry if purchased *a la carte*.

The target market for VoDSL service are the customers who can't afford T-1 service and find ISDN's meager bandwidth inadequate, but still need some kind of high speed data transport, Internet access and multiple voice lines. Until VoDSL, this valuable, but under-served community of voice and data users has been stuck with using dial-up Internet service and multiple POTS lines to fill their needs. Installing VoDSL service puts an end to such inequitable treatment, while at the same time delivering capabilities that go far beyond what is offered by legacy telephone systems.

This undervalued market represents a tremendous revenue and earnings opportunity for any VoDSL service provider that can cost-effectively bundle voice services with a range of applications enhanced by DSL technology. In other words, VoDSL service providers aren't selling voice *per se*, they're selling services and value-added applications along with voice. With VoDSL on board, anyone can get the bandwidth of a T-1 *and* multiple voice lines *and* enhanced services at a cost-effective price.

Benefits of VoDSL

VoDSL's appeal is simple — it's inexpensive and convenient. VoDSL splits the bandwidth of a DSL connection into anywhere from two to 24 virtual phone numbers. As we've already seen, it offers many benefits not only for the customer, but also for the service provider.

According to a report issued in mid-2001 by Research First Consulting (Birmingham, AL — 205-995-8866, www.researchfirst.com) revenue from SME-based VoDSL services will soon explode at a compound annual growth rate of 156%, from $169 million in 2001 to $7.2 billion in 2005. The report, entitled "Enriching the Broadband Offering: the Small Business Voice over DSL Opportunity," was based on interviews with more than 900 owners of small businesses (5 to 50 employees) across a range of industries. This is the fastest growing business segment in the U.S. and one that spends more than $70 billion annually on telecommunications services.

The principal finding of the report was that Internet-centric small businesses recognize the powerful competitive advantage of VoDSL, provided that the service provider offers cost-competitive bundled services. Other findings of the report were:

- By 2005, 11.5% of all U.S. small business telephone stations will be served by VoDSL.
- Companies dissatisfied with current dial-up customer service are particularly interested in VoDSL.
- Companies with on-location manufacturing activities are more likely to use VoDSL.
- VoDSL delivers cost-effective services that enable small businesses to better compete with their larger counterparts.

But the U.S. isn't the only country where VoDSL is beginning to blossom. There has also been activity in Europe, although it has occurred at a slower pace than in the U.S. Some European CLECs are pushing VoDSL as a telecommunications alternative for Europe's SMEs who are trying to wiggle out from under the grip that the ILECs have heretofore had on voice services, ISDN services and leased lines such as E-1. These CLECs include COLT Telecom Group plc (London, UK — www.colt.net) which currently has operations in 32 European cities in 12 countries, and Versatel Telecom International N.V. (Amsterdam, Netherlands — www.carrier.versatel.com) which is based in the Netherlands but is focused on the Benelux countries and Germany. But, unlike the U.S., where most VoDSL customers find lower costs a buying incentive, European (and Japanese) VoDSL customers may initially find that the impetus toward VoDSL will be the enticement of the feature rich services VoDSL technology can provide; not less expensive telecommunication services.

Although ISDN has been successful in the SME marketplace in both Europe and Japan, the paltry 64/128 Kbps feed is quite anemic when compared to the data stream offered by most forms of VoDSL service. Hence, Versatel expects that VoDSL will grow as it finds acceptance the European marketplace. The VoDSL service provider will just need to breach the beachhead that ISDN has established in Europe, possibly through the use of a marketing approach that pits VoDSL's "single monthly bill for bundled services" against ISDN's "metered bill for usage whether voice or data."

Advantages for the Business Community

All telco customers feel the crunch from telecommunications costs: local telephone service, long distance service, Internet access, even wireless service, not to mention

VPN, web hosting and costs associated with telework programs.

Service providers know that VoDSL can help mitigate these costs. For example, Mpower Communications (Pittsford, NY — 716-218-6550 , www.mpowercom.com), a VoDSL service provider with operations in California, Florida, Georgia, Illinois and Nevada, is already delivering bundled VoDSL services specifically aimed at the SME market and as the reader will soon learn, many more services providers are jumping on the bandwagon — even the "big honchos" like MCI Worldcom (Clinton, MS — 800-465-7187, www.wcom.com), AT&T (Basking Ridge, NJ — 908-221-2000, www.ATT.com) and SBC Communications (San Antonio, TX — 210-821-4105, www.sbc.com) have either begun testing the technology, have VoDSL trials in progress or have rolled out their own brand of VoDSL services. This is just the beginning. Communications Industry Researchers, Inc. (Charlottesville, VA — 434-984-0245, www.cir-inc.com) projects that there will be more than 25 million VoDSL lines in service worldwide by 2005.

VoDSL is attractive to the SME community because it allows them to have Internet access and up to 24 voice lines with customized features at a relatively low price. As pointed out previously, the typical small business doesn't need and can't afford a T-1/E-1 line, which can cost anywhere from $500 to $1000 per month and many SMEs find that ISDN service doesn't meet their needs. But, VoDSL can easily deliver, at an attractive price, six, eight, ten, 16 or even 24 phone lines as well as an Internet connection, which fits perfectly the needs of the average small business. And as VoDSL becomes more widely available, a whole host of new pricing plans and bundled services designed with a specific niche market in mind (SME business offices, satellite offices, call centers, SOHOs or residential customers) will quickly become available.

Even the customers that claim they are doing just fine with their 56 Kbps analog modems or 128 Kbps ISDN connections must admit that dealing with a relatively slow connection and high telecommunications costs can sometimes be frustrating. Factor in an inability to benefit from such applications as video conferencing, e-learning and VPNs and you can begin to see why many customers would find VoDSL attractive. These are just some of the many ways in which VoDSL can address the disparity that exists between the SME market and the enterprise market. Most of all, VoDSL enables *anyone* to utilize bandwidth-intensive applications, thus allowing *everyone* to benefit from improved communications, productivity and efficiency.

VoDSL offers a number of benefits that many telco customers find irresistible, such as:

• Lower costs due to the ability to use the existing telcom infrastructure. A single DSL-equipped local loop can feed up to 24 POTS lines, plus Web and data streams through one VoDSL connection. Due to dynamic bandwidth allocation, circuit usage is maximized at all times, with voice taking the precedence.

- Lower prices due to bundled offerings. There will be VoDSL service providers that also offer local and long distance voice services priced by the "bucket of minutes," replacing the fixed-price paradigm.
- Voice features and feature access remain available to the customer. Yet, there is none of the technical problems associated with pure VoIP.
- No need to replace legacy customer premise equipment, such as telephones, fax machines, analog modems.
- Convenience since the customer only needs to deal with one service provider for all of their telecom needs.
- Flexibility due to simplified equipment, which makes it easy to add new services, including self-provisioning of new services.
- The delivery of IP-based Centrex features.

Advantages to the Service Provider

Yet, when it comes to the convergence of voice and data over a high-speed transmission mode, the SME market is still underserved (even counting the multitude of ISDN installations in Europe and Japan) and the residential market has been almost totally overlooked. Thus, VoDSL has room to grow if the service provider makes a simple service offering with a clear message at a competitive price and bundles its services.

ILECs

At first glance, deployment of VoDSL doesn't seem like a great proposition for many service providers, especially ILECs; although upon closer examination, ILECs find that VoDSL solves a number of near and long-term ILEC challenges. For example, VoDSL can:

- Provide an answer to the "copper exhaust" issue (too much demand for phone lines which then overloads a local loop).
- Assist with the shrinking revenue puzzle by helping to grow the average revenue generated per customer.
- Help with the inevitable migration of their networks to the more efficient pure packet-based switching architecture.
- Mitigate the overall impact of encroaching competitors.
- Leverage the investment made in their existing DSL access networks and Class 5 digital switches.
- Increase the overall operational efficiency of the ILEC's local networks.

The business community isn't the only market ordering additional telephone lines, residential customers, too, have been adding to the copper wire shortage condundrum. This market needs extra phone lines to accommodate roommate situations, teenagers, home offices, fax access and Internet access. VoDSL offers a cheaper and better way, for both the customer and the service provider to meet this demand.

With many telcos running into severe copper "exhaustion," there's been a flurry of activity on the installation front — copper loops can't be installed fast enough to keep up with the demand. This approach is capital intensive, eating into the telcos' already stretched maintenance budget, and significantly impacting the overall profitability of POTS. But with VoDSL in their service package, telcos can liberate up to 23 copper pairs per customer served. Do the math — an impressive amount of an ILEC's copper plant can be freed up just by converting only a small portion of its customers to VoDSL service. In other words, an ILEC that proactively offers VoDSL can cap or even reduce the size of its copper plant.

Another advantage of VoDSL is that service providers can use "lineless" Class 5 switches since many forms of VoDSL service don't require line cards. VoDSL gateways can terminate directly into the GR-303 interface of the Class 5 switch. Consequently, since VoDSL enables the use of not only fewer Class 5 switches (due to the prospect of capping the size of the copper plant), but also lineless switches, it can dramatically reduce an ILEC's overall costs. Moreover, a new-generation softswitch device such as Lucent's PathStar Access server doesn't even need a voice gateway, since it integrates local loop termination, a Class 5 telephony switch, a Voice-over-IP gateway and an edge router into a single network element, thereby saving equipment costs and precious central office space.

The new generation of VoDSL technology also holds promise for the ILECs that want to reap more revenue from their residential customers.

Service Providers Other than ILECs

VoDSL is also quietly making headway among independent service providers, which are extending their VoDSL services in areas that have fallen off the ILEC's' radar screen. Some of the service providers in this group are actually building out their own fiber to underserved areas, instead of using the existing unbundled copper platform of the ILEC. But the providers who are leasing ILEC lines find that they can generate $500 per month revenue on a line they lease from the ILEC for $20 or less — a compelling set of numbers. Thus, you can see why it's mostly the non-ILEC service providers who are rolling out VoDSL bundled services.

Chapter Nine

VoDSL represents the only profitable method for delivering a compelling bundle of services to the ever-expanding SME market, the SOHO customers and the upscale residential market. Reduced copper and installation costs are key to VoDSL's cost savings; but service providers also save money because they pay access fees to the ILEC based on the number of twisted copper pairs, not each virtual phone number. By enabling the delivery of up to 24 telephone lines and high-speed Internet access over a single copper twisted pair and a centralized Class 5 switch, the economic drivers behind VoDSL deployment is very clear for the non-ILEC service provider.

Let's take another look at Mpower Communications. Mpower's primary marketing focus is the bundling of its VoDSL services; offering a comprehensive package of services — local voice, long distance voice and high-speed Internet access. In many ways Mpower is a typical example of how service providers can best take advantage of VoDSL. By bundling services using VoDSL technology, any service provider can offer its customers more for less: multiple voice lines (each with its own set of services), high-speed Internet access, and so on.

A research and strategic consulting firm, the Yankee Group (Boston, MA — 617-956-5000, www.yankeegroup.com), estimates that U.S. revenue for VoDSL services reached $567 million in 2001 and will rise to $3.2 billion by 2004. In addition to the VoDSL service providers mentioned previously (Mpower, Colt and Versatel), there are other pioneers offering a wide variety of VoDSL services:

- Broadview Networks (New York, NY — 212-269-4147, www.broadviewnet.com).
- Focal Communications Corp. (Chicago, IL — 312-895-8400, www.focal.com).
- Network Plus (Randoph, MA — 800-552-5254, www.networkplus.com)
- Network Telephone Corp. (Pensacola, FL — 850-469-1508, www.networktelephone.net).
- Panhandle Telecommunication Systems Inc. (Guymon, OK — 580-338-7525, www.ptsi.net).
- Picus Communications Inc. (Hampton Roads, VA — 800-290-0461, www. picus.com).
- Rio Communications (Bend, OR — 541-762- 4357, www.rio.com).
- Transbean (New York, NY — 212-683-8330, www.transbeam.com)

Let's examine in detail at a few of these service providers. Broadview Networks began widely offering VoDSL services in the New York City metropolitan area in July 2001. According to a company spokesperson, the public reception has been tremendous — sales have doubled almost every month. Since Broadview bundled service packages are designed specifically for the SME market and have a 10 to 30% lower cost than comparable service from Verizon, it's easy to understand why a business might call Broadview for its telecommuncation needs.

Network Telephone Corp. is using VoDSL technology to deliver local and long distance voice services, business class calling features, high-speed Internet access, web hosting and website development services to SMEs throughout the southeastern U.S. Network Telephone credits its success to the VoDSL technology that enables it to provide high-quality bundled services.

AT&T's VoDSL isn't quite at the stage where a prospective customer can say, "let's call AT&T and order their VoDSL," but AT&T clearly indicated when it purchased NorthPoint Communications that its plans were to offer VoDSL. For example, Mark Siegel, an AT&T spokesperson, has been quoted as saying, "We see DSL in a fundamentally different way than most of the rest of the world. Everybody seems to associate DSL with high-speed access. While that's obviously very important, we see DSL as a platform for a variety of different services that includes telephony."

Economics

A DSL access network can be upgraded to provide VoDSL at a small incremental cost to the service provider, and service providers have proven to be more than amenable to passing on some of the price and service benefits to their customers.

By combining voice gateway technology with DSL access technology, voice and data service can be integrated onto a single copper line infrastructure, meaning the customer's legacy copper twisted-pair wiring can be used exclusively for all telecommunications needs. All that's needed is an integrated access device (IAD) at the customer location and a gateway device on the service provider's end. The lower cost of service delivery translates into lower costs to the customer. Also, the customer doesn't need to incur any costs of re-wiring or purchasing new telephones or replacing other legacy equipment.

The VoDSL industry analysts claim that service providers will be able to provision integrated voice and data service to customers at a cost that will run somewhere from 30% to 50% less than traditional POTS.

The Value Proposition

Now that VoDSL is ready for prime time, service providers and customers must find benefit in this service. It must meet a minimum set of expectations in both the marketplace and the service provider community.

The customer's perspective is discussed throughout this chapter with an emphasis on the overall lower communications costs through VoDSL's bundled services. But, there is still confusion in the marketplace; the potential customer base needs to be educated about the advantages of VoDSL. According to a recent poll conducted by Mori (London, U.K. — 020-7347-3000, www.mori.com), the largest independently-owned market research company in Great Britain, close to 80% of SMEs don't see the value proposition of VoDSL. Of course, part of the problem may be that in Europe, unlike the U.S., ISDN has a stable customer base in this market group. Still, this poll does alert the service provider community that they need to put a great deal of effort into ensuring that their customers understand the cost-savings VoDSL offers.

For instance, high costs have often prevented many SMEs from connecting satellite or branch locations. Until the availability of DSL and VPNs there was really no other option to T-1/E-1 service — a technology certainly beyond the budget of most SMEs.

From the service provider's perspective, VoDSL enables them to:

• Deliver competitively priced services to the business community using the only ubiquitous communication architecture available for now and the foreseeable future — the copper loop.

• Cover the ILEC's charges for the local loop with revenues from the bundled services that VoDSL allows them to deliver over that copper twisted pair, thus substantially reducing the fixed recurring cost for each line.

• Market a variety of bundled services resulting in differentiation from competitors.

• Create new revenue streams.

• Create customer loyalty — attractively priced services makes it unlikely the customer will switch providers.

Attractive Bundled Services

By uncoupling voice and data services from the physical circuitry, the service provider can give its customers a wide range of offers and price points. Bundled services are an integral part of most service providers' portfolio. Several service providers have published pricing packages for their VoDSL services. A comparison of these prices with the traditional telecom costs of the average business or residential user provides a telling financial story. It's clear that by going with a service provider that bundles voice and data services together in a single package, the customer enjoys one-stop shopping for voice and data needs, as well as consolidated billing and lower costs.

Bundled services can include not only multiple voice lines and Internet access, but also video conferencing, VPNs, Centrex, unified messaging and more. Since VoDSL extends the packet-switched network of the PSTN to the customer's premise, it provides the perfect foundation for such services. And, the value of bundled service packages accrues to both the VoDSL service provider and the customer.

The service provider can increase its revenue potential through a higher ratio of sales of value-added services. Once the customers do a bit of comparison shopping, they will see the advantage of purchasing a VoDSL service bundle.

With the right marketing campaign, a VoDSL service provider can significantly grow its overall customer base and, thus, its profit margins. For instance, a small business provisioned with a DSL line having Internet access and six phone lines with call waiting can be converted to a VoDSL customer with the same Internet access, the same six phone channels, but besides call waiting, other feature rich options can be easily added, such as caller ID, conferencing, voice mail and unified messaging. Factor in 500 minutes of long distance at a set monthly rate and the business customer (and the service provider) have a very attractive, accountant-pleasing proposition.

This is just one of many VoDSL service packages that could be put together for the business community. Other packages could include VPN services, video conferencing, Centrex services, web hosting and more. Bundled services aren't just for the business community, a popular residential bundle might include four voice lines with caller ID, call waiting, voice mail, high-speed Internet access and with the purchase of a long distance package, free long distance on the weekends.

The theory behind bundling is that items in the bundle will be offered at an effective price that is lower than the *a la carte* price, while at the same time keeping both the customers and the service providers happy. The customers are thrilled with the bundled services because they address their cost concerns. The service provider is delighted with its ability to offer low cost packages because it's selling services that have a low cost provisioning rate, and which it wouldn't have been able to sell if the same services were offered unbundled at a non-discounted price.

As illustration: A customer currently has three POTS lines (3 x $16 = $48), call waiting and caller ID on two lines ($6 + 8 = $14), long distance at 7 cents per minute with the average bill being $35 (300 minutes plus the infamous "extra" charges), and DSL service with Internet access ($39). The total monthly payment for that customer's telecommuncations services would be $136, which would typically be spread among two or three service providers. However, if a VoDSL service provider offered a bundled service of: three lines, *all* with call waiting and caller ID, 300 minutes of long distance service, and high-speed Internet access for $110, everyone wins.

Recent research has shown a strong desire on the customers' behalf to sign up for service bundles that include both voice and data services.

According to Research First Corporation, there are close to 7.5 million small businesses in the U.S. that make use of almost 40 million telephone handsets and about 75% of these handsets are served off of key telephone systems (KTS) since traditional POTS-based Centrex is out of the reach of many SMEs, mainly because of the high costs associated with provisioning traditional Centrex services. Thus, the local exchange carriers have been unable to lower the price of their full-featured Centrex services to the point where they are cost competitive with small KTSs and hybrid PBXs having fewer capabilities.

However, with VoDSL in the picture, more reliable and feature-rich Centrex services can be brought onboard. A VoDSL customer can order telephone service with the same terms (but more features) offered by KTSs and PBXs. For example, a VoDSL service provider could offer 4x10 Centrex service (10 Centrex lines are provided, but only 4 can be active at the same time) and those 10 lines are standard Centrex Direct Inward Dial lines running off a Class 5 switch, supporting all available Custom Local Area Signaling Services (otherwise known as "CLASS"), as well as custom calling features, voice mail and more.

Caveat Emptor — Let the Buyer Beware

The author believes VoDSL to be the optimum technology for specific target markets and has provided numerous compelling reasons to install VoDSL on the reader's premise; nonetheless, we should review the disadvantages of VoDSL.

Backup: In case of power outages, the customer may need battery backup or a separate landline or wireless access. However, one VoDSL service provider (Network Telephone) includes a small Uninterruptible Power Supply (UPS) in the packages it offers to the small business market, which addresses the backup power issue. But note that both ISDN and T-1 service also lack inherent backup, so many businesses that use these services tend to buy multiple "pipes" from multiple service providers so as to provide the necessary redundancy. The same can be done with VoDSL services, at a lesser cost; or else the customer can choose to leave a few POTS lines in place as backup.

Reliability: Some service providers don't rank repairs on par with other services in importance. For instance, if your VoDSL service provider is dependent upon an ILEC for any repairs, then they may run into some difficulties. While ILECs normally have a 24-hour mean-time repair objective to solve problems, DSL and VoDSL repairs generally take a back seat to other kinds of service repairs. However, it might be possible to subscribe to a service agreement (such as those offered with Frame Relay and T-1/E-1 services), which will guarantee quick repair service within a designated time frame.

Even then, however, unless the customer negotiates *Committed Information Rates (CIR)* — which will add to the VoDSL service's overall costs — the Mean Time To Repair (MTTR) on an unbundled DSL loop will be anywhere from 24 to 48 hours. That's nowhere near the standard four-hour repair time enjoyed by customers of traditional leased services such as T-1/E-1 and Frame Relay. Also be advised that service providers, such as WorldCom, charges from $200 to $1300 per month for CIR on its DSL service.

Quality of Service (QoS): Although VoDSL vendors and service providers are working feverishly on the issue of quality of service (QoS), there's still room for improvement. At present, most non-ATM VoDSL service providers offer a "best-effort" service, which isn't as stringent as service guarantees for something like T-1/E-1 or Frame Relay. Yet, for VoDSL to really take off as a business option, the service providers must offer VoDSL with solid offer *service level agreement (SLA)* guarantees. During 2002, however, service providers such as Verizon and BellSouth began to put in place equipment that gives them to ability to provide QoS and, thus, the wherewithal to support SLA guarantees.

Dynamic Bandwidth Issues: Then there's the bandwidth issue to consider — the worst case scenario is that each telephone session without compression eats up 64 Kbps so the more virtual telephones you have installed and in use at the same time, the less data bandwidth you will have available. Voice will always take precedence. Fortunately, most DSL operators provide for dynamic bandwidth allocation (but ask yours to be certain), so when a call has ended, that bandwidth is automatically made available for data.

The Prospects are Bright

William Rodey, chairman of the DSL Forum, an all-embracing worldwide body covering all the varieties of DSL (www.adsl.com), stated at the March 2002 DSL Forum meeting in Rome that as of January 1, 2002, DSL deployment had reached 18.7 million installations worldwide. He went on to state that "with 188% global growth in 2001 and accelerating deployment in many markets around the world, DSL broadband is well on the way to a global mass market." While, today, most DSL installations are residential, thanks to G.SHDSL (a standard that was specifically designed for the SME market), DSL installations are growing in leaps and bounds in the SME community. The next logical step is to integrate voice and data services, i.e. VoDSL. This will allow service providers to sell to a broader range of customers, while at the same time gaining additional profit from each DSL line, meaning that VoDSL represents an industry trend that can truly cement the continued success for DSL.

By using VoDSL technology, SMEs will, for the first time, be able to afford not only telephony applications with features and functions that heretofore were only available to the enterprise market due to costs, but also have access to applications that require intensive bendwidth, such as conferencing and e-learning tools. No wonder the Cahners In-Stat Group, which is now Reed Business Information (Scottsdale, AZ

— 480-483-4440, www.reedbusiness.com) forecasts 10 million VoDSL lines to be up and running by 2004.

Another report, this one from Allied Business Intelligence, Inc., commonly referred to as ABI (Oyster Bay, NY — 516-624-3113, www.alliedworld.com), indicates that the telecom industry downturn during the 2001-2002 timeframe undoubtedly helped to delay deployments of VoDSL solutions. However, the study revealed that as early as the fourth quarter of 2001, VoDSL installations were starting to accelerate. Carriers (ILECs in particular) initially avoided the whole subject of VoDSL, fearing both the technical difficulties involved in deploying it and the potential cannibalization of existing telephony revenues. Now, however, interoperability issues have almost been fully resolved, test trials have been successful, and VoDSL deployment is now seen as an attractive opportunity. And although the first wave of CLECs may have floundered, CLECs worldwide are now experiencing strong subscriber growth for VoDSL service, and ABI expects this trend to continue, with CLECs in North America, Europe, and South Korea leading in VoDSL deployments.

"The revival of VoDSL is taking place. A weathered and experienced core group of CLECs has endured and these players are learning to scale their deployment of VoDSL services in a much more realistic regional manner; in comparison to the wide-deployment model of their failed predecessors," says John W. Chang, ABI Senior Analyst and report author. "In the longer-term the ILECs remain critical for a broader rollout of VoDSL services."

ABI's research findings include the following:

• Revenue generated by both service providers and equipment vendors of VoDSL for year-end 2002 will be $159.7 million dollars, representing over a 90% increase from 2001.

• By year-end 2002, approximately 96,550 VoDSL subscriber lines will be in use versus 14,000 in 2001.

• By 2003, South Korea will be the leader of VoDSL deployment in Asia.

• All incumbent carriers have tested VoDSL technology, and some incumbent carriers will have field deployments in 2002.

The Softswitch's Pivotal Role

Service providers hope to boost sales through new technology, hybrid services and enterprise management tools. To this end, the future of VoDSL includes network-hosted telephony applications made possible through the use of application-enabled softswitches.

VoDSL can be seen as a "data ready" solution. When a customer wants access to a specific feature, function or application, the service provider (or even the customer) can quickly and easily "turn on" the service via soft provisioning.

Advanced technologies are making possible powerful innovative deployment and business models that are emerging daily. For example, service providers have implemented softswitches that can combine voice via VoDSL technology with a variety of new services, making it possible for the VoDSL industry to offer more powerful solutions. Many of these softswitches are "applications-enabled," which means the service provider can roll out services that give their business customers the option of immediately replacing legacy PBXs and KTSs, and/or Centrex services, with a hosted model of telephony management. Among the possibilities: SMEs can outsource their phone systems, saving anywhere from $20,000 to $50,000 up front by avoiding the cost of a PBX, while, at the same time, retaining control of the management of their telephony needs.

With softswitch technology, business customers can manage their telephony services via a business phone's LCD display and/or an intuitive browser, which makes the tasks much easier than programming a PBX, and quicker than waiting for the telecom provider to make moves, adds or changes to the customer's service. And, the applications go well beyond Centrex and PBX features. Examples of what service features can be implemented include the following:

- Call logs to track missed calls and inbound and outbound calls.
- A feature that lets the customer simply mouse click on a button to return calls or to e-mail a response.
- Personalized call treatments, i.e. the ability to tailor the handling of individual incoming calls, allowing for sophisticated call handling. One example: "VIP" callers can be forwarded to the called party wherever he or she might be, while other calls go automatically to voice mail.
- Categorize incoming calls or allow certain incoming calls to be forwarded to a specific number during a specific time period, such as during weekends, during a travel period or when a person is on vacation.

Softswitches can also simplify maintenance and support, since a PBX or key system is no longer necessary. Also, the installation of VoDSL via an applications-enabled softswitch is easier.

Applications

Powerful new applications enabled by VoDSL technology seem to be emerging daily. Thanks to softswitches, these applications provide voice with new services that can

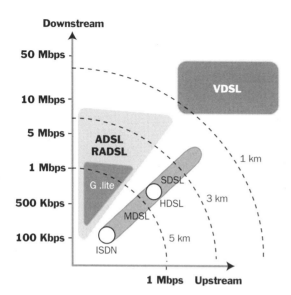

Fig. 9.1. With all the DSLs available, every application or customer can be served by a particular technology. This diagram illustrates bandwidth and "reaches" of various DSLs.

leverage the current DSL infrastructure, enabling the VoDSL service provider to offer more powerful communications solutions to their customers.

Adoption of the H.323 and SIP standards opens the door for the deployment of compatible voice applications running under a common protocol, enabling the rapid deployment of intranet voice services within the SME market and their teleworkers.

Moreover, DSL and VoDSL make VPNs not only possible but practical for the SME, which can now pass their private data safely over the Internet, connect remote and branch offices, build corporate intranets and extranets and take advantage of exciting applications such as interactive distance learning (e-learning), video conferencing, telemedicine, video e-mail and video streaming. SMEs will be amazed at the cost savings and productivity gains that applications like video conferencing offer. With VoDSL technology in place, it's possible to organize and conduct online meetings and communicate with remote workers, customers, partners and suppliers with the same speed and efficiency as larger competitors.

In Summary

VoDSL has moved from theory to reality. The benefits are substantial: better bandwidth usage, conservation of the legacy copper plant, leveraging usage of existing legacy equipment for both the customer and the service provider, and cost-effective integrated services allowing the customer to scale and customize their voice and data services more efficiently. Thus, there's no doubt that VoDSL with its bundled services will be an integral part of most service providers' portfolios. DSL provides a good foundation for services such as multiple voice lines, Internet access, conferencing, e-learning, and more.

Who doubts that VoDSL is ready for prime time? As more and more VoDSL service providers come online and compete with each other to provide innovative and cost-effective services, the benefits will become even more pronounced for both the service provider and their customers. There will not only be lower costs for everyone, but also multi-service products that accommodate a wide range of broadband access technologies and bundled service offerings that can accommodate almost any customer's needs.

We've seen how the benefits to the customer, whether business or residential, includes reduced costs, more and better applications, and the ability to share the same resources for voice and data. But the cost savings alone make VoDSL worth a closer look.

VoDSL wasn't brought to market in a heated rush and as a result it had the opportunity to be "done right." And as new Customer Premise Equipment (CPE) comes on the market, they will allow VoDSL service providers to offer a rich set of communications features, making the value proposition for VoDSL even more compelling. Potential features include remote call centers, bandwidth prioritization, universal messaging and customer self-provisioning on the Web, which means that the customer can easily add and subtract virtual lines and services as needed.

Voice is the one application that every business uses, with millions of telephone calls being made every day. VoDSL promises to revolutionize the way SMEs use telephone or voice services. But for VoDSL to enter the mainstream, the service provider will need to implement a campaign to educate its potential customers in how VoDSL technology works and to reassure its customers that quality and reliability issues have been addressed through prioritization mechanisms that have been built into the technology. The VoDSL service provider will also need to tout how VoDSL modernizes and improves the installation process, whether for data or voice related services. This includes explaining such benefits as: why the customer no longer needs to wait for a truck roll; that a simple call to the service provider (or a visit to a configuration web site) will result in extra capacity or services being added almost immediately.

CHAPTER TEN

The Future

There are two basic themes to the future of VoDSL. One is technological and the other could best be described as "economic / bureaucratic."

VoDSL Transformed by Evolving Technology

Perhaps the chief problem with nearly all of the DSLs we have looked at is that the form of "broadband" we call DSL may not be considered to be very "broad" in the future as the per capita consumption of bandwidth continues to rise. In an era when most users were accustomed to narrowband 56 Kbps dial up modems, the appearance of ADSL (and even its diminutive brother, G.lite) over the analog local loop was considered to be a near-miraculous phenomenon. Today, the speed of such technologies no longer impress those customers demanding advanced services, though that fact hasn't stopped the approximately 3.3 million people who currently subscribe to some type of xDSL service (it's still a great improvement over dial up modem access). Indeed, some projections conclude that around 10.5 million people will subscribe to xDSL services by 2005, this in spite of the fact that the deployment of xDSL broadband actually *dropped* in the second quarter of 2001 owing to the collapse of the communications economic sector during 2000-2002

The Future Dominance of G.SHDSL

Peering a short ways into the murky future, however, one can be reasonably certain that while G.lite will certainly be popular with less demanding residential and SOHO users, both ADSL and G.lite will be sharing DSLAM space with the Symmetric (or Single-line) High Bit Rate Digital Subscriber Line (or Loop), or G.SHDSL. This actually won't be too difficult a technical feat, since G.SHDSL works in an ADSL DSLAM chassis, so G.SHDSL cards can easily be placed into existing DSLAMs to service small-to-medium businesses.

G.SHDSL combines elements from both SDSL and HDSL2. G.SHDSL is the first true, internationally recognized, DSL standard and it will probably usurp all competing DSLs — with the exception of Very-high-speed DSL commonly referred to as VDSL or VHDSL. G.SHDSL technology will operate over ordinary telephone copper wires at speeds ranging from 192 kilobits per second to slightly over 2.3 megabits per second on one pair of wires and 384 Kbps to 4.624 Mbps on two pairs. G.SHDSL will transport T-1, E-1, ISDN, ATM and IP signals, as well as channelized VoDSL. When handling the equivalent 1.544 Mbps bandwidth of a T-1, G.SHDSL can reach over 2.4 km. (7,874 feet) on 0.4-mm (26 AWG) cable and under self-near-end-crosstalk (self-NEXT) conditions. By comparison, single-pair HDSL merely reaches 1.8 km (5905 feet) under similar conditions. Moreover, like HDSL, G.SHDSL will support range extenders. Further, the G.SHDSL modems and range extenders can be powered from the central office over the copper link.

As early as 1999, equipment manufacturers began preparing for G.SHDSL's arrival. In August 1999, Cisco committed several programs to integrate G.SHDSL into its carrier-class portfolio of CO and CPE products, the first being a multi-port G.SHDSL line card option in its 6000 Series IP DSL Switch family, followed by support for the technology in its Cisco 67x series SOHO / Teleworker routers.

G.SHDSL has become wildly popular in Europe as a replacement for E-1, since E-1 lines tend to be five to 10 times more expensive than T-1 lines in the U.S. and the European Economic Community has even mandated that future E-1 service will be done over G.SHDSL. G.SHDSL has the ability to maintain the 2.048-Mbps data rate of an E-1 up to 2.4 km (7874 feet) despite worse-case line conditions. The U.S. is only just now catching on to G.SHDSL, partly because of the languishing economic situation that existed when the ITU made G.SHDSL a global standard (G991.2) in February 2001.

Perhaps the most interesting part about G.SHDSL is its ability to support a plethora of services — such as native IP, Frame Relay and ATM — without running a processor hungry (and latency-inducing) ATM or Frame Relay stack. This is totally unlike ADSL which is pretty much tied into using ATM as a transport for both voice and data.

The probable ascendancy of G.SHDSL will be a boon for all small and some medium-sized businesses, as its symmetrical nature makes it suitable to support an equivalent number of incoming and outgoing calls. A surge in the number of small call centers should take place since the economics for establishing them will now be so favorable. For example, according to a study done by the Yankee Group, a 24-line VoDSL configuration should be very attractive to smaller customers, since it is roughly 75% cheaper than a T-1 line and 40% cheaper than DSL and the same number of POTS lines (even a 16-line voice offering should be enticing, since 95% of small businesses currently use 12 or fewer telephone lines). Higher voice usage should come as a relief to service providers, since although data traffic is growing each year at triple-digit rates, 75% of revenues from small to medium-sized businesses still derive from voice services.

A major technical limitation with many DSLs, however, shall remain: unless special extenders and remote DSLAMs are installed in the access network, the customer premises must be situated within about a three-mile radius of a central office. Even so, according to Probe Research Inc.(Cedar Knolls,NJ — 973-285-1500, www.proberesearch.com), Verizon already has 79% of their lines DSL capable, BellSouth has 70% and SBC and Qwest have 60% ready to go. The incumbent carriers have invested over $100 billion and competitive carriers have invested over $56 billion in deploying new facilities and upgrading their existing facilities. As a result of these investments as well as the investments of cable companies, approximately 85% of U.S. households have access to broadband.

"Next-Gen" Networks and the Rise of the Applications Server

We've seen how early ATM-based "loop replacement" VoDSL systems were adroitly overlaid onto legacy voice access networks and GR-303 gateways, making the whole system still dependent on the central office's ancient Class 5 switch for calling features, switching, dial tone, and interconnection to the PSTN.

In the IP-centric world of the future, softswitches and media servers take the features and functionality of the gateway and Class 5 switch and separate these functions into a fully distributed architecture in which switching (physically directing voice from end to end), signaling (ensuring that a call is routed to the right destination along with the requested advanced features), and value-added applications are divided into separate components placed at optimum points in a carrier network. Such a widely decomposed and distributed system nevertheless appears to the outside world as a single VoIP gateway, thanks to various protocols (such as

Megaco/H.248 and SIP) that dart back and forth across the network among the components, creating a single functioning entity.

This "next generation" architecture is composed of a media gateway controller, which itself can be distributed over several computer platforms, a signaling gateway that communicates with the SS7 network, and several media gateways, that perform the conversion of media signals between PSTN circuits and packet pipes. In a prototypical configuration, this distributed gateway system interfaces on one side with one or more telephony switches (circuit switches), and on the other side with a SIP or H.323 compliant system.

This "next-gen" network will provide the functions of both Class 5 and Class 4 tandem switches in a distributed architecture, offering all the benefits of packet networking while providing traditional voice service functionality in a flexible, scalable framework. Such a simplification of today's system will increase flexibility to introduce new services to customers at low cost.

Under the new architecture, voice transport will be handled by media and access gateways, the call control will be handled by a media gateway controller, and applications development will be handled by many independent application providers. The rapid development of more and more increasingly sophisticated applications will have an escalating impact on the next-generation network.

In Chapter 5, Figure 5.17 illustrates how a media gateway controller not only connects to a media gateway and signaling gateway controller, but also uses the SIP protocol to communicate with a media server (and SIP services servers) embedded in the IP / Internet Services layer. This is because, although a typical media gateway has powerful media encoding/decoding capabilities and may provide media-related functions such as playing announcements, it is unlikely to record messages or support media-intensive applications which require functions such as speech recognition. These functions are more optimally performed in the media server, which in turn preserves the simplicity, reliability, and price performance of the media gateway.

The media server (always an IP media server, unless one believes that ATM will one day take over the whole world's telecom infrastructure) provides a platform for adding enhanced telephony services to this new network. It optimizes service delivery, facilitates growth, and supports the interconnectivity advantages of IP. While the IP media server platform supplies traditional enhanced services such as conferencing, messaging, and announcements, it also supports packet voice and all of the new enhanced services such as integration with web-enabled environments, wireless technologies, and mobile users. Optionally, a media server can be designed to provide services to both IP and PSTN end points simultaneously. The activities of such a media server still focus on intensive media processing, but now with the flexibility to accept either PSTN or packet-based inputs and outputs.

The following are the core media capabilities required in an IP media server:
- Automatic speech recognition (ASR)
- Conferencing
- DTMF ("touch-tone") handling
- Fax messaging
- Interactive Voice Response (IVR)
- Media streaming
- Text-to-Speech (TTS)
- Transcoding and compression

One "phantom" network component which doesn't appear in the previously referenced Chapter diagram but will one day take its place in importance among the other softswitch components is the *application server*. The application server communicates with the media gateway, the softswitch and the media server. It may use MGCP, Megaco/H.248 or SIP to communicate with these distributed softswitch constituents (indeed, like everything else in the next-gen architecture, applications servers themselves can be aggregated or distributed in the network, and SIP can be used to communicate between the various "engines" making up the application server, thus allowing it to be broken into separate components which can be distributed throughout the network). The application server is responsible for coordinating with the media server to provide more media intensive functions than the gateway can support. The application server generally provides the processing logic for IP media functions, while the media server provides the media processing. For example, the application server may contain the database information for a credit card calling application, and use the media server to provide audio prompts and gather account information (account number, PIN, etc.) using voice recognition. The application server could include directory servers, policy servers, or other servers that provide applications to the converged network.

This is where the future finally becomes "futuristic" for both business and residential customers. One could, for example, provision and transfer voice calls through a web brower, or perhaps select services from a service provider and activate them simply by clicking on a mouse, or selectively forwarding your calls to your cell phone, home phone, or voice mail based on the time of day (the so-called "follow-me" services). Applications that had formerly been available to wireless users (inbound/outbound call logs, speed dial, directory services) can now be brought to landline telephones, and unified messaging will finally become possible.

Businesses will become so entranced by these new applications that they will fail to take notice that the old IP / ATM debate will have abated, thanks to flexible, versatile platforms that are genuinely multiservice in nature.

For example, in 2002, Advanced Fibre Communications (Petaluma, CA — 707-794-7700, www.afc.com) a provider of multiservice broadband solutions, introduced the Telliant 5000, a carrier-class, scalable multiservice edge switching platform that combines ATM switching and DSL capabilities with future IP, *Multi-Protocol Label Switching (MPLS)* and emerging optical access technologies in a device that takes up a single rack's single shelf. AFC's Telliant 5000 leverages a high-performance switching fabric to deliver 15 Gbps of non-blocking switching capacity that combines the native cell/packet switching fabric with sophisticated traffic management and protocol inter-networking functions to deliver a single access platform that can provide concurrent layer-2 and layer-3 service models.

AFC feels that the two trends highlighting the deployment of broadband services in the wireline network will be "packetization" (the transition of analog circuits to digital packets) and "fiberization" (the transition from copper to fiber-optic transport). Collectively, these trends should reduce costs, increase bandwidth and provide additional revenue for carriers. The impact of these trends has already been realized throughout most of the network core, but the edge of the network where users, the communication endpoints and the concentrator equipment dwells, will still need significant carrier investment to be fully upgraded. This current deficiency at the edge separates carriers from the promise of new broadband revenues.

Once packetization and fiberization are fully realized at the edge of the network, carriers will be able to simultaneously and cost-effectively transport multiple services such as voice, data and video over a converged network. The Telliant 5000 was designed with such network evolution in mind, providing current high-performance ATM switching but with an easy migration to future IP capabilities.

AFC believes that carriers will migrate their legacy networks to an all-IP-based network because of the capabilities of new equipment such as their Telliant 5000.

As this book went to press, CyberPath (Piscataway, NJ — 732-463-7700, www.cyberpathinc.com), also announced a device along these same lines. Its CyberPath ZX2000-L Multi-Service Gateway, a next-generation voice gateway, allows Competitive Local Exchange Carriers (CLECs), Local Exchange Carriers (LECs), Integrated Communications Providers (ICPs), and Interexchange Carriers (IXCs) to provide integrated toll-quality voice, data, and video services to small business, SOHO, and residential customers.

The CyberPath ZX2000-L Multi-Service Gateway can supply toll-quality broadband voice over various forms of DSL, ATM, cable, or channelized T-1/T-3. It can route voice traffic over the PSTN or IP network, terminating calls on either a GR-303 Class 5 switch or an H.323 managed IP network.

The ZX2000-L can integrate with a SONET ring architecture, providing added network reliability. It can even implement "smart routing" policies based on time of day or traffic congestion or other network conditions, for more efficient use of backbone facilities and maintenance of service levels. Service profiles can be configured on a per-subscriber basis, to offer a choice of service levels and pricing plans, and to enforce service level agreements.

The device even offers media gateway functionality in a softswitch environment.

By integrating *Time Division Multiplexing (TDM)*, ATM switching, and SONET interfaces on a single platform, various schemes can be achieved for added network reliability and for performing load balancing among voice gateways.

Next-generation equipment such as those from CyberPath and AFC will completely transform the world of communications, and will allow customers to enjoy data and voice applications previously unimagined, much of it delivered over DSL and VoDSL.

According to a March 2002 report by the high-tech market research firm Cahners In-Stat/MDR (now called Reed Business Information, www.reedbusiness.com), the softswitch market is in the very early stages of what the high-tech market research firm believes will be a long growth period, as the PSTN is transformed into a packet-switched network capable of supporting voice, data and video.

They predict that the softswitch will be a key component of this transformation as it controls the media gateways, IP phones and Integrated Access Devices (IADs) in the packet-switched network and provides interworking with the traditional circuit-switched network.

Norm Bogen, a Director with InStat/Reed stated, "The reduced service provider capital equipment budgets in 2002 dampened growth in the softswitch market. With reduced budgets, service providers will limit their spending to products that reduce costs or increase revenues and deliver an appropriate return on investment (ROI). Softswitches and related technologies can reduce costs and provide for increased service revenue, which leads us to conclude that they will continue to be deployed."

In-Stat/Reed predicts that it will take at least ten years, and probably closer to twenty years, to convert from a circuit-switched telephone network to a converged packet-switched network. Many service providers are already evaluating, and in some cases deploying, softswitch technology.

In-Stat/Reed has also found that:

- The worldwide softswitch market is expected to reach $1.32 billion in 2006, representing a compound annual growth rate of 60.2% over the forecast period (2001-2006). The market will continue to experience strong growth beyond 2006.
- Companies competing in the softswitch market include traditional telecommunications equipment and software suppliers like Lucent Technologies (Basking Ridge, NJ — 908-719-7657, www.lucent.com), Nortel Networks (Research Triangle Park, NC — 800-466-7835, www.nortel.com), Tekelec (Calabasas, CA — 818-880-5656, www.tekelec.com) and Siemens (Munich, Germany — +49-89-636-00, www.siemens.com), as well as start-ups such as NexVerse Networks (San Jose, CA — 408-750-9400, www.nexverse.com) and Sonus Networks (Westford, MA — 978-392-8100).
- In the short term, vendors that supply both a media gateway and softswitch, will be in a stronger position than those than supply only the softswitch.

The ultimate question, of course, is: After all the capital expenditure to create the next generation network, will new services (and their delivery via broadband services such as DSL) really be accepted with open arms by business, simply because new technology makes them realizable and of bearable cost? Not all of the services made possible by the next-generation network may be suitable for all types of business, partly because of the unrelenting pace of evolving technology and partly because of varying economic conditions.

Managed VPNs

To illustrate how complicated the situation can be, take, for example, the idea of outsourcing secure *Virtual Private Networks (VPNs)*, which involves IP VPN services being delivered from the "network cloud" by managed service providers or carriers. In 1999 there was a flurry of *IP Service Switches (IPSSs)* developed by hardware and software vendors such as Check Point Software Technologies (Redwood City, CA — 650-628-2000, www.checkpoint.com), CoSine Communications (Redwood City, CA — 877-436-7463, www.cosinecom.com), Netscreen (Sunnyvale, CA — 408-730-6000, www.netscreen.com), Quarry Technologies (Burlington, MA — 781-505-8300, www.quarrytech.com), Shasta — now part of Nortel, RapidStream (San Jose, CA — 408-519-4888, www.rapidstream.com) and SpringTide — now part of Lucent Technologies. The hardware and software from these companies promise to deliver value-added security services from the provider's edge.

Service providers have enjoyed slow but steady success in deploying IPSSs and offering secure managed VPNs to customers. These providers include such well-known names as SAVVIS Communications Corporation (Herndon, VA — 703-234-8000, www.savvis.com), Broadwing (Cincinnati, OH — 513-397-9900, www.broadwing.com), and Qwest Communications (Denver, CO — 303-992-1400, www.qwest.com). Still, acceptance for managed IP VPNs remains slow, and the market is still said to be maturing.

While some service providers have found it easy to offer VPN as a service, some weren't initially motivated to offer IPsec services because IPsec VPNs and their related security technologies (firewalls, policy management, etc.) require considerable attention to detail. During the economic downturn of 2000-2002, some service providers didn't have sufficient trained staff on hand to manage these services in large-scale rollouts. Other providers have come up with an amusing solution — just as a business outsources its IPSec VPN and its associated encryption / security functions to a service provider, so the provider in turn can outsource such network services to a specialist services company that manages the IPSec VPN over the service provider's infrastructure, thus freeing the provider to stick to its core business competency.

As America emerged from the economic downturn in 2002-2003, the most active segment of the IPSec VPN market turned out to be larger enterprises. Although enterprises are in a better position to pay a modest premium for a managed VPN service, they are nevertheless reluctant to outsource their security infrastructure (IPsec VPNs, firewalls, etc.) to third parties such as managed service providers. Most of the large enterprise dealings with service providers have centered on the ability of the service providers' equipment to aggregate a large amount of voice and data access lines and pipe them into the network core, not to specifically use the service providers' IP VPN services. The attitude of the average large enterprise appears to be one of: "You can furnish us with the pipes, but we prefer to do our own traffic encryption for the VPN." Moreover, the expense of maintaining a reasonably skilled staff (once a principal driver of outsourcing) is less than it used to be thanks in part to the influx of Asian and other foreign nationals in both the U.S. and Europe.

Small and Medium Enterprises (SMEs) may have fewer security concerns about managed VPNs than vast multinational enterprises, but even in the case of SMEs, an ever-increasing flow of increasingly automated, inexpensive technology may make "in-house" VPNs over DSL an easier and less expensive proposition than buying it as a managed service. Tom Nolle, president of CIMI Corporation (Voorhees, NJ — 856-753-0004, www.cimicorp.com) was quoted in the June 2002 issue of *Network* magazine as saying, "Even for a very small company, firewalling an Internet connection is relatively trivial, and the CPE is a hundred bucks. There's no business model to promote the survival of the providers that offer Internet tunneling. Could a carrier do antivirus and firewall effectively? Not [if they want to] make money, which is all that counts."

Carriers are indeed wary and slow to offer VPNs as a service using their own resources. Carriers tend to be not so good at managing CPE for customers; their legacy management systems don't scale, and they must deal with all sorts of difficult installation, redundancy, software management, and inventory issues. For these reasons, some of the more technically adept managed service providers are pondering business models involving partnerships with the big carriers. The service provider would simply manage these VPN services to businesses on the behalf of the carriers, enjoying the respectable "name brand" of the carrier in the bargain.

In the meantime, *Multiservice Edge Router (MER)* vendors such as Juniper Networks (Sunnyvale, CA — 408-745-2000, www.juniper.net), Unisphere Networks (Westford, MA — 978-589-5800, www.unispherenetworks.com), Laurel Networks (Pittsburgh, PA — 412-809-4200), and Cisco Systems (San Jose, CA 408-526-4000, www.cisco.com) have taken advantage of the service provider market's confusion by adding IP VPN technology to their boxes. These devices are situated at the network edge, but closer to the core than the customer premises or the service provider's first Point of Presence (POP). Unisphere has experienced success in the edge aggregation space, focusing on Layer 2 and Layer 3 internetworking, so providers can run ATM and Frame Relay-enabled VPNs with Multiprotocol Label Switching (MPLS) for a gradual, controlled migration to pure IP VPNs (presumably delivered over the last mile to customers via some form of DSL).

These new MERs do not have as many value-added features as high-end IPSS devices, and most of the IPSS vendors plan to support tiered security services, such as network-based firewalls or intrusion detection. Indeed, aside from their yeoman duty serving up VPNs to businesses as a managed service, smaller IPSSs play a role in mass-market security solutions such as SME broadband, and supporting the remote or off-network sites of enterprises. They also are ideal for the static "private" networks of large enterprises. It's just another, more secure, way to deliver point-to-point connections; but it's being done via IP and MPLS, IPSec, or both.

Although the MER market appears to be quite vibrant, many experts predict a gradual two or three year trend that will result in both IPSS and MER functionality being integrated into a single kind of device. The resulting platform will have a comprehensive feature set and should be popular among those who would offer IP VPNs as a managed service to businesses.

Of course, without a solid value proposition, economic argument and customer acceptance that managed VPN traffic encryption will not occur until it hits the car-

Chapter Ten

rier edge, there would not be much sense in considering carrier or service provider-based managed VPNs. In theory, success should come to any service provider who can offer a decent, inexpensive, *Operations Support System (OSS)* package that can do the following:
- Provision and manage firewalls and IPsec VPNs.
- Do policy management.
- Manage equipment from multiple vendors.
- Scale up to meet high levels of service provider or even carrier traffic.

Despite the naysayers, one can be certain that network-based IP VPNs will, to some degree, serve businesses of all sizes in the future. Customers will be able to select from among several VPN architectures. So the question really isn't whether to use network- or CPE-based gear for VPNs, but rather, which provider or vendor combination can provide the most flexible choices for deploying VPNs.

However, there is an even larger "fly in the ointment" involving advanced services offered by next generation networks. Due to a lack of promotion / education by service providers and carriers, users may not be aware of all the advanced services that may be available to them, hence a lack of demand for them. This phenomenon is not unique to futuristic voice services — Microsoft Outlook has a feature that lets people with modems use them to dial voice calls from their address book. Almost no one uses the feature, because almost everyone is unaware of its existence.

Similarly, the phrase "next-generation voice services" generally brings to mind Voice-over-IP (VoIP) and some foggy notion of value-added applications resulting from the melding of voice and data, but few particular services come to mind.

This is probably because the softswitch market is still attempting to gather momentum. Since the PSTN did not lay down and die when the idea of the next-generation packet-based network was first proposed, softswitches have had more success at performing Internet off-load and Class 4 switch functionality than as replacements for Class 5 switches. All experts and vendors are certain that a great transition to packet-switched networks and converged services is in the offing, but because of the economic doldrums of the 2000-2002, RBOCs and carriers will move to converged services slowly, at least until a sudden surge around 2004.

In the meantime, many start-ups focused on convergent applications are and will continue to bring IP PBX and other advanced features, such as unified messaging, to businesses. These vendors either sell directly or partner with a softswitch vendor such as Sonus to reach providers such as Touch America (Butte, MT — 406-497-5100, www.tamerica.com), Masergy Communications (Irving, TX — 866-588-5885, www.masergy.com), or the voice Application Service Provider (ASP) TalkingNets

(Wilmington, NC — 877-661-8255, www.talkingnets.com).

Sonus, for example, has many partners that only do enhanced converged applications, such as BayPackets (Fremont, CA — 510-743-2500, www.baypackets.com) for Internet call waiting, voice and unified messaging, and voice VPN; NetCentrex (San Jose, CA — 408-521-7400, www.netcentrex.net) for call center and Automatic Call Distribution (ACD) features; and Voyant Technologies (Westminster, CO — 303-223-5000, www.voyanttech.com) and Pactolus Communication Software (Westborough, MA — 866-722-8658, www.pactolus.com) for conferencing services.

Sonus and other softswitch vendors or provider customers have also partnered with Broadsoft (Gaithersburg, MD — 301-977-9440, www.broadsoft.com), which along with Sylantro (Campbell, CA — 408-626-2300, www.sylantro.com) is leading the IP Centrex charge. Other key players in this space include LongBoard (Santa Clara, CA — 408-571-3300, www.longboard.com) and VocalData (Richardson, TX — 972-354-2100, www.vocaldata.com). These companies also compete with pure IP PBX vendors, such as the more established Shoreline Communications (Sunnyvale, CA — 408-331-3300, www.shoretel.com), and august veterans such as Lucent and Nortel.

"Up and coming" perhaps best describes these companies, each of which have scored wins in the U.S. (and a few internationally), with customers ranging from residences to vertical markets such as real estate, banking, travel, and construction. The size of the implementations ranges from very small offices to distributed enterprises of 400 or more people. Service-oriented industries have been among the strongest adopters, because of features that facilitate their handling of customer calls — such as advanced find me/follow me and rules-based call forwarding. Shoreline's sweet spot is the multisite enterprise with 100 to 3,000 employees. It has racked up many customers, most of whom are deploying VoIP equipment at new locations.

Ironically, most customers use convergent applications for services they have been comfortable with for years; services having features similar to those derived from the time-honored Class 5 switch, albeit with a web *Graphical User Interface (GUI)* that simplifies provisioning and management via a point-and-click user environment. The reason for adopting a next-generation network, however, is not solely to mimic services found in the PSTN, but to boldly go where no voice service has gone before. The applications made possible by such advanced packet networks should be so novel that they act as a differentiator for the service provider, thus building customer loyalty and reducing the dreaded customer "churn," which actually means "market share."

Like the sales of advanced next-generation softswitches, there has been a slow but steady growth in IP Centrex solutions for SMEs. As convergent applications appear, they will be easy to install (existing phones can be plugged into the IAD instead of throwing them away and using IP phones) and contribute in part to a smooth, controlled, and inexpensive migration path to an IP-centric environment.

VoIP services will be customizable via the GUI, which has abilities ranging from an end user to a telecom manager or network administrator, such as centralized moves, adds, and changes. One such change might grant an employee long distance calling privileges to only specific area codes at certain times of the day, week, or month, and then revoke the privilege, based on business rules. The GUI should be able to display relevant billing data to that employee or manager, as well as missed/inbound/outbound call logs and online directories.

When "computer telephony" (the control of phone calls by computer intelligence) appeared in the early 1990s, one of its selling points was that users no longer had to remember long sequences of DTMF touchtones ("star codes") to program and control their phone system. Command shortcuts and clicking on icons in a GUI could be used instead. These ideas have been carried over to end user interfaces to the next-generation network, so that icons can be linked to a specific function command, such as moving extensions, setting up new users, or implementing a speed dial or follow-me schedule.

Sylantro's platform can be modified via the web, Microsoft Outlook, a Wireless Application Protocol (WAP)-enabled cell phone, or the LCD display on a desk phone. Shoreline's system relies on applications in compliance with Microsoft's Telephony API (TAPI), which means that the features can be distributed throughout all sites in an enterprise voice network. End users can take control of any standard phone, IP or analog, at any company office by simply having a PIN number available; then all communications — voice, e-mail, voice mail, and so on, which shift to that desktop.

If these apps work, the result should be something like a VoIP intranet, providing all linked company employees with a uniform, feature-rich set of free services. Dave Passmore, research director of the Burton Group (Midvale, UT — 801-566-2880, www.tbg.com), likes to call it "a single virtual PBX." Specific vertical markets will find this useful, such as medicine and finance, which can use toll-saving VoIP apps for call forwarding.

Those companies plugged into a next generation network, but feel comfortable with the PBX paradigm, will like the concept of the IP PBX, a Customer Premises Equipment (CPE), IP data-enabled version of a traditional PBX and its functionality (though this kind of PBX is programmable with a modern GUI instead of an old command line interface).

What's Next?

Generally, when a discussion of any advanced technology gets going, one is almost always forced to speculate about "the next big thing." During the entirety of the 1990s, "the next big thing" was unified messaging (it still is!).

Lately, the "next big thing" made easy to deploy by the softswitch-powered next generation network is expected to be what's called "presence detection" or "presence management" applications, where users can see if the person they want to call is available, on the phone, in a meeting, or what-not. Some platforms will even tell you when a person you're trying to reach gets off the phone, then automatically dials the number. The goal of this technology is not to hound users or invade their privacy, but for the system to obtain enough intelligence so that the network can determine the best way for you to receive information based on your availability. For example, an executive going into a meeting can opt to block cell phone calls, and automatically redirect calls to voice mail and even convert messages to Instant Messaging (IM) text.

Indeed, these presence-management collaboration technologies are a sort of value-added cousin to the Instant Messaging (IM) services of America Online, MSN and Yahoo!, and some of them have already appeared in the cellular phone sector. For example, phone makers Motorola, Nokia, and Ericsson formed the Wireless Village initiative April 2001 to define and promote a set of universal specifications for mobile instant messaging and presence services called the Wireless Village Instant Messaging and Presence Services (IMPS) standard. This led in February 2003 to the IM company, Personity (Pittsburgh, PA — 412-325-1800, www.personity.com), and Motorola (Arlington Heights, IL — 847-632-2560, www.mot.com) launching the Motorola Messenger system powered by Personity — a messaging and presence solution which supports IM with presence capability and mobile chat rooms based on the standard *Short Messaging Service (SMS)*. Amazingly, all of this can be made to work with the present wireless infrastructure. Related systems have been devised by companies such as Bantu (Washington, D.C. — 202-822-3999, www.bantu.com), which gives a taste of what will be possible when the next-generation network really gets going.

Several convergence vendors are tackling aspects of the provisioning process by focusing on OSS issues. LongBoard, for example, is focusing on next-generation apps that link Microsoft Outlook with other voice features, while including hooks that talk to telcos' back office systems for billing, provisioning, and so on. LongBoard's platform is essentially a service-creation tool environment enabling developers to create customized applications, a much-hyped but little-realized advantage of an open system.

LongBoard's technology mixes control call management with Microsoft Outlook and presence features to support new hybrid apps composed of Web pages, VoIP, and

IM and presence services. One could, for example, list all of one's fellow users currently logged onto the network at any particular time, or perhaps set up a call to a specified group of users at a specific time, when they're all simultaneously available to take the call.

Nortel envisions even more sophisticated collaborative messaging technology that allows co-workers to share files, Web pages, and interactive whiteboard sessions in real time. To simulate the experience of actually extending one's presence into cyberspace, users would need to know who's on the call, who's speaking, whether someone's raising his or her hand to speak, and be able to have side conversations via text chats, and so on. Such an everything-and-the-kitchen sink application will no doubt use copious amounts of DSL bandwidth and will take several years to bring to the market.

A simple logical outgrowth of many of these features is "session level control, such that within a given session you can move among various media. Worker #1 could initiate an interaction with Worker #2 by sending an instant message asking if Worker #2 has time to discuss a matter of importance. They converse over the phone for a few minutes before Worker #1 sends a graphic file of a project. The two workers then agree to meet, but before they do so they escalate the session to a video conferencing call, so each knows what the other person looks like.

Hopefully, through a blizzard of advertising or some mass-educational effort, the public will realize that the next-generation network is easily customizable and programmable.

But according to Tom Nolle of CIMI Corporation (Voorhees, NJ — 856-753-0004, www.cimicorp.com), the conundrum faced by the entire industry is that, "Buyers [from a focus group] could not, without prompting, come up with any suggestions for new apps." Our conclusion: These apps could be developed with marketing effort, but probably can't pull through spontaneous changes. Users don't plan for things that aren't offered in the carrier service set, so they're bad at conceptualizing what they might want."

Part of the problem, says Nolle, is that buyers can't assess the Return on Investment (ROI) of a given app, contrary to several vendors' claims. He goes on to state, "Our surveys, going back to the early 1980s, have consistently shown that buyers aren't a reliable guide for what they might buy if it were available, only on what they will buy of the collection of currently available services."

But also as Probe Research's Christine Hartman so succinctly puts it, "Often the people developing and funding the next generation are not representative of the general population. Many technical people thrive on resolving complex problems, so the services they tend to create are complex, when what the larger market wants is simplicity."

As Doug Allen, the senior editor of *Network* magazine, wrote in a 2002 issue:
> Often, VoIP buyers will invest in new gear for the cost savings. They're certainly aware of a migration path to converged apps, but are likely to begin using them slowly. The features alone don't necessarily motivate the sale. In contrast, most prospective buyers won't buy simply to lower costs — they want advanced functionality to justify the purchase, too. This leads to a catch-22: Which comes first, the features or the cost savings? The jury's still out, if only because so many of the benefits of enhanced features remain unproven.

The success of the next generation network voice and data services (as delivered to customers via DSL) will hinge on whether applications can be developed that actually solve business problems and are easy to use.

Beyond G.SHDSL: VDSL and Ethernet Hybrids

VoDSL will evolve beyond simply a lower cost delivery mechanism for bundled services, to become a major enabler in the transition to converged multimedia services and applications; and in the transition — from circuit-switched to next-generation networks.

This is because, as bandwidth hungry applications such as video conferencing, streaming video and Video-on-Demand rise in popularity, currently deployed forms of xDSL may not be up to the task. Both businesses and some residential customers will be looking to technologies that offer advanced broadband speeds of 5 to 10 Mbps in the near-term, and 100 Mpbs or more in the longer-term. Such higher broadband technologies can be used not just in a local loop scenario, but can have an impact in *Metropolitan Area Networks (MANs)* and perhaps in the core itself.

Even today in Multi-Tenant Units, a combination of Ethernet and Very-High Speed DSL (VHDSL or VDSL) is used to provide bandwidths far in excess of what something like, say, G.Lite is capable of. Indeed, Ethernet itself is becoming the new video backbone. Interconnection costs for Ethernet can be 10 times lower than traditional analog video routing and switching.

Ethernet may either be teamed with DSL (DSL / VoDSL in buildings, with Etherent as the backhaul link to the central office) or it may become a ubiquitous competitor of DSL. In June 2002, Atrica Inc. (Santa Clara, CA — 408-562-9400, www.atrica.com), a provider of optical Ethernet equipment for the MAN market, announced that it had teamed with FlexLight Networks (Kennesaw GA — 678-290-4801, www.flexlight-networks.com), a new company delivering a *Gigabit Passive Optical Networking (GPON)* solution for the last mile. GPON solution delivers the benefits of greater bandwidth per customer, integrated voice and data solutions on a single fiber, longer reach per CO served and a greatly reduced payback period. When used in conjunction with Atrica's Optical Ethernet System, carriers can devise cost-effective Metro Ethernets that can deliver Ethernet data as well as TDM voice solutions to customers located many miles from the metro ring.

Atrica's Optical Ethernet System combines standard 10 Gigabit Ethernet with optical switching and traffic engineering and management capabilities to give carriers up to ten times better price/performance as compared to other Metro networking solutions including SONET and next-generation SONET. Comprised of the A-2100 Optical Ethernet Edge Switch, the A-8000 Series Optical Ethernet Core Switches and Atrica's Service Platform for Ethernet Networks (ASPEN), Atrica's Optical Ethernet System delivers the carrier-class functions necessary to enable deployment of Ethernet in the Metro including sub-50 millisecond resiliency, guaranteed SLAs, TDM support and comprehensive service management.

FlexLight's state-of-the-art optical access platform serves as the backbone of the Access Layer. The company's Optimate solution enables transport of various existing, as well as emerging, protocols over complex topologies. The company's solution provides 2.5 Gbps per optical wavelength across a *Passive Optical Network (PON)* network using *Coarse-Wave Division Multiplexing (CWDM)*. The Optimate family provides Fast and Gigabit Ethernet services for data as well as T-1 / OC-3 services for TDM traffic, and efficiently carries that traffic through the access network in their native formats with Quality of Service/Class of Service (QoS/CoS) guarantees, which results in a cost-effective, highly-efficient network to provision services to and from the metro ring.

Of course, deploying a fiber ring in the core of some metropolitan areas can be vastly expensive.

DSL research is not standing still, either. Although various forms of "pure" (non-Ethernet-related) VDSLs exist that can supply bandwidths up to and exceeding 52 Mbps, VDSL, like G.SHDSL, appears to be better received in Europe than in the U.S., although some independent operating telecom companies in America have conducted trials. However, with the exception of Qwest, the other Regional Bell Operating Companies (RBOCs) have largely taken a wait-and-see attitude toward VDSL. This is partly because VDSL requires very short loop links, 4000 feet (1220 meters) or less, and sufficient bandwidth to achieve good video performance occurs only if the link is shortened to 2000 feet (610 meters) or less. That is possible if neighborhood gateways exist, thus allowing short runs to each residence or business. But since fiber (such as Ethernet fiber) is now being deployed too, telcos have been confused over whether it makes more sense to run fiber all the way to the user or to a neighborhood VDSL gateway instead.

In the meantime, the xDSLs' biggest competitor shall remain what is still the most common type of broadband delivery service, the cable modem, offered by cable TV companies such as Comcast, Cox, Time Warner, AT&T, and Charter. According to the Yankee Group, nearly 7.6 million people subscribe to cable modem service and the number is expected to grow to over 15 million by the year 2005. Such users within a neighborhood node access the Internet at speeds of up to 5 Mbps, although service can be slowed dramatically when a large number of users are online at the same time. Providing telephony services over such an architecture presents some technological challenges, though there are cable operators who are already offering state-of-the-art cable modem services which deliver cable TV and phone service (Voice over Cable, or VoCable) over the same coaxial cable.

When VDSL finally appears it may also have to compete with Wireless Fidelity (Wi-Fi) a new kind of wireless connection — essentially a particular form of well-known Wireless LAN (WLAN) technology — which allows users to connect an inexpensive ($175 or so) base station to the high-speed Internet backbone and share that connection with others in a building, small neighborhood, or office park. The connection, from base station to computer, is made through a $50 antenna module snapped into the user's computer. Networks are popping up across the country in airports, highway rest areas, cafes, and parks. Wi-Fi network speeds can reach 11 Mbps. Future wireless LAN technology will achieve approximately 50 Mbps. Security on the systems remains an issue and there have been many news reports detailing how some aspiring hackers can drive cars into parking lots of companies and, using a laptop and a Wi-Fi card/antenna, surreptitiously log onto an unsuspecting company's network. Also the 802.11 unlicensed spectrum that Wi-Fi uses could force user crowding and so the ultimate spectrum rights will have to be worked out.

VoDSL Usage Transformed by Government Legislation

The United States lags far behind other countries in terms of broadband use. An October 2001 report on the development of broadband access in Organization for Economic Cooperation and Development (OECD) countries ranks the United States fourth, behind South Korea, Canada, and Sweden, in broadband proliferation. In Canada, the percentage of high-speed Internet users is twice that of U.S. subscribers, while Korea is more than quadruple that. In 2001 the international research firm NetValue (Neuilly Sur Seine Cedex, France, www.netvalue.com) claimed that more than half of Korean households

(57.3%) use broadband, and so it is not surprising that 73.9% of the Korean Internet population used audio or video at least once a month, and over half (54.1 per cent) used a gaming protocol. Over a third of Internet users in Hong Kong used audio/video and Spain had the highest audio/video usage in Europe (33.8%). In contrast, only 29.5% of U.S. Internet users and only 23.8 % of U.K. Internet users viewed video or used audio. Broadband and audio/video trends emerging in Korea are good indicators for what the future may hold for Europe and the U.S.

Whereas European and Asian broadband industries are encouraged by robust government strategies for broadband deployment, in the U.S. there appears to be a multi-way "battle royal" among ILECs, CLECs, the FCC and Congress.

Currently, four competing technologies provide consumers with high-speed Internet access: cable modem, xDSL, fixed wireless and satellite. Only xDSL is highly regulated — when provided by ILECs — and must satisfy federal and state requirements. Moreover, service providers complain that they need tax breaks and other financial relief to help establish a broadband infrastructure, since the lower retail pricing of xDSL creates an illusion that the cost structure is more favorable than dedicated T-1s. In fact, the two are much closer than people realize with xDSL requiring a higher upfront investment in capital equipment. With dedicated T-1 access, service providers can aggregate T-1 local loops served from multiple central offices (COs) into a single point of presence (POP). At each CO, the T-1 circuits can be "groomed" into channelized T-3s that are presented to the service provider's POP. The access equipment requirements are minimal — a T-3 aggregation switch at the service provider's POP and a T-1 router or integrated access device (IAD) at each subscriber's customer premises.

In contrast, when deploying DSL services, to enjoy a similar footprint, a service provider needs to co-locate DSLAMs in multiple COs, even before there are subscribers local to that CO, or at least when the first subscriber signs up. There are an estimated 12,000 COs in the U.S. and 1100 cities with a population over 25,000 people. A typical city with 25,000 to 50,000 people may have only two COs, while large cities such as New York may have upward of 20 COs. To establish a footprint in an average sized *Metropolitan Service Area (MSA)*, a service provider may need to co-locate DSLAMs in as many as 10 or 12 COs. On top of that, there are monthly co-location facilities charges at roughly $1000 per month per CO, which leads to a "burn rate" of up to $12,000 per month, whether or not customers have been secured to pay service subscription fees.

Controversy continues to stem from the U.S. Telecommunications Act of 1996. For example, the Act's Section 271 prohibits the former Regional Bell Operating

Companies (RBOCs) from offering interLATA services, which include both long-distance telephony and Internet transmission services, in those states where they also provide local telephone service, until they have satisfied certain market-opening requirements. Therefore, while RBOCs may operate dial-up and broadband Internet Service Providers (ISPs), customers can only connect to the Internet through a regional or national ISP operated by another company. The situation is further complicated by the fact that almost all Internet communications occur across state lines and may or may not include voice transmissions.

Also the U.S. Telecommunications Act of 1996 required the ILECs to "unbundle" their circuits to CLECs ("unbundling" refers to giving competitors the opportunities to use an incumbent's phone lines and hardware to deliver services). The unbundling phenomenon led to a new class of CLECs offering data via xDSL. CLEC investment in DSL led to a roughly equivalent expenditure by ILECs, and thus appears to have driven the overall rate of DSL deployment, as has the recent appearance of Voice-over-DSL (VoDSL) technology.

When incumbent providers offer DSL, the service comes under the historical purview of telecommunications regulation. When the incumbent providers sell an enhanced, non-regulated service over a basic service, the incumbent must provide the basic service to others too. DSL is seen as a basic service, thus the ILECs must unbundle their service at two levels. They must unbundle their physical capabilities so competitive DSL providers can implement DSL, and they must unbundle their DSL service so that competitive ISPs can sell Internet access over the incumbents' DSL service.

Although the Telecommunications Act of 1996 mandated that incumbents offer CLECs access to unbundled network elements at reasonable rates, the ILECs continue to control over 90% of local market revenues and customers, and they remain subject to federal and state price regulation. The diminutive CLECs are not subject to such price regulation, but most of these went out of business during the 2000-2002 time frame.

At the same time, horizontal consolidation has occurred among telephone companies, plus vertical integration of such companies: Qwest acquired US West; NYNEX merged with Bell Atlantic, which then merged with GTE to become Verizon; SBC acquired Pacific Telesis and Ameritech; MCI merged with WorldCom, which also merged with UUNet; and AT&T acquired TCI and other cable interests.

Thus, although the Telecommunications Act of 1996 eliminated legal barriers to entry in those states where barriers had existed, economic and technical barriers are coming down much more slowly if at all. Despite this, competitors have made some inroads among business customers in urban markets, and issues posed by open access in broadband have prompted some FCC initiatives.

The Telecommunications Act of 1996 calls for the FCC and the states to encourage the deployment of advanced technologies for all Americans on a reasonable and timely basis. But what qualifies as "advanced," and who represents "all," and what are "reasonable" and "timely," are all a matter of debate. Indeed, the Act calls for access to advanced telecommunications and information services in rural and in high-cost areas to be "reasonably comparable" to that in urban areas in terms of price and quality. The wording appears to join unregulated information services with regulated telecommunications services, and what that implies for future policy remains yet to be seen.

At the time this book went to press, a series of U.S. legislative initiates involving broadband began to appear in Congress. For example, Senators Lieberman and McCain are intending to weigh in on the debate with their own individual bills to encourage broadband deployment. In 2002 Sen. Lieberman issued a 60-page white paper on the need for a national plan for broadband deployment and is introducing the National Broadband Strategy Act of 2002, a bill that highlights the need for a coherent and comprehensive national strategy for providing widespread availability of broadband and for motivating research and advances in broadband applications and content. Because broadband implementation has been piecemeal, and stalled in significant part because numerous government agencies have failed to act quickly in deciding a wide range of broadband issues now pending before them, the bill calls upon the Administration to recommend a coherent, cross-agency national broadband strategy in a series of key government policy areas.

Parallel to that, and focusing on how the U.S. will get to truly advanced broadband speeds (in the range of 10 Mbps and 100 Mbps), Senator Lieberman plans to introduce a series of substantive pieces of legislation addressing four key elements integral to a national strategy for advanced broadband deployment. The key elements are:

1. FCC regulatory framework: Direct the FCC to explore all of the broadband deployment and delivery technology options to enable us to reach advanced broadband speeds. Retaining technological neutrality, the FCC will be asked to develop the regulatory framework to enable and implement a plan to deploy this advanced Internet capability.

2. Tax credits: Establish tax credits and incentives for a range of advanced broadband deployment and broadband utilization efforts. These could include credits for infrastructure deployment, equipment implementation, employee utilization, installation in atypical settings, and innovative applications.

3. Advanced infrastructure R&D: Ensure that fundamental R&D issues are tackled in a coordinated manner to overcome the scientific and technological barriers to advanced widespread broadband deployment. The U.S. has already established successful interagency and interdisciplinary initiatives under the National Information Technology Research & Development Program to advance critical IT technologies. We must leverage our existing expertise in these programs to resolve fundamental obstacles to effective broadband deployment and hasten the next generation of technologies. A cooperative R&D program, including government, industry and universities, will be critical to advanced broadband.

4. Application R&D and deployment: Require federal agencies to undertake R&D and promote the development and availability of major applications in areas where government plays a central role, including e-education, e-medicine, e-government, e-science and homeland security. This could stimulate demand for broadband and promote bridging of the digital divide consistent with the missions of government agencies. And the government should lead by example in moving to expand opportunities for broadband-based e-commerce in federal procurement, bidding, and contracting.

Certainly the benefits that will arise from the development and implementation of a National Broadband Strategy fully justify making it a high priority. Indeed, Sen. Lieberman's plan compares it in importance to President Kennedy's 1961 challenge to America to land a man on the moon before the end of the decade. As Lieberman's report states: "Like putting a person on the moon, the deployment of broadband to Americans should be a national mission and a national priority. It is the ultimate economic stimulus, the next superhighway system for our next generation of leaders, our children, and grandchildren. With the right policies and leadership, industry and policy makers can work together to accomplish this imperative. Otherwise, millions of Americans will miss out on the personal growth, higher wage jobs, knowledge, and quality of life opportunities that broadband can deliver."

Another controversial debate in the U.S. Congress that centers on the telephone delivery of broadband involves the Tauzin-Dingell Internet Freedom and Broadband Deployment Act (H.R. 1542), which passed the House of Representatives on February 27, 2002.

The bill asks that the Regional Bell Operating Companies (RBOCs) provide high-speed data services (defined as 384 Kbps in at least one direction) to their customers over the next five years by upgrading their networks or providing alternative broadband services from another source if they do not. They must provide service only within 15,000 feet of their central offices and can affiliate with another provider to serve customers beyond that distance. In return, they would receive the ability to block entry to their Internet facilities.

The Tauzin-Dingell bill attempts to provide for greater competition by relaxing operating restrictions on ILECs. The Bell companies argue that their delivery of broadband is regulated, while cable delivery is not. In addition, they claim that there is no real incentive for them to build broadband infrastructure while regulations require them to open that infrastructure to their competitors.

On the other hand, long-distance carriers and consumer organizations oppose the bill, raising questions regarding the potential harm that Tauzin-Dingell could do by actually reducing competition within the telephony sector, and the possible weakening of consumer protection aspects of current U.S. telecommunications law. They further believe that the Tauzin-Dingell bill would create a duopoly between cable and the Bells in the delivery of broadband services.

During the spring and summer of 2002, the broadband debate began to move to the U.S. Senate, when the Broadband Regulatory Parity Act of 2002 (S.2430) was introduced by Senators Breaux and Nickles in May 2002. The proposed Act seeks to provide parity in regulatory treatment of broadband services and of broadband access services providers, among other things. It attempts to provide a Tauzin-Dingell compromise by separating the voice and data components in order to maintain regulation of voice services while freeing data services of the regulations imposed by the Telecommunications Act of 1996. The main thrust of the legislation calls for deregulation of DSL technology at both the federal and state levels, since out of the four identified technologies it is the only one subject to strict regulatory requirements. The bill requires the FCC to determine which regulatory requirements — if any — should be retained and which should be eliminated within 120 days of the bill's enactment.

The "Baby Bells" support the Breaux and Nickles legislation, since they view it as a way of providing the incentives they say they need. Consumer choice of ISP would be preserved when using DSL provided by telephone companies by requiring them to provide all ISPs with access to their networks. In addition, the legislation retains existing requirements that incumbent telephone companies provide competitors with access to their voice networks.

Senator Hollings, however, strongly opposes the Breaux bill on the grounds that it does not create parity within telephony and would effectively squash competitive carriers. Senator Hollings offered his own legislation, the Broadband Telecommunications Act of 2002 (S.2448) that focuses on deployment in rural and underserved areas (as well as schools and libraries) through loan and grant programs. Money for these programs will come from the establishment of a trust fund, to be appropriated by Congress from FY 2003 to 2007, and administered by the National Telecommunications and Information Administration (NTIA).

As Hollings has said: "I most certainly support the deployment of broadband nationwide and it can be accomplished without compromising competition. In fact, I believe it is through a combination of policies such as — competition, loan programs, tax credits, consumer privacy protections, and addressing the 'demand' problem — that broadband can be achieved. There is no silver bullet here, and an approach that destroys competition will undoubtedly undermine the deployment of broadband and other innovations. Such an outcome would set communications policy back for decades."

Telecommunication entities in the United States are regulated at the federal level by the FCC an independent government agency. The FCC's mission is to encourage competition and to protect the public interest. Congress — via legislation — directs the FCC to develop and to implement policy concerning interstate and international communications by radio, television, wire, wireless, satellite, and cable.

During 2002, the FCC had five proceedings in progress to examine the broadband issue, all promising relief in some way, none of which appeared to be leading to any definitive (or tentative) conclusions.

In Summary

In the not-too-distant future, a home or business owner will walk to a store, buy an IAD, sign up for VoDSL service via a web page, plug the IAD into a single copper phone line, then plug all premises telephony devices into the IAD. Within moments, after a remote provisioning and self-configuration process, a vast array of inexpensive but powerfully useful voice and data services will be available to the customer.

This is what VoDSL is ultimately all about. It will transform the way we do business, making it more interesting, yet more economical. It represents not only the future of the "last mile" and a way of delivering high-bandwidth bundled services, but it will also help facilitate the world's transition to the packet-based "next generation" networks of tomorrow.

Although VoDSL suffered somewhat during the economic doldrums of the early 2000s, it will rise like the proverbial phoenix to emerge as the small and medium businessman's best friend.

APPENDIX A

The Open Systems Interconnection (OSI) Model

The OSI is a standard description or reference model for how information from one point in a network (such as a voice software application running in a computer, an IAD or an IP phone) is transmitted through a telecommunications network to another endpoint (another software application in another computer).

OSI was the first worldwide effort to standardize the entire field of computer communications, or data networking, in the form of a networking framework for implementing hardware and protocols. The OSI model was developed by the International Organization for Standardization (ISO) and was originally supposed to be a detailed specification of computer internetworking interfaces formulated by representatives of major computer and telecommunication companies during committee meetings held beginning in 1983. Instead, the committee decided to establish a common reference model for which others could develop interfaces, which in turn could become standards.

The OSI model was completed in 1984, and it is still considered the chief architectural model for intercomputer communications. OSI continues to be administered by the ISO, so any new standard that seeks validation as an ISO standard for computer communications must be compatible with the OSI reference model. The model also can be used to guide product developers so that their products will consistently interoperate with other communications products. Finally, since the OSI reference model is a common point of reference for categorizing and describing network devices, protocols, and issues, it has value as a recognized, single view of communications that gives everyone a common reference point for education and discussion about communications.

The Open Systems Interconnection Model

The OSI reference model is purely a conceptual model, it does not do any "communicating" itself. The OSI model is composed of an architecture or framework of seven layers, each specifying particular network functions. Everything from a cable to a web browser fits into this layered framework.

The tasks that move information between networked computers or communicating devices are divided into seven task groups, with each task or group of tasks then assigned to the appropriate OSI layer. The layers are in two categories: The upper layers, sometimes called the *Application Layers*, and the lower layers, or *Data Transport Layers*. The upper three or four layers are used whenever a message passes from or to a user and are generally implemented only in software. The lower three layers (up to the network layer) handle data transport issues and are used when messages travel through the host computer or device. The bottom two layers, the *Physical Layer* and *Data Link Layer*, are implemented in hardware and software, though with IP networks only the bottom layer (the Physical Layer) need actually be hardware, since it is closest to the physical network medium and is responsible for actually putting information on it.

Each layer has its own function and is basically self-contained, so that the tasks assigned to each layer can be implemented independently. Data going to and from the network is passed layer to layer. Each layer is able to communicate with the layer immediately above it and the layer immediately below it. This way, each layer is written as an efficient, streamlined software component. When two computers or other devices communicate on a network, the software at each layer on one device assumes it is communicating with the same layer on the other device. For example, the Transport Layer of one computer communicates with the Transport Layer on the other computer. The Transport Layer on the first computer doesn't care how the communication actually passes through the lower layers of the first computer, across the physical media, and then up through the lower layers of the second computer — whatever layers are working below function as a transparent, background process. This can occur because, when a layer receives a packet of information, it checks the destination address, and if its own address is not there, it passes the packet to the next layer. One layer's functionality can thus be updated without affecting adjacent layers.

Although manufacturers and telecom / datacom product developers do not always strictly adhere to OSI in terms of keeping related functions together in a well-defined layer, practically all communications products are described in relation to the OSI model. Different network devices are designed to operate at certain protocol levels, and each network protocol can be mapped to the OSI reference model.

The OSI model assumes that each communicating user is at a computer or device equipped with hardware and software adhering to these seven functional layers. When one person sends a message to another, the data at the sender's end will pass down through each layer in that device to the bottom layer, then over the channel, and at the other end, when the message arrives, data will flow back up through the layer hierarchy in the receiving device and through the application to the end user.

The seven layers are (from the top, downward):

Layer 7. The Application Layer: This highest of layers in the OSI architecture ultimately leads the outside world (e.g. the user). However, the application layer is not itself the user application; it only provides the system independent interface to the actual data communications application and its own user interface (though applications may indeed perform Application Layer functions).

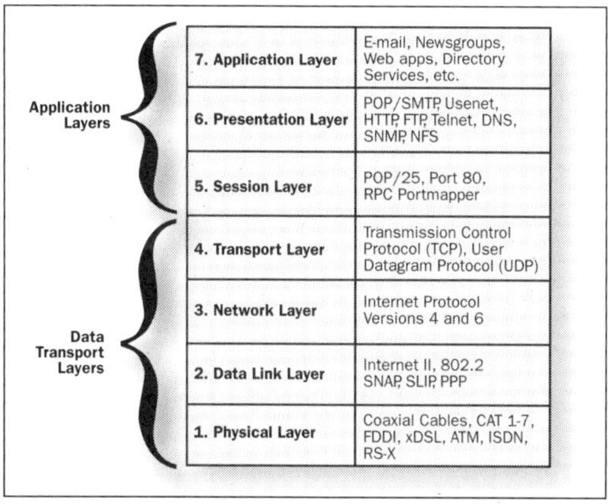

Although the highest and seemingly the most abstract layer, the application layer actually consists of a complex set of standards and protocols. Application Layer Protocols are classified into Common Application Specific Elements (CASE) and Specific Application Specific Elements (SASE).

Layer 7 is where communication partners are identified, quality of service is identified, and user authentication and privacy are determined. This layer contains functions for applications services, such as file transfer, database processing, e-mail, remote file access and virtual terminals.

Fig. A.1 OSI Model.

	Layer	Examples
Application Layers	7. Application Layer	E-mail, Newsgroups, Web apps, Directory Services, etc.
	6. Presentation Layer	POP/SMTP, Usenet, HTTP, FTP, Telnet, DNS, SNMP, NFS
	5. Session Layer	POP/25, Port 80, RPC Portmapper
Data Transport Layers	4. Transport Layer	Transmission Control Protocol (TCP), User Datagram Protocol (UDP)
	3. Network Layer	Internet Protocol Versions 4 and 6
	2. Data Link Layer	Internet II, 802.2 SNAP, SLIP, PPP
	1. Physical Layer	Coaxial Cables, CAT 1-7, FDDI, xDSL, ATM, ISDN, RS-X

Layer 6. The Presentation Layer: Sometimes called the *Syntax Layer*, this layer is usually that part of an operating system that is concerned with the representation (syntax) of messages' data associated with an application during the transfer between two application processes. Applications routing data are simply routing binary streams, which have no meaning without a definition as to how it is to be formatted. A raw binary representation alone isn't good enough, since two computers communicating with each other may have totally different configurations: One could be using a 16 bit word size, the other 32; a PC could be using an ASCII character set, while an IBM mainframe could be using EBCDIC, etc.

The presentation layer must therefore do its part to provide transparent communications services by masking the differences of varying data formats between dissimilar systems and, in general, converting incoming and outgoing data from one presentation format to another (for example, from a text stream into a popup window).

The presentation layer is also concerned with methods of data encryption and data security, and compression algorithms that may have also changed the data format.

Here's how the presentation layer works: The Presentation Layer in one computer will attempt to establish a "transfer syntax" with the other Presentation Layer in the other computer by negotiating a common syntax that both applications can use. Failure to do so results in a non-connection. A widely used standard for the Presentation Layer is ISO 8824 and 8825

Layer 5. The Session Layer: This is a sort of interface between the hardware (the bottom four layers) and the software (the top two layers). This layer is the "dialog manager." The Transport Layer handles how a data stream is directed, but it is the Session Layer that says when data can flow. It negotiates and creates a connection between two Presentation Layers when requested, then controls the data flow. More technically, it establishes, maintains and otherwise controls the use of the basic communications channels provided by the Transport Layer, by setting up, coordinating, and terminating conversations, exchanges, and dialogs between the applications at each end. It handles session and connection coordination. A session establishes the connection between communicating devices, providing synchronization, security authentication, and network naming. Protocols that function at this layer include RPC, XNS and LDAP. This layer is often combined with the Transport Layer.

Layer 4. The Transport Layer: This layer defines the rules for information exchange and manages the end-to-end flow control and delivery (for example, determining whether all packets have arrived) and error-checking / recovery within and between networks. It ensures complete data transfer.

This is the highest of the lower layer protocols in the OSI protocol stack. It provides the means to establish, maintain, and release transport connections on behalf of session entities. It provides a connection-oriented or connectionless service. In a connection-oriented session, a circuit is established through which packets flow to the destination (most protocols at this layer are connection-oriented). Error and flow control are dealt with at this layer (ACK). And most gateway functions are found at this level.

The Transport Layer is the first of what's called the *"peer-to-peer"* layers, which means that once the lower layers are implemented, a Transport Layer can transparently communicate directly with the Transport Layer of another data entity. It provides an idealized full duplex bit pipe to the upper layers in which the binary stream sent from one end, makes it, in order, to the other end. The result is that the upper layers get an idealized bit pipe and need only be concerned with what the data is, not how it arrived. Some protocols that reside at this level are TCP, UDP, SPX and TFTP.

Layer 3. The Network Layer: This layer handles how data is routed from the source computer to the destination computer, sending it in the right direction over the correct intermediate nodes to the proper destination, and receiving incoming transmissions at the packet level. The Network Layer does routing and forwarding within and between individual networks and can provide a *Connection-Oriented Network Service (CONS)* or a *ConnectionLess Network Service (CLNS)*. No matter what the route of the actual connection, the data arrives at the destination as if the two data entities were directly connected. It is at this layer that network traffic problems such as *Quality of Service (QoS)* and *Type of Service TOS)* are managed. Devices operating at this layer include routers, brouters (a combination bridge and router), and Layer-3 switches. Protocols working in this layer include ARP, DLC, ICMP, IP, IPX, NetBEUI and RARP. Routing Protocols at this level include BGP, EGP, EIGRP, IGMP, IGRP, OSPF, and RIP.

Layer 2. The Data Link Layer: This layer concerns itself with the procedures and protocols for operating the communications channels (transmission protocol knowledge and management), and provides error detection / correction and synchronization for the Physical Level. Devices at this layer include switches, bridges and intelligent hubs. The Data Link layer consists of two sublayers: The *Logical Link Control (LLC)* and the *Media Access Control (MAC)*. The LLC sublayer is in charge of establishing and maintaining links between communicating devices. Ethernet and Token Ring protocols operate in the Data Link Layer, as do protocols such as ATM, CSMA/CD, FDDI, PPP and SLIP. The MAC sublayer is responsible for framing data. Computers or other communicating devices identify themselves via MAC addresses and Network Interface Cards (NICs), while the Data Link Layer at the same time organizes information from the higher layers into "blocks of bits," called frames (such as the "Synchronous Data Link Control" frame), for orderly transfer and error control. Sometimes this involves bit-stuffing to pad out strings of 1's.

Layer 1. The Physical Layer: This bottom layer is responsible for activating, maintaining, and deactivating the physical connection for bit transfers between "Data Link Entities." It provides the hardware means of sending and receiving a bit stream (a "bit pipe") through the network at the most basic electrical, mechanical, functional and procedural levels. An example of a Physical Layer ISO standard is the RS-232 interface. Devices at this level include cables, connectors, hubs, multiplexers, repeaters, receivers, switches, and other hardware.

A simple mnemonic phrase exists to remember the correct order of the layers:

Application — **A**ll
Presentation — **P**eople
Session — **S**eem
Transport — **T**o
Network — **N**eed
Data Link — **D**ata
Physical — **P**rocessing

APPENDIX B

A Plethora of Protocols

A protocol is an agreed-upon set of rules governing the format of messages that are exchanged between computers or other networked devices. "Agreed-upon" means that the members of some standards body — such as the International Telecommunication Union (ITU) or the Internet Engineering Task Force (IETF) — have formed working groups and committees and have spent a long time formulating a protocol, eventually ratifying it as a "standard." Some "protocols" such as TCP/IP aren't just one protocol, but several. That's why you'll often hear it referred to as a protocol suite, of which TCP and IP are the two principal protocols.

It may come as a surprise that something as well-established as high-bandwidth (45 Mbps) T-3 transmission technology lacks a protocol standard and so DS-3 signals are proprietary and differ among vendors. Since these DS-3 methodologies are not compatible with each other, users must ensure that a particular vendor's equipment exists at both ends of a link, or else the varying spans must be connected with something such as an optical SONET link, which is itself a standard and, therefore, can interconnect equipment from various vendors.

There are so many protocols in the telecommunications and networking industries that it is impossible to list them all here. Early packet (datagram) networks were based on X.25 and other statistical multiplexing protocols running over primitive analog modems or low-bandwidth digital circuits. They were subject to considerable processing overhead and the packets of data they transferred were often delayed because of network idiosyncrasies. In the 1990s, Frame Relay and ATM protocols allowed for high bandwidth networks to carry packet data with great accuracy and with few incidents of data retransmission. These days, the Internet Protocol (IP) has become the most popular of all protocols, primarily because of its use in the most popular of all networks, the Internet.

Different protocols have different sizes and capabilities in the network. In addition to well-known protocols for transport and routing, there are also protocols for interoperability (H.323), fixing line echo impairments (G.131), softswitch component control (MGCP, Megaco/H.248), voice call control (SIP), store-and-forward fax (T.37), real-time fax (T.38), POTS-based videoconferencing via modems (H.324), and maintaining the quality of service of packet networks (DiffServ, MPLS). There are even protocols such as the Internet Group Management Protocol (IGMP) and Distance Vector Multicast Routing Protocol (DVMRP) that enable a virtual network to be layered on top of the Internet so that multicast sessions can take place.

No single protocol or protocol suite can do everything, though some try, such as ATM. ATM can do LAN emulation and it supports multiprotocol over ATM services, and it has a signaling protocol in its suite that can, when needed, automatically set up channels called Switched Virtual Connections (SVCs). Such comprehensive features hints at the fact that ATM is a complex series of related protocols — so complex in fact, that it has its own layered system, just like the OSI Model in Appendix A. ATM deployment thus demands a very complex and intensive integration of software and the protocol infrastructure.

In many cases various different kinds of protocols or protocol suites must be used in concert, each with their own dedicated purposes, which would lead to a network disaster if it were not for the fact that the interactions between the protocols can be clearly defined. Appendix A of this book explains how the International Standards

Vovida.org For Open Source Protocols

Aside from the vendors and standards bodies that devise new protocols, there is Vovida.org (www.vovida.org), a communications community site sponsored by Cisco Systems, Inc., which is dedicated to providing a forum for open source software used in datacom and telecom environments. It is said to be the only site to provide a Linux- and Solaris-based open sourced communications system (VOCAL), developmental protocols including MGCP, RTP, RTSP, SIP, COPS, RADIUS, OpenOSP and TRIP, and it has links to other Open Source information sites.

Organization (ISO) essentially created a kind of "meta-protocol" that defines communications between protocols, which takes the form of a seven-layered structure called OSI that keeps each protocol in its place. OSI defines a protocol as "a set of rules and formats (semantic and syntactic), which determines the communication behavior of N-entities in the performance of N-functions." In OSI nomenclature "N" represents a layer, and an entity is a service component of a layer. The data sent between layers is called a Service Data Unit (SDU), and OSI defines the analogous data transfer between two communicating devices as a Protocol Data Unit (PDU). The information flow is controlled by a set of actions that define the state machine for the protocol, which OSI defines as Protocol Control Information (PCI).

In Chapter 5 we saw how IP packets, which themselves have no Quality of Service (QoS) standards for real-time delivery demanded by voice applications, are often broken down into smaller sections and packaged into ATM cells before they are sent over an xDSL line, since ATM does have rigorous QoS standards. This is called "encapsulation" or the packaging of one protocol and it's payload inside another. For example, the document RFC 1483 defines an encapsulation method for supporting multi-protocol encapsulation over ATM AAL5. The tunneling technique used by Virtual Private Networks (VPNs) is essentially a form of encapsulation, wherein the PDU is encapsulated into a new PDU at the entry to the network boundary and "decapsulated" upon exiting the network, with the packet payloads extracted and reconstituted at the destination.

Under the OSI model, *segmentation* is the process of breaking an N-service data unit (N-SDU) into several N-protocol data units (N-PDUs), while *reassembly* performs the reverse operation, bringing together several N-PDUs into one N-SDU. At a more abstract level, *blocking* is the aggregation of several SDUs (which might be from different services) into a larger PDU within the layer in which the SDUs originated, while *unblocking* involves the dicing up of a PDU into several SDUs in the same layer.

Taking things to yet a higher level of abstraction, *concatenation* is the process of one layer combining several N-PDUs from the next higher layer into one SDU, which is similar to blocking except that it occurs across a layer boundary. The reverse of concatenation is *separation*, in which a layer breaks a single SDU into several PDUs for the next higher layer — this is similar to unblocking except that it occurs across a layer boundary.

THE "G" SERIES

Some "protocols" are actually compression/decompression algorithms (codecs). In particular, certain codecs are designed specifically to compress and uncompress human speech. Since a stable set of sounds called phonemes occur in human speech, these can be "tokenized," resulting in great compression of voice packets and allowing for many separate voice channels to be established over a packet network. These voice compression protocols are known as the *G Series* protocols, which come under the provenance of the ITU.

Ironically, the earliest of these, G.711 (used in VoATM-based VoDSL, T-1 lines and even some IP telephony installations) doesn't really do "compression" at all. G.711 is a Pulse Code Modulation (PCM) scheme operating at a 8 kHz sample rate, with 8 bits per sample. Since a signal must be sampled at twice its highest frequency (as dictated by the Nyquist theorem), G.711 can thus encode frequencies between 0 and 4 kHz (there is more than enough bandwidth, since analog channels in the U.S. are clipped at 3.3 kHz and European analog systems only reach to 3.4 kHz). The algorithm allows for the transmission and reception of A-law and mu-law voice by converting linear PCM input signals (13 bits for the international A-law standard and 14 bits for mu-Law) sampled at an 8 kHz sampling rate into an 8-bit compressed floating-point PCM representation. The actual technique of converting between linear PCM and G.711 PCM is known as "companding" (compressing/ expanding). The analog-speech waveform, once having been encoded as binary words, is then transmitted serially, at digital bit rates of 48, 56, or 64 Kbps. ISDN channels and digital phone sets on digital PBXs use G.711. Support for this algorithm is required for ITU compliant videoconferencing (the H.320 / H.323 standard).

Although considered to be a popular "codec" or "vocoder," G.711 excessive bandwidth output at 64 Kbps is pretty much uncompressed, and is thus used as a reference against which the speech quality of voice compression algorithms with high compression and lower bit rates are measured.

VoATM-based VoDSL systems typically utilize either the standard 64 Kbps version of G.711 or G.726 (which simply takes the 64 Kbps G.711 PCM and resamples it at 32 Kbps). G.726 can also compress a 64 Kbps PCM channel to a 40, 24 or 16 Kbps channel using what's called an "adaptive differential" form of PCM.

When IP-centric VoDSL systems are used (or when the VoDSL system makes a Voice over IP (VoIP) call over an extensive IP network such as the Internet), more channels can be freed up, and better efficiency can be achieved on the Wide Area Network (WAN), by employing G Series vocoder algorithms with higher compression rates than G.711.

Interestingly, use of G.711 underwent a bit of a revival during the economic crunch years of 2000-2002. One reason is that there was a glut of bandwidth and so using high levels of compression were not as important as they used to be. Instead, people wanted toll-quality audio. Also the cost per port on IP telephony systems had declined so much that the intellectual property costs for G.723 and G.729a were starting to appear significant. Fortunately, G.711 doesn't have any intellectual property costs, so it's not just an easier vocoder to implement and a high density vocoder, but it's also a cheaper vocoder (technically speaking, of course, it actually isn't a vocoder at all). Moreover, many high compression coders can't handle compressing and transmitting fax signals. When they run into problems sending faxes, the back-off default codec is inevitably G.711.

However, when network congestion rises and voice packets start to get lost, G.711 degrades quickly — at a 2% packet loss some listeners actually find G.711 "less than good." G.729, on the other hand, has a packet loss signal correction scheme that compensates for lost packets.

Other G Series codecs include the following:

G.721 — An old ADPCM speech coding standard at 32 Kbps, but it has now been replaced by G.726.

G.722 — ITU-T Recommendation: This algorithm produces digital audio through a wideband speech coder operating at 64, 56 and 48 Kbps. The sampling rate is 16 kHz. All the other ITU-T speech coding standards use a sampling rate of 8 kHz.

G.723 and G.723.1 — These are ITU-T standardized dual rate speech coders for multimedia communications transmitting at either 5.3 and 6.3 Kbps. G.723 was designed for sending compressed digital audio over ordinary analog lines as part of the H.324 videoconferencing standard.

Once there was a different G.723 coder (an old ADPCM speech coder than could compress 64 Kbps speech to 40, 32 and 24 Kbps), but it was ultimately folded into G.726. To avoid confusion, the ITU changed the name of the currently adopted G.723 coder to G.723.1. Thus, there is no real distinction between G.723.1 and G.723 with reference to the currently adopted G.723.1 standard.

G.723.1 encodes speech or other audio signals in frames using linear predictive analysis-by-synthesis coding. Speech signal input to the G.723.1 coder are 16-bit linear PCM samples, sampled at 8 kHz. G.723.1 encodes 240 sample frames (30 ms) of 16-bit linear PCM data into either twelve 16-bit code words or twenty-four 8-bit code words for the 6.3 Kbps rate and ten 16-bit code words or twenty 8-bit code words for the 5.3 kbps rate. Total algorithmic delay is 37.5 ms (which includes 7.5 ms of look ahead). The 6.3 Kbps rate uses as its excitation signal Multi-Pulse Maximum Likelihood Quantization (MP-MLQ) "code book" search, while the 6.3 Kbps version is based upon the Algebraic-code-excited Linear-Prediction (ACELP) system. It's possible to switch the G.723.1 frame stream between the two rates on the fly, at any 30 ms frame boundary. There's also an option for variable rate operation using discontinuous transmission and Voice Activity Detection (VAD), which removes the bandwidth-wasting silent pauses between words prior to transmission, filling in the non-speech intervals later with low-level white noise.

The result of the algorithm is approximately 4 kHz of near-toll quality speech bandwidth under clean channel conditions. Indeed, the quality of G.723.1 measured by the standard Mean Opinion Score (MOS), normally used to rate speech codec quality, is 3.98. Since a rating of 4.0 is considered toll quality, this means that the G.723.1 vocoder has 99.5% the audio quality of analog telephony.

Since it provides high quality speech voice compression and decompression over a narrow digital band, G.723.1 is used extensively in IP telephony and it is the default voice coder for H.323 and is actually required for H.323 compliance. Relative to the G.729 / G.729A coders, the G.723 speech coders pass DTMF "touch tones" through with less distortion, though music and sound effects suffer a bit.

Four companies collaborated closely towards the development of the G.723.1 speech technology: The Audiocodes Ltd., the DSP Group, France Telecom, and the University of Sherbrooke.

G.726 — This defines an ADPCM voice coder operating at 40, 32, 24, and 16 Kbps. This has replaced the old ADPCM standards of G.721 and G.723.

G.727 — Defines an embedded ADPCM with voice encoded at 40, 32, 24 and 16 Kbps.

G.728 — ITU-T Recommendation: Encoding/decoding of speech at 16 Kbps using low-delay code excited linear predictive (LD-CELP) methods. Like G.722, it is optional for H.320 compliance.

G.729, G.729 Annex A (G.729A) and G.729 Annex B (G.729B) — G.729 is a Conjugate-Structure Algebraic-Code-Excited Linear Prediction (CS-ACELP) speech compression algorithm (an audio codec, or "vocoder") approved by ITU-T that supports 3.4 kHz speech at a mere 8 Kbps. G.729A is a reduced complexity version of the G.729 coder that processes signals with 10 ms frames and has a 5 ms look-ahead resulting in a total algorithmic delay of 15 ms. The input/output of the G.729A algorithm is 16 bit linear PCM samples converted from/to the 8 Kbps compressed data stream.

G.729A is quickly becoming a widely accepted standard for voice compression. Being only an 8 Kbps codec, it offers opportunities for significant increases in bandwidth utilization to existing telephony and wireless applications and codec offers functionality such as toll quality, low complexity (MIPS, RAM, ROM) and low delay (10 ms).

G.729A was developed for use in multimedia and it is also used in the analog voice modem V.70 and is the ITU-T standard for Digital Simultaneous Voice and Data (DSVD) modems. DSVD allows the simultaneous transmission of data and digitally-encoded voice signals over a single dial-up analog phone line. DSVD modems use V.34 modulation (up to 33.6 Kbps), but may also use V.32 bis modulation (14,400 Kbps). Indeed, the DSVD voice/data multiplexing scheme is an extension of the V.42 error correction protocol widely used in modems today. DSVD also specifies fallbacks that enable DSVD modems to communicate with standard data modems (i.e. V.34, V.32 bis, V.32 and V.22).

G.729B provides for the addition of a Voice Activity Detector (VAD) and a Comfort Noise Generator (CNG) to G.729 which reduces the transmission rate during periods of silence that occur during speech. Digital Signal Processor (DSP) software exists that incorporates all of this functionality into one flexible G.729AB package that is a complete implementation of the G.729 Annex A and Annex B specifications. Voice services utilizing G.729A/B would require approximately 24 Kbps for three voice channels as opposed to 96 Kbps with ADPCM. Even with six G.729AB voice channels, there would still be a considerable amount of spare bandwidth that could be allocated for data traffic.

TRANSCODING

Just because voice is compressed doesn't mean that it isn't any good — with a managed, IP backbone, where the network operator has control over packet loss, latency, and jitter, existing codecs are more than capable of coding and decoding voice very quickly — on the order of 10 to 15 milliseconds, which, so long as there are no other network problems, allows for PSTN-like toll-grade quality voice calls. Indeed, the variants of G.729 can produce what many listeners perceive to be CD quality, which is even higher than toll-grade quality.

However, as different G Series codecs are encapsulated in a QoS preserving protocol such as the Real-time Transport Protocol (RTP) and sent over the network to various kinds of endpoint devices, often DSP resources must be brought to bear to "transcode" or convert between different codec types in an RTP stream in case the codecs don't match. For example, a compressed G.729 media RTP stream across the Internet might need to terminate on a device that supports only G.711. The transcoding DSP resource would terminate the G.729 media stream and convert it to G.711. Since conferencing systems generally use G.711 exclusively for voice, if a would-be conferencing participant is using an IP phone equipped with only a low-bit-rate codec, the call must be transcoded to G.711 before the user can join in the conference.

The old, high-bandwidth G.711 codec keeps showing up as a standard for transcoders because of G.711's lack of compression and multiple coding ability. As complex multiservice networks continue to expand in size, multiple codings are often necessary when transporting voice signals between PSTN and VoIP networks (or between either one and a wireless network). For example, voice mail systems use high compression vocoders, and so receiving or forwarding voice messages from such a technology can result in multiple codings.

Say That Again? Quality of Service (QoS) Protocols

Ironically, whereas currently ATM dominates the backbone, IP has begun to slowly encroach onto ATM's "turf." ATM has been displaced somewhat from the core to the edge of the network and the local loop (where it's currently found working with xDSL to guarantee some sort of QoS for voice and data calls made over xDSL lines). This means that at the current time, when voice packets comprising a customer's call leaves the secure realm of VoATM over xDSL within an access network and reaches the network core, it may not be possible to guarantee QoS — particularly if the call must traverse a large IP network — until the packets reach the local loop at the other end of the call.

As the PSTN becomes integrated with — and may ultimately be replaced by — next generation IP-centric networks, the crystal-clear QoS enjoyed by generations of phone users no longer remains a simple affair. Without a dedicated channel nailed up for their exclusive benefit, real-time communications such as voice and video have a tough time on packet networks, which are merely "best effort" services.

If you have too many real-time communications going on in a conventional Time Division Multiplexed (TDM) network such as a regular analog local loop leading from your home or business to an ATM backbone, any attempts to add a call exceeding the total maximum possible bandwidth results in some sort of busy signal or SS7 announcement ("all circuits are busy"). A busy, overburdened vanilla IP "best effort" network, however, will actually let you make that extra call, but your packets will get delayed (or dropped) and the voice quality of all phone calls in the network will start to degrade.

And even if a service is running perfectly at exactly maximum capacity, unpredictable aggregation or surges of packets caused by user activity can result in some users getting poor or non-existent service, unless "traffic shaping" is done at borders. (Traffic shaping is simply using whatever technique is necessary to force your data traffic to conform to a certain specified behavior, such as: "This network can tolerate 100 Mbps of data for a maximum burst of a second seconds and its average over any 10 second interval will be no more than 50 Mbps.").

And it doesn't help matters that IP traffic is exploding in popularity. Esmeralda Swartz, Director of Strategic Marketing at Avici Systems (North Billerica, MA — 978-

964-2000, www.avici.com) tells the author that "Figures taken from the FCC, AT&T and IDC indicate that IP back-haul traffic is growing at the rate of 110 percent compounded annually. Today, IP traffic is about 30 percent of all network traffic, 21 percent of the traffic is still private line service, 8 percent is switched non-IP, and 41 percent is conventional long distance voice. By the year 2004, 79 percent of the long-haul traffic will be IP, 11 percent will be private line, 3 percent will be switched non-IP traffic, and 8 percent of traffic will be conventional long distance voice."

Clearly, to achieve quality on a network not originally designed for it, one needs to impose a QoS architecture, which is based on — you guessed it — more protocols.

But what exactly is QoS? Amusingly, everyone, like blind men examining a camel, defines it differently.

From an engineering perspective, QoS is a study of delays in IP packet arrivals (some say anything above 150 to 200 milliseconds is unacceptable), packet loss (an infinite delay), and jitter (variations in packet delay). Men in white coats look at bandwidth and tally the performance of routers and switches.

From a corporate IT manager's perspective, QoS means that the network meets a certain standard of availability (such as no more than an hour or two of downtime per week), that users and applications will be allocated the bandwidth they need, and that the network's performance will meet some specific requirements favored by the business.

And from a user's perspective, a phone call either sounds good or it doesn't.

In reality, QoS encompasses many things: Operating systems (real-time scheduling, threads), communications protocols, data networks, scheduling and traffic management issues, etc.

Indeed, a true QoS must encompass the workings of all seven layers of the OSI Model, as well as every network element from end-to-end. Any QoS "guarantees" are only as good as the weakest link in the transmission "chain" between a source of voice / data traffic and its destination. As applications become more distributed, QoS support becomes important at the lower network layers. Since most operating systems don't support real-time delivery, things become ever more problematic.

A buggy application or weak endpoint system resources (too little memory, slow CPU, faulty network interface card) can ruin QoS for a user, no matter how tremendous the network bandwidth or clear the signaling. Ultimately, then, QoS is simply when a user subjectively judges the workings of an overall system to be satisfactory.

Unfortunately, it's difficult for a customer to use his or her subjective notion of QoS to wrangle a Service Level Agreement (SLA) with a service provider. For a QoS architecture to be successful, there must be some quantifiable (not to mention verifiable) level of assurance that the traffic and service requirements of each network component (e.g. an application, host or router) can be satisfied, even though each component must interface with other components, each with its own set of traffic and service requirements!

Carriers and service providers selling IP minutes would like to come up with some kind of universally accepted "number" to describe QoS, since studies of cellular networks show that as voice quality increases, the time users spend on the network also increases, which means that billable hours increase too. QoS thus becomes a differentiator that can be used to hold existing customers and entice new ones.

Still, to be an economically viable service, customers have to know what they're buying and they must be confident that their service will be stable.

Thus, so far we've seen that QoS is an umbrella term for a number of techniques that match the needs of specific applications to available network resources. This is done by first identifying the applications that will run on the network and then by allocating an appropriate amount of network resources such as bandwidth and relative priority.

Of course, some applications are more stringent about their QoS requirements than others, and for this reason (among others) we have two basic types of QoS available.

Bandwidth vs. Prioritization

With all the voice compression, packet loss and jitter going on, the two fundamental issues in QoS have always been bandwidth vs. traffic engineering or prioritization. If you have unlimited bandwidth, then you have near-perfect QoS for all traffic flows. (I use the description "near-perfect" because, as some technicians cryptically remark, "drinking water out of a fire hose can occasionally lead to problems.")

Considering the growth rate of IP networks, however, no service provider relishes the thought of ripping out his routers and replacing them with new ones every eight months or so. Also, much of the equipment out there hasn't really been optimized to handle the increased IP traffic and the new applications that now, and in the future, will be putting increased pressure on carriers.

Fortunately, scalable core routers have appeared on the market. But always having unlimited bandwidth is an expensive proposition; most networks cannot have unlimited bandwidth and overprovision all routers and switches. The industry has concluded that you also need to do some traffic engineering to achieve a good QoS.

Thus we return to a solution based upon clever protocols. Protocols for QoS fall into these two categories:

- Integrated Services (resource reservation), where network resources are apportioned according to an application's QoS request, and subject to bandwidth management policy.
- Differentiated Services (prioritization), where network traffic is classified and network resources apportioned according to bandwidth management policy criteria. To enable QoS in this case, network elements give preferential treatment to classifications identified as having more stringent requirements. Thus is born the concept of *Class of Service (CoS)*. Different applications are associated with two or three or more grades of service.

Some argue that CoS is fiction, since we're not talking about real, established services but simply arbitrarily defined ways of forwarding packets. ("If the packet header looks like X then hold on to it for a millisecond. But if a Class bit in the header is set to 1, then send it first.")

These types of QoS can be applied to individual application "flows" or to flow aggregates, so there are two other ways to describe types of QoS:

- "Per Flow," a flow being defined as a stream of packets — an individual, unidirectional, data stream between two applications (sender and receiver), with the packets all uniquely identified by a "five-tuple" — they share the same transport protocol, source address, source port number, destination address, and destination port number.
- "Per Aggregate," which is two or more flows. Each flow has something in common, such as any one or more of the five-tuple parameters, a label or a priority number, or authentication information.

Different kinds of applications and network topologies impose which type of QoS is most appropriate for individual flows or aggregates. Some QoS protocols and algorithms are as follows:

ReSerVation Protocol (RSVP): Provides the signaling to enable network resource reservation (Integrated Services). Generally used on a per-flow basis; though it's also used to reserve resources for aggregates. RSVP makes a great effort to provide the closest thing to circuit emulation on IP networks. It's been used in videoconferencing for years. But in order for integrated services to work well, the router must maintain state information on each flow; the router determines what flows get what resources based on available capacity. RSVP doesn't have much scalability (processing must be done on every individual flow on the core Internet routers) and it lacks good policy control mechanisms.

RSVP, if completely implemented, can in theory deliver QoS over any media, even if the media itself has no provisions for providing QoS. However, the RSVP definition deals mostly with routers, since when one router wants to talk to another, RSVP is used to request a certain allocation of bandwidth. But the immense Internet encompasses many kinds of routers, most of which do not support RSVP.

Differentiated Services (DiffServ): A coarse and simple way to categorize and prioritize network traffic (flow) aggregates. Instead of maintaining individual flows on all routers, the flows are aggregated into an aggregate flow that receives "treatment" (per class or per service state). Service classes are identified, and each packet is marked to identify it as belonging to a particular service, then sent through the network. Each router in the path must examine the packet's header to determine how it will be treated (hold it in favor of a high priority packet or let it go now).

A minimal DiffServ system needs a number of fundamental features:

- Admission Control — the ability of the network to refuse customers when demand exceeds capacity.
- Packet Scheduling — the packet scheduler is a sort of "traffic control module" that regulates how much data an application (or flow) is allowed, and can thus be used for treating data from different customers in accordance with their service level agreements. For example, some customers have paid for premium high QoS, high bandwidth service.
- Traffic Classification — the ability for the system to sort packet streams into "substreams" (each of which gets a different priority) and "policies" and "rules" (for allocating the network's resources). Thus, important real-time traffic such as voice gets a higher priority than, say, e-mail data.

Multi-Protocol Labeling Switching (MPLS): Favored by companies such as Cisco, Juniper and Avici, MLPS provides bandwidth management for aggregates via network routing control according to labels inside (encapsulating) packet headers. MPLS routing establishes "fixed bandwidth pipes" similar to ATM or Frame Relay virtual circuits, but with MPLS you're dealing with "coarse levels" of QoS — it's not actually as fine grained as the virtual circuits and provisioning of, say, ATM.

MPLS resides only in routers and is multi-protocol so it can be used with network protocols other than IP (ATM, IPX, PPP or Frame-Relay) or even directly over datalink layer.

MPLS is now a sort of *de facto* standard, since various equipment manufacturers have gotten together and demonstrated interoperability of their MPLS-enabled equipment through a third party, George Mason University's Internet Lab. The equipment from these manufacturers has now been certified as being interoperable.

MPLS may in fact be the "the winner" in the QoS protocol contest if only because of the huge companies lining up behind it — but while QoS protocols vary, they are not mutually exclusive of one another. On the contrary, some of them (MPLS and DiffServ) are said to complement each other.

Protocols Alone Can't Cure Bloated Networks

For all this talk of exotic QoS protocols, sometimes service providers and carrier have just got to reengineer their network. VoDSL service on the access network should be rock-solid right to the gateway of the core network, but if the core network happens to be an IP-based next gen network instead of the PSTN, then you may find yourself experiencing one of those "objectionable" 150 to 200 millesecond delays in Voice over IP (VoIP) conversations. If you have ten router hops between the two endpoints, and each router contributes 50 milliseconds of delay, then you end up with a 500 millisecond delay. Neither you nor your carrier can throw bandwidth at this kind of problem, nor solve everything by simply marking the packets for "special delivery" and expect any kind of improvement.

Instead, for a next-generation network to live up to its promise, both you and your service provider may have to swap out any slow software-based IADs or routers and install new high speed switches, or reduce the number of hops in the network somehow. Making backbone routers faster speeds performance but at the expense of intelligence and flexibility. Give the backbone routers too many QoS abilities, such as priority queuing, and latency and administrative overhead will rise, bringing down router performance.

All in all, however, support for more and more protocols is a good thing for VoDSL and the next generation multiservices network, since softswitches and gateways can only benefit from having the flexibility of supporting all the main communication links, such as ATM, Frame Relay, IP and the Internet, as they may exist between any two sites. With the flexibility to convey high-quality voice over all of these kinds of links, legacy equipment can be leveraged and investments can be protected.

Glossary

μ-law. Also called mu-law. A standard analog signal compression algorithm, used in digital communications systems of the North American digital hierarchy, to optimize the dynamic range of an analog signal prior to digitizing.

1000Base-T. Gigabit Ethernet for high bandwidth LAN and WAN applications.

100Base-T. A 100 Mbps LAN that maintains backward compatibility with 10Base-T networks running at 10 Mbps. Competitor to 100VG AnyLAN.

10Base-T. A 10 Mbps Ethernet LAN that runs over twisted pair wiring. This network interface was originally designed to run over ordinary twisted pair (phone wiring) but is predominantly used with Category 3 or 5 cabling.

23B + D. Way of representing the Primary Rate Interface (PRI) in ISDN; a circuit that is divided into twenty-three 64 "bearer" (B) channels for carrying voice, data, video, or other information simultaneously and one "delta" (D) channel for telephony signaling and other data data. See also Primary Rate Interface.

25 Pair AMP Champ Amphenol. 25 Pair indoor connector used in high cable density applications.

2B + D. Way of representing the Basic Rate Interface (BRI) in ISDN. A single ISDN circuit divided into two 64 Kbps digital B channels for voice and data and one 16 Kbps channel for low-speed data and signaling. Either one or both of the 16 Kbps channels may be used for voice or data. In ISDN, 2B + D is carried on one or two pairs of wires (depending on the interface). See also Basic Rate Interface.

2B1Q. Two Binary, One Quaternary. A line coding technique that compresses two binary bits of data into one time state as a four-level code. Used by ISDN and for multiple versions of symmetric DSL.

2W. A 2-Wire circuit (analog or digital).

4W. A 4-Wire circuit (analog or digital).

5ESS. A digital Central Office switching system originally made by AT&T, then Lucent Technologies.

A

A and B Bits. Bits used for signaling information in robbed bit signaling.

A and B Leads. Term used in many countries to represent tip and ring.

AAL. See ATM Adaptation Layer.

AAL1 (ATM Adaptation Layer Type 1). Addresses

constant bit rate (CBR) ATM traffic such as digital voice and video and is used for applications that are sensitive to both cell loss and delay. See CES.

AAL2 (ATM Adaptation Layer Type 2). This AAL supports time-sensitive, connection-oriented, variable bit rate (VBR) isochronous traffic such as compressed, packetized voice and video.

AAL3/4 (ATM Adaptation Layer Type 3/4). Handles bursty connection-oriented traffic, such as variable-rate connectionless traffic, like LAN file transfers. It is designed for traffic that can tolerate some delay but not the loss of a cell.

AAL5 (ATM Adaptation Layer Type 5). Accommodates bursty LAN data traffic with less overhead than AAL 3/4. Is used by many VoDSL systems to transport IP packets over the local loop. AAL5 has been adapted by the ATM Forum for a Class of Service called High Speed Data Transfer.

ABCD Parameters. The transfer characteristics of a two-port network describing the input voltage and current to the output voltage and current.

ABR. See Available Bit Rate.

Access Code. A short sequence of digits that allows a user to access a specific facility, service, feature or function of a telecom network.

Access Control. Methods, such as login passwords and time and computer restrictions, for controlling user access to network resources.

Access Control List. Database that describes the type of access each user has to a service.

Access Device. A physical device that terminates the local loop, such as a CSU/DSU for T-1.

Access Latency. Maximum time that a port will take to either successfully transmit a packet or discard it as measured from the time the packet is presented to the MAC for transmission.

Access Line. The physical telecommunications circuit connecting an end user location with the serving central office in a local network environment. Also called the local loop or "last mile." See also Local Loop.

Access Method. The method by which networked stations determine when they can transmit data on a shared transmission medium. Also, the software within an SNA processor that controls the flow of information through a network.

Access Network. That portion of a public switched network that connects access nodes to individual subscribers. The access network today is predominantly passive twisted pair copper wiring.

Access Nodes. Points on the edge of the Access Network that concentrate individual access lines into a smaller number of feeder lines. Access Nodes may also perform various forms of protocol conversion. Access Nodes can include Digital Loop Carrier (DLC) systems concentrating individual voice lines to T-1 lines, cellular antenna sites, PBXs, and Optical Network Units (ONUs).

Access Rate. The transmission speed of the physical access circuit between the end-user location and the local network. This is generally measured in bits per second; also called Access Speed.

Access Security System. Remote access security software that works with network-based security servers.

Active Hub. A multiport device that amplifies LAN transmission signals.

A-D Conversion. Analog-to-digital conversion. A-D conversion is a process in which an analog signal is modified into a digital signal. A-D conversion takes place, for example, when an analog modem call reaches a digital modem.

Adapter. Usually a NIC (Network Interface Card). See also Adapter Card.

Adapter Card. Circuit board or other hardware that provides the physical interface to a communications network; an electronics board installed in a computer which provides network communication capabilities to and from that computer; a card that connects the DTE to the network. Also called a Network Interface Card (NIC). See also Data Termination Equipment and Network Interface Card.

Adaptive Clocking. A source CES IWF sends data to a destination CES IWF. The destination CES IWF writes the received data to the segmentation and re-assembly buffer. Then the data is read with a local T-1/E-1 service clock. It is from the received CBR data that the local service clock is determined.

Adaptive Differential Pulse-Code Modulation (ADPCM). A voice compression technique that compresses a traditional 64 Kbps PCM channel to 32 Kbps or even less bandwidth. ADPCM is used in VoDSL to carry multiple voice calls over a single DSL circuit.

Address Resolution Protocol (ARP). TCP/IP Interior Gateway Protocol for dynamically mapping Internet addresses to physical hardware addresses on LANs; limited to LANs that support hardware broadcast.

ADM (Add/Drop Multiplexer). A SONET network node that combines and splits signals.

ADPCM. See Adaptive Differential Pulse-Code Modulation.

ADSL. See Asymmetric Digital Subscriber Line; Asynchronous Digital Subscriber Line.

ADSL Forum. The organization developing and defining xDSL standards, including those affecting ADSL, SDSL, HDSL, and VDSL. Members participate in committees to vote on ADSL specifications; auditing members receive marketing and technical documentation.

ADSL Lite. Also called G.lite. A sub-rated ADSL system, with a data rate limited to 1.5 Mbps or less downstream and 500 Kbps upstream. ADSL Lite systems offer lower data throughput, and reduced digital signal processing requirements, than full-rate ADSL systems. ITU-T G. G.992.2 is the standard describing this.

ADT (ADSL Digital Transceiver). In an architecture with a separate passive splitter, the ADT consists of electronics which supply the high-rate and low-rate terminations at the customer premises.

Advanced Distributed Recovery Intelligence. A function that tracks down a network problem and automatically isolates and resolves hard error connections already present on a Token Ring.

Advanced Peer-to-Peer Internetworking (APPI). Open-standard IP architecture for SNA peer-to-peer networking.

Advanced Peer-to-Peer Networking (APPN). SNA protocol that allows network nodes to interact without a host computer.

AFE (Analog Front End). Functions including the analog-digital conversation, analog filter, and line driver.

AGC (Automatic Gain Control). Receiver adaptation to the received signal level so as to reduce dynamic range of the signal input to the analog-to-digital converter.

Agent. Intelligent management software embedded in a network device. In network management systems, agents reside in all managed devices and report the values of specified variables to management stations. An agent can be an SNMP agent wherein SNMP-enabled devices can respond to SNMP requests issued by one or more NMSs; the agent can also issue unsolicited SNMP traps (event messages) to one or more NMSs.

AIS. See Alarm Indication Signal.

Alarm Indication Signal (AIS). Also known as a blue alarm. In T-1, this is an unframed all ones signal.

A-law. A standard compression algorithm (the PCM coding and companding standard defined in CCITT G.711) used in digital communications systems of the European digital hierarchy, to optimize the dynamic range of an analog signal for digitizing.

A-link. An SS7 term. A-links connect an end office or signal point to a mated pair of signal transfer points. They may also connect signal transfer points and signal control points at the regional level with the A-links assigned in a quad arrangement.

Always On. Current analog dial-up services require the user to "make a call" to the ISP. The connection is only active during the duration of the call. Most xDSL implementations (including ADSL, UADSL, and SDSL) enable the connection to be always on in a fashion similar to a LAN.

AM (Amplitude Modulation). Modulation of amplitude of a carrier signal by the amplitude of a payload signal. See also FM, PM.

American National Standards Institute (ANSI). U.S. coordinating body for industry standards groups. Member of International Standards Organization (ISO).

American Standard Code for Information Interchange (ASCII). Seven- or eight-bit computer code used for data communications.

Amplitude. Measure of the distance between the high and low points of a waveform.

AN (Access Node). A point on the edge of the access network that concentrates individual access lines into a smaller number of feeder lines. Access nodes may also perform various forms of protocol conversion. Typical access nodes are DLC systems concentrating individual voice lines to T-1 lines, cellular antenna sites and PBXs.

Analog. An electrical signal or waveform in which the amplitude and/or frequency vary continuously. The current basis for most residential telephone service.

Analog Front End. The analog front ends are responsible for converting the digital signal to analog and forcing the signal onto the twisted pair line.

Analog Line. A line that transmits data by means of an analog signal.

Analog Transmission. Current method of voice transmission used in telephone systems. Analog converts voice to electronic signals and amplifies them, allowing the signals to be sent over long distances.

Analog-to-Digital Converter (ADC). Device that samples incoming analog voltage waveforms, rendering them as sequences of binary digital numbers; passing waveforms through an ADC introduces quantization noise.

ANSI (American National Standards Institute). The primary standards organization for the U.S. The ANSI accredits standards bodies, such as Committee T1 for telecommunications. One of the key standards bodies involved with DSL. Member of the ISO.

API. See Application Programming Interface.

APON (ATM Passive Optical Network). A passive optical network running ATM.

AppleTalk. An Apple Computer networking system that operates over STP wire at 230 Kbps.

Application. Software that provides a set of services such as electronic mail, spreadsheets, word documents, etc.

Application Layer. Layer 7, the highest layer of the OSI Reference Model; defines the way applications interact with the network. Implemented by various network applications, including electronic mail, file transfer, and terminal emulation.

Application Service Provider (ASP). A company that offers individuals or enterprises access over the Internet to applications and related services that would otherwise have to be located in their own personal or enterprise computers.

Applications Server. Used in a next generation network along with a softswitch and media server to provide enhanced telephony services to xDSL and other broadband subscribers.

Architecture. Network architecture refers to the total design and implementation of the network. It includes the network's topology, transmission technologies and communications protocols, management and security systems, and any other attributes that give a network a particular set of capabilities and functionalities.

ASIC (Application Specific Integrated Circuit). A chip designed for a specific purpose or application. Examples of an ASIC application can be G.SHDSL or other broadband solutions.

ASP. See Application Service Provider.

Asymmetric Digital Subscriber Line (ADSL). A technology that allows more data to be sent over existing copper telephone lines (POTS). ADSL supports data rates from 1.5 to 9 Mbps when receiving data (known as the downstream rate) and from 16 kbps to 640 Kbps when sending data (known as the upstream rate). Asymmetrical variations include: ADSL, G.lite ADSL (or simply G.lite), RADSL and VDSL. The standard forms of ADSL (ITU G.992.1, G.992.2, and ANSI T1.413-1998) are all built upon the same technical foundation. The

Glossary

suite of ADSL standards allows for interoperation between all standard forms of ADSL.

Asynchronous. Data communication in which transmission is sent by individual bytes at different time intervals.

Asynchronous Transfer Mode (ATM). A high-speed, connection-oriented ITU standard protocol that packs digital information into small fixed-sized cells that are switched throughout a network over broadband virtual circuits. Data is transmitted / received in 53-byte frames including 48 data bytes (payload) and 5 control bytes. ATM was originally designed to carry priority-rated, video, and data transmissions over high-bandwidth networks. ATM scales easily and speeds typically range from 25 Mbps to OC-192 (10 Gigabits per second). Today ATM is a high-speed networking standard used in WANs and often used to connect DSL lines to your Internet backbone.

Asynchronous Transmission. Data transmission of characters in succession to a receiving device with intervals of varying lengths between transmittals and with start bits at the beginning and stop bits at the end of each character to control the transmission. In xDSL and in most dial-up modem communications, asynchronous communications are often found in Internet access and remote office applications. See also Synchronous Transmission.

ATIS (Alliance for Telecommunications Industry Solutions). Sponsors Standards Committee T-1.

ATM Adaptation Layer (AAL). Each AAL consists of two sub layers: the segmentation and reassembly (SAR) sub layer and the convergence sub layer. AAL is a set of four standard protocols that translate user traffic from higher layers of the protocol stack into a standard size and format contained in the ATM cell and return it to its original form at the destination.

ATM Cell. An ATM cell is 53 bytes long containing a 5-byte header and a 48-byte payload. The header of an ATM cell contains all necessary information for data to reach the appropriate endpoint. The payload portion of an ATM cell can contain any type of information, be it voice, video, or data.

ATM Connection. An ATM connection is actually one physical connection between two endpoints that contains multiple virtual circuits or connections (VCs). Furthermore; multiple VCs can be grouped to traverse a VP. See also Permanent Virtual Circuit, Switched Virtual Circuit, Virtual Channel Identifier, and Virtual Path Identifier.

ATM Forum. The organization tasked with developing and defining ATM standards. See http://www.atmforum.com for more information.

ATM Switch. A hardware device that directs an incoming ATM cell to one or more output interfaces.

ATM Traffic Descriptor. A generic list of traffic parameters that can be used to capture/define the traffic characteristics of a specified ATM connection.

Attachment Unit Interface (AUI). An IEEE 802.3 cable connecting the media access unit (MAU) to a networked device; AUI also may refer to the host back panel connector to which an AUI cable attaches. AUI connections adapt between two different cabling types.

Attenuation. Signal loss resulting from traversing the transmission medium. The decrease in magnitude of the power of a signal is measured in decibels. As attenuation increases, signal power decreases.

ATU (ADSL Transceiver Unit). The ADSL Forum uses terminology for DSL equipment based on the ADSL model for which the Forum was originally created. The DSL endpoint is known as the ATU-R and the CO unit is known as the ATU-C. These terms have since come to be used for other types of DSL services, including G.SHDSL, RADSL, MSDSL and SDSL. ATU now represents xDSL services.

ATU-C (ADSL Transceiver Unit-Central Office). The ADSL modem or line card that physically terminates an ADSL connection at the telephone service provider's serving central office.

ATU-R (ADSL Transceiver Unit-Remote). The ADSL modem or PC card that physically terminates an ADSL connection at the end-user's location.

Authentication. A process for validating a user's login information; it usually involves comparing the username and passwords with a list of authorized users. If the username and password matches an entry on the list, the user is able to log in and access the system, limited by the permissions assigned to that user account.

Authority Format Identifier. Specifies the format of the initial domain identifier (IDI) in the initial domain part (IDP) of an OSI network service access point (NSAP).

Autonomous System (AS). In Internet (TCP/IP) terminology, a series of gateways or routers that fall under a single administrative entity and cooperate using the same Interior Gateway Protocol (IGP).

Auto-Partitioning. A function of all repeaters, whereby a faulty segment is automatically isolated to prevent the fault affecting the entire network; the segment is automatically reconnected by the repeater when the fault condition is rectified.

Autosensing. For a LAN, automatically determining which network users have 10 or 100 Mbps capabilities and matching them to the bandwidth they require.

Available Bit Rate (ABR). An ATM service category that is a "best effort service" that allows the setting of both minimum and peak cell rates. Thus, it provides a guaranteed minimum capacity but allows data to be bursted at higher capacities when the network is free. The ABR service category is similar to nrt-VBR, since it is also used for connections that transport variable bit rate traffic for which there is no reliance on time synchronization between the traffic source and destination, and for which no required guarantees of bandwidth or latency exist. ABR has flow-control mechanisms to dynamically adjust the amount of bandwidth available to the originator of the data traffic. The ABR service category is meant for any type of traffic that is not time sensitive and expects no guarantees of service. ABR is often used with TCP/IP traffic and other LAN protocols.

AWG (American Wire Gauge). A measure of the thickness of copper, aluminum, and other wiring in the U.S. and elsewhere. Copper cabling typically varies from 18 to 26 AWG, the higher the number, the thinner the wire. The thicker the wire, the less susceptible it is to interference. Thin wire cannot carry the same amount of electrical current the same distance that thicker wire can, unless perhaps it is made of a special alloy and is cooled to superconducting temperatures.

B

B Channel. A "bearer" channel is the information-bearing channel for an ISDN circuit and is thus a fundamental component of ISDN interfaces. Carries 64 Kbps in either direction (full duplex), is circuit-switched, and can carry either voice or data. See also Basic Rate Interface (BRI), Primary Rate Interface (PRI), and Integrated Services Digital Network (ISDN).

B2C. Business-to-consumer e-commerce; marketing to consumers over the Internet.

B8ZS. See Binary 8 Zero Substitution.

Back Door. A networking security term referring to a hole in a compromised system that allows continued access to the system by an intruder even if the original attack is discovered.

Backbone. Although often associated with large intercontinental optical fiber carrier networks, a backbone can be any LAN or WAN and its concomitant infrastructure that interconnects intermediate systems (bridges, switches, and/or routers), providing connectivity for users of the distributed network. The term is used interchangeably with Backbone Connection.

Backbone Connection. Connection that groups multiple LANs into a single LAN, usually between network devices; for example, between routers and routers, switches and switches, or routers and switches. Distances covered by backbones formed in this way can extend up to several miles. A building backbone is one that interconnects all the LAN segments within a building, while a campus backbone interconnects two or more building backbones. Common backbone technologies include Fast Ethernet, Gigabit Ethernet, and ATM.

Backward Error Correction (BEC). The sender retransmits any data found to be in error by the receiver upon feedback from the receiver.

Balanced Line. A transmission line with two identical conductors with the same electromagnetic characteristics in relation to other conductors, ground references, and the external environment. See Coax, Flat Pair, Twisted Pair, Unbalanced Line.

Balun (balanced-unbalanced). An impedance-matching device that connects a balanced line (such as a twisted-pair line) with an unbalanced line (such as a coaxial cable).

Bandwidth. As related to frequencies, bandwidth can be considered the difference between the highest and lowest frequencies of a band that can be passed by a transmission medium without undo distortion. As a measure of information carrying capacity, bandwidth indicates how many bits per second (bps) a link can carry, but does not take into account any of the impairments (such as packet loss, delay and jitter) that lower the actual deliverable quantity of data.

Bandwidth Bound. An application that can only run properly with a minimum amount of bandwidth at its disposal, but which will not necessarily benefit from lower delay in a network, such as a bulk transfer file application.

Bandwidth Reservation. Request made by protocol (or by some other method) to the network to allocate a specific amount of bandwidth for a data flow.

Bandwidth on Demand Interoperability Group (BONDing). Also called Bandwidth On Demand; sometimes written as BOND-ing or BONDING. An international standard for the automatic aggregation of multiple 64 Kbps B channels into a single larger, logical "pipe." Often used in videoconferencing. Similar to Dynamic Bandwidth Allocation.

Baseband. Transmission scheme whereby the entire bandwidth, or data-carrying capacity, of a transmission medium such as a coaxial cable is used to carry a single digital data signal between users. Note that this limits such transmission to a single form of data transmission, since digital signals are not modulated. The earliest electrical baseband transmission device was the telegraph. Contrast with Broadband.

Basic Encoding Rate (BER). See Bit Error Rate

Basic Encoding Rule. Rule for encoding data units described in ANS.1.

Basic Rate Interface (BRI). An ISDN interface typically used by smaller sites and customers. This interface consists of a single 16 Kbps data (or "D") channel plus two 64 Kbps bearer (or "B") channels for voice and/or data for a rate of 144 Kbps. Also known as Basic Rate Access, or BRA.

Baud. Transmission rate of a multilevel-coded system when symbols replace multiple bits. Baud rate is always less than bit rate in such systems, where the Baud Rate indicates the actual symbol frequency being used to transmit data. Often used incorrectly as an equivalent to bits per second (bps). For example, both ITU-T V.22bis (2400 bps) and V.22 (1200 bps) modems transmit data at 600 baud, but V.22 bis modems use four bits per symbol and V.22 modems use two.

BDSL. Broadband DSL, used interchangeable with VHDSL or VDSL.

Bearer Services. A communication connection's capacity to carry voice, circuit, or packet data. The two B channels in an ISDN BRI connection are bearer channels and serve as the foundation for the feature set of bearer services offered over ISDN. See B Channel, Basic Rate Interface.

BEC. See Backward Error Correction.

Bell System. Before 1984, the local telephone companies that belonged to AT&T were commonly grouped together as the Bell system. All others were independents. After 1984, it became common to speak of the entire telephone network as the public switched telephone network (PTSN). See Independent Telephone Company (ITC), PTSN, and Regional Bell Operating Company (RBOC).

Bellcore (Bell Communications Research). See Telcordia.

BELR. See Binary Equivalent Line Rate.

BER. See Bit Error Rate.

BERT. See Bit Error Rate Test.

BICC (Bearer-Independent Call Control). Used by some media gateway controllers to communicate with each other.

Binary 8 Zero Substitution (B8ZS). Also called Bipolar Eight Zero Suppression. A code technique used in T-1 / DS-1 that modifies the AMI encoding to ensure minimum pulse density without altering customer data. When eight zeros in a row are detected, a pattern with intentional bipolar violations is substituted. These violations enable the receiving end to detect the pattern and replace the zeros.

Binary Equivalent Line Rate (BELR). Data rate in bits/second before line coding/modulation or after decoding/demodulation. Transmitted symbol/baud rates and analog bandwidths (in kHz) on loop depend on modulation and coding schemes.

Binder Group. Cable pairs are typically arranged under the cable sheath in binder groups. The binder is a spirally wound colored thread or plastic ribbon used to separate and identify cable pairs by means of color-coding. The enclosed pair group is called a binder group. The groups are composed of insulated twisted copper pairs that are also twisted within each binder. Typically they are wrapped in 25 pair bundles. For example, pairs 1-25 might be in one binder group and pairs 26-50 in another. In xDSL, one often hears discussions of signal interference between adjacent pairs within a binder group. The best of all worlds is to keep a data pair separated from another data pair by assigning it to an adjacent binder group. If the data pairs are too close to each other they create what telcos call "disturbers" (i.e., crosstalk). If a "disturber" exists in the binder group serving your SNI, NID, MPOE, etc., you may not "qualify" for DSL service.

Biphase. A baseband line code, also known as the Manchester line code.

Bipolar Return to Zero. A bipolar signal in which each pulse returns to zero amplitude before its time period ends. This prevents the buildup of DC current on the signal line.

Bipolar Violation (BPV). For an AMI-coded signal, it is the occurrence of a pulse of the same polarity as the previous pulse. For a B8ZS or HDB3-coded signal, it is the occurrence of a pulse of the same polarity as the previous pulse without being a part of the zero substitution code.

B-ISDN or BISDN. See Broadband Integrated Digital Network.

Bisynchronous communication ("bisync"). Character-oriented data-link protocol for applications.

Bit. A contraction of "binary digit." A bit is the smallest element of information in the digital system. The binary units, 0 or 1, are used in the binary numbering system. Eight bits are needed to create one byte or character.

Bit Error Rate (BER). Also called the Basic Encoding Rate. It's a measure of transmission quality, the ratio of the number of bits incorrectly transmitted in a given bit stream compared to the total number of bits transmitted during a given time interval; also, a rule for encoding data units described in ANSI. See Bit Error Rate Test.

Bit Error Rate Test (BERT). A test that reflects the ratio of errored bits to the total number transmitted. Usually shown in exponential form to indicate that one out of a certain number of bits are in error.

Bit Rate. The number of bits of information that can be transmitted over a channel in a given second. Typically expressed in bits per second (bps).

Bit Robbing. A technique in T-1 multiplexing in which the least significant bit (bit 8) of each byte in selected frames is robbed from being used to carry message information and instead is used to carry signaling information.

Bit Stuffing. Extra bit(s) that are conditionally inserted into the frame to adjust the transmitted bit rate.

Bits Per Second (bps). Measurement unit for transmission speed over data communications lines. The bps rate may be equal to or greater than the baud rate depending on the modulation technique used to encode bits into each baud interval. The correct term to use when describing modem data transfer speeds.

BLEC. See Broadband Local Exchange Carrier.

BLES. See Broadband Loop Emulation Services.

Blocking. Whenever bits cannot make their way from an input port to an output port in a network node, they are considered to be blocked. In the voice network, the call will not go through. In a data network, the bits may be stored in a buffer or discarded, depending on the situation.

BNC Connector. Standard connector to link IEEE 802.3 10BASE2 coaxial cable to a transceiver. Also used on Network Interface Card to connect a PC to a LAN or certain routing equipment.

BONDing. See Bandwidth on Demand Interoperability Group.

Bottleneck. Occurs when data passes through a port at a slower speed than the actual data transmission. Thus, there can be traffic slowdowns when too many network nodes try to access a single node, often a server node, at once.

bps. See Bits Per Second.

Bps. See Bytes Per Second.

BPV. See Bipolar Violation.

BRA (Basic Rate Access). See Basic Rate Interface.

BRI. See Basic Rate Interface.

Bridge. A device that connects two networks as a seamless single network using the same networking protocol, such as TCP/IP. ADSL modems are typically bridges. Strickly speaking, a bridge is a device that connects two networks of the same type (using the same protocol), though newer ones are sophisticated enough so that they can be used to interconnect local or remote networks. Bridges form a single logical network, centralizing network administration. They operate at the physical and link layers of the OSI Reference Model. Contrast this with Router and Gateway.

Bridge Protocol Data Units. A packet to initiate communications between devices under a spanning-tree protocol. Compare PDU.

Bridge Tap (or Bridged Tap). A connection of another local loop to the primary local loop. Generally it behaves as an open circuit at DC, but becomes a transmission line stub with adverse effects at high frequency. It is generally harmful to DSL connections and should be removed. Extra phone wiring within one's house is a combination of short bridge taps. A POTS splitter isolates the house wiring and provides a direct path for the DSL signal to pass unimpaired to the ATU-R modem.

Broadband. A data transmission scheme in which multiple signals share bandwidth, allowing transmission of voice, data, and video signals over a single medium. Strictly speaking, broadband refers to a telecommunications link that runs at more than 1.5 Mbps in the U.S. and more than 2 Mbps everywhere else. This is about equal to Primary Rate of ISDN, or HDSL, SDSL and G.SHDSL. Broadband differs from Baseband which carries a single digital data signal.

Broadband Integrated Digital Network (B-ISDN, or BISDN). Somewhat archaic term for a digital network with ATM or STM switching operating at data rates in excess of 1.544 or 2.048 Mbps, designed for high-bandwidth multimedia applications and the integration of voice, data, and video.

Broadband Loop Emulation Services (BLES). Defined by DSL Forum to describe specific means of accomplishing VoDSL, based heavily on ATM Forum and DSL Forum standards. A term often used interchangeably with Loop Emulation Service (LES).

Broadcast. A message forwarded to all network destinations.

Broadcast Storm. Multiple simultaneous broadcasts that absorb available network bandwidth and can cause network time-outs.

Brouter. A device that can provide the functions of a bridge, router, or both concurrently; a brouter can route one or more protocols, such as TCP/IP and/or XNS, and bridge all other traffic. Also called a Bridge/Router.

Brownout. In the context of DSL, a situation that occurs when a CO (or local exchange) cannot handle all of the calls attempted and even disrupts calls in progress. Also called a "brown down."

BT (Burst Tolerance). Parameter defined by the ATM Forum for ATM traffic management. For VBR connections, BT determines the size of the maximum burst of contiguous cells that can be transmitted.

Buffer. A storage area in a computer or other processor's memory dedicated for telecommunications purposes. The whole art of network design is a balancing of the need to buffer bits in order to store and process them and the need for adequate bandwidth and delay to actually send the bits somewhere. Larger buffers can compensate for slower links and network nodes that experience blocking, but at the risk of offending waiting users.

Building Local Exchange Carrier (BLEC). A LEC (local exchange carrier) is the term for a public telephone company in the U.S. that provides local service. A BLEC is a LEC that provides broadband local service as an in-building service. Not to be confused with a Broadband Local Exchange Carrier, which simply supplies broadband to anyone as would a CLEC.

Bus Networks. A bus network is a multiple access medium for small networks and usually only consists of one cable and the devices that are attached to it.

Bypass Mode. Operating mode on ring networks such as FDDI and Token Ring in which an interface has de-inserted from the ring.

Byte. A group of bits, normally eight, which represent one data character.

Bytes Per Second (Bps). The speed at which bytes are transmitted across a data connection; although most data rates are actually given in bits per secton (bps).

C

Cable Binder. A cable binder is used to bundle multiple insulated copper pairs together in the telephone network.

Cable Modem. A modem designed to operate over cable TV lines; used to achieve extremely fast access to the Internet. In theory capable of speeds up to 10 Mbps for downloading, though most speeds are restricted to about 1.5 Mbps, less if too many users are active on a particular cable segment. Cable modems are the primary competitor to DSL.

Cable Modem Termination System (CMTS). A CMTS is the head-end box in a high-speed cable network that aggregates multiple cable channels. A CMTS that terminates local subscriber cable lines is analogous to a DSLAM in a high-speed DSL network.

Call Agent. See Media Gateway Controller.

Caller ID. See Calling Number Identification.

Calling Number Identification (Caller ID). A telephone company service that delivers the calling party's telephone number to the called party. The number can appear on an ISDN telephone, an LCD screen, a computer screen, or another device.

Call-Progress Tones (CP). Call-Progress tones are generated by the telephone network to provide feedback on call status to the user. Examples include dial tones, busy signals and fast-busy signals.

Campus Area Network (CAN). A network that involves interconnectivity between buildings in a set geographic area, such as a campus, an industrial park, or other such private environment.

CAP. See Carrierless Amplitude & Phase Modulation; Competitive Access Provider.

Carrier Failure Alarm (CFA). Alarm state on a T-1/T-3 that refers to a loss of signal/loss of frame.

Carrier Service Area (CSA). Also called Carrier Serving Area. The area served by a LEC, RBOC, or telco, using Class switches often used in conjunction with Digital Loop Carrier (DLC) technology.

Carrierless Amplitude & Phase Modulation (CAP). A transmission technology for implementing a DSL connection. Transmit and receive signals are modulated into two wide-frequency bands using passband modulation techniques. Licensed by Globe Span Technologies, Inc. CAP-modulated ADSL can reach data speeds of 7.1 Mbps. CAP is a competitor to DMT modulation, though most DSL sytems use DMT.

CAS (Channel-Associated Signaling). Signaling

in which the signals necessary to switch a given circuit are transmitted via the circuit itself or via a signaling channel permanently associated with it.

Category 3 Cabling (CAT3 or CAT-3). A rating for twisted pair copper cabling that is tested to handle 16 MHz of communications. Handles 10 Mbps of LAN traffic and is often used as telephone wiring.

Category 5 Cabling (CAT5 or CAT-5). A rating for twisted pair copper cabling that is tested to handle 100 MHz of communications. CAT-5 cable is generally required for higher-speed data communications, such as Ethernet LANs and possibly low-speed ATM.

CATV (Community Access Television). Also known as Cable TV.

CPE (Customer Premises Equipment). Refers to that portion of equipment that resides within a customer's premises.

CBR (Constant Bit Rate). Delay sensitive applications such as video and voice require a continuous bit stream. CBR traffic requires guaranteed levels of service and throughput. The BW for this class of service is in constant use.

CCITT See Comité Consultatif International de Télégraphie.

CCS. See Common Channel Signaling.

CDDI. See Copper Distributed Data Interface.

CDM. See Code Division Multiplex.

CDSL. See Consumer DSL.

CDV (Cell Delay Variation). The actual measurement of the variation in inter-cell spacing introduced by the multiplexing process of a cell switch.

CDVT (Cell Delay Variation Tolerance). ATM layer functions may alter the traffic characteristics of ATM connections by introducing Cell Delay Variation. When cells from two or more ATM connections are multiplexed, cells of a given ATM connection may be delayed while cells of another ATM connection are being inserted at the output of the multiplexer. The upper limit tolerable cell delay is the CDVT.

Cell. A fixed-length protocol unit used in a data link layer protocol. Since it is fixed-length, a cell requires no special delimiting symbols. ATM is the international standard for building networks that employ cells. Contrast this with Packet and Frame.

Cell Loss Priority (CLP). A bit in the Payload Type (PT) field of the ATM cell header that conveys two levels of priority of an ATM cell: CLP=0 cells have higher priority than CLP=1.

Cell Transfer Delay (CTD). The elapsed time for a cell to traverse the ATM network over a given virtual channel from source to destination.

Central Office. Formerly called a Local Exchange office or an "end office." A local telephone company office where the local loops of all customers in a given serving area connect to the Public Switched Telephone Network (PSTN) and where circuit switching of customer lines occurs.

Central Site. See Central Office.

Centrex. A type of business telephone service that resembles having a private branch exchange (PBX) located in your local central office. A single-line telephone service delivered to individual station sets with additional features.

CES. See Circuit Emulation Service.

CES-IWF. CES Internetworking Function. This is the physical hardware (or software) at the Central Office that runs the ATM Circuit Emulation Service (AAL1) between two constant bit rate (CBR) devices.

CEV. See Controlled Environmental Vault.

CFA. See Carrier Failure Alarm.

CFTC. Copper From The Curb, See Copper To The Curb.

CGI. See Common Gateway Interface.

Challenge Handshake Authentication Protocol (CHAP). Security feature that prevents unauthorized access to devices; supported only on lines using PPP encapsulation. CHAP is essentially a security protocol that arranges an exchange of random numbers between computers. The machine receiving the number from the first computer performs calculations on that number using a previously agreed-upon string of characters as a secret encryption key.

Channel. A generic term for a communications

path on a given medium; multiplexing techniques allow providers to put multiple channels over a single medium. See also Multiplexer.

Channel Aggregation. Combines multiple physical channels into one logical channel of greater bandwidth.

Channel Service Unit (CSU). A device on the Digital Trunk Interface that is the termination point of the T-1 lines from the T-1 provider. The CSU collects statistics on the quality of the T-1 signal. The CSU ensures network compliance with FCC rules and protects the network from harmful signals or voltages.

Channel Service Unit (CSU) / Data Service Unit (DSU). Connects an external digital circuit to a digital circuit on the customer's premises. The DSU converts data into the correct format, and the CSU terminates the line, conditions the signal, and participates in remote testing of the connection.

Channelized VoDSL (CVoDSL). The least expensive form of DSL and is targeted at the residential market and perhaps very small businesses. CVoDSL does not carry voice as a packet service (such as ATM, IP or Ethernet). Instead, CVoDSL sends voice directly over the xDSL circuit using TDM. Originally championed by Aware Inc. (www.aware.com) under the name "Voice enabled DSL."

CHAP. See Challenge Handshake Authentication Protocol.

CI (Customer Installation). Customer wiring and CPE at customer premises.

CID. AAL2 Channel Identifier.

Circuit. A communications path through a network between two or more devices, usually carrying electrical current.

Circuit Emulation Service (CES). The ATM Forum CES specification supports emulation of existing TDM circuits over ATM networks, including support for CBR traffic over ATM networks (that comply with other ATM Forum interoperability agreements). "Structured" CES is how T-1/E1 channels may transported across different ATM PVCs. "Unstructured CES" is how all T-1/E-1 channels are transported over the same ATM PVC.

Circuit Switching. A switching system that establishes a dedicated physical communications connection between endpoints, through the network, for the duration of the communications session; this is most often contrasted with packet switching in data communications transmissions. See also Packet Switching.

Circuit-Level Gateway. A specialized function that relays TCP connections without performing any additional packet processing or filtering.

Circuit-Switched Data (CSD). A circuit-switched call for data in which a transmission path between two users is assigned for the duration of a call at a constant, fixed rate.

Circuit-Switched Network. A circuit-switched network connects calls from source to destination along a dedicated route through the network. This route is "permanent" in that it is allocated to that call for the entire duration of the call, allowing the transmission of a continuous (non-packetized) voice or data stream. Also called a Switched-Circuit Network (SCN).

Circuit-Switched Voice (CSV). A circuit-switched call for voice in which a transmission path between two users is assigned for the duration of a call at a constant, fixed rate.

CLASS. See Custom Local Area Signaling Services.

Class of Service (CoS). A way of managing traffic in a network by grouping similar types of traffic (for example, e-mail, streaming video, voice, large document file transfer) together and treating each type as a class with its own level of service priority. This allows relative priority between types of traffic, but does not guarantee true QoS.

CLE. See Customer Located Equipment.

CLEC. See Competitive Local Exchange Carrier.

Competitive Local Exchange Carrier. A service provider that competes with the local RBOC or Incumbent Local Exchange Service Provider (ILEC). To start one, you file with your State public utility commission to be a competitive carrier. The company then negotiates an interconnection agreement with the ILEC.

CLEI. See Common Language Equipment Identifier.

CLI. See Command Line Interface.

Client. (1) A software program designed to fetch information from a server in a "client-server" environment. Internet browsers are examples of client software. (2) An intelligent workstation that makes requests to other computers known as servers. PC computers on a LAN can be clients.

Client/server. A distributed system model for computer interactions. Usually, a user runs a client process to obtain and process information present on a remote server. That is, the client "talks" and the server "listens" and responds to client requests. The interaction between client and server is usually asymmetrical, with more bits flowing from server to client than vice versa.

Clock. Any of the sources of timing signals used in isochronous data transmission.

Cloud. A commonly used term that describes any large network.

CLP. See Cell Loss Priority.

CMTS. See Cable Modem Termination System.

CNG. See Comfort Noise Generation.

CO. See Central Office

Coaxial ("coax"). High-capacity networking cable consisting of a hollow outer braided copper or foil shield surrounding a single inner copper conductor with plastic insulation between the two conducting layers. Used for broadband and baseband communications networks and cable TV; usually free from external interference and capable of high transmission rates over long distances

Code Division Multiplex (CDM). A spread spectrum technique of sharing a media by greatly expanding the bandwidth of each channel by multiplying the data by a unique wideband code. See also FDM, TDM.

Codec (Coder/Decoder or Compression/Decompression). A hardware device or software program that converts analog information streams into digital signals, and vice versa; generally used in audio and video communications where compression and other functions may be necessary and provided by the Codec as well. In IADs, codecs are specified in voice gateway profiles. The various codecs available may specify straight voice (PCM) or compressed voice (ADPCM).

Collapsed Backbone. Network architecture under which the backplane of a device such as a hub performs the function of a network backbone; the backplane routes traffic between desktop nodes and between other hubs serving multiple LANs.

Collocation cage. A cage in a CO that is erected by the ILEC and rented to a CLEC. CLEC personnel can access and maintain the equipment in the cage.

Comfort Noise Generation (CNG). CNG is employed at the far-end of a voice conversation to produce background noise intended to emulate the "silence" at the near-end.

Comité Consultatif International de Télégraphie, or International Telephone and Telegraph Consultative Committee (CCITT). The former name for an international standards body that developed telecommunications standards; now called the International Telecommunications Union (ITU). The ITU's Telecommunication Standardization Sector, or ITU-TSS, develops communications standards, known as "Recommendations", for all internally controlled forms of analog and digital communication. Recommendation X.25 is an example of an ITU-TSS standard. See ITU.

Command Line Interface (CLI). A user interface to a computer's operating system or an application program in which the user responds to a visual prompt by typing in a command on a specified line, receives a response back from the system, and then enters another command, and so forth.

Commercial End-user. See Service user.

Committed Bit Rate (CBR). Absolutely guaranteed specified bandwidth with specific range of QoS parameters, appears in Service Level Agreements (SLAs) between customer and service provider.

Common Carrier. Telephone companies that

provide long-distance telecommunications services at government-regulated rates. Also known in U.S. as XC (InterExchange Carrier).

Common Channel Signaling CCS). A communications system in which one channel is used for signaling and different channels are used for voice/data transmission. Signaling System 7 (SS7) is a CCS system.

Common Gateway Interface (CGI). A standard interface between a Web server and an external program, such as a Web browser. Its purpose is to add interactivity to a website by responding to user input. Most CGI scripts are written in Perl or MacPerl.

Communications Protocols. Fundamental mechanisms for network communications. They specify the software attributes of data communications, including the structure of a packet and the information contained in it. Protocols may also prescribe all or some of the operational characteristics of the hardware on which they will run. Popular network protocols include IPX, TCP/IP, and AppleTalk.

Community Antenna Television. Also known as Cable TV.

Companding. Compressing the dynamic range of a signal prior to transmission, with matching expansion at the receiver to regain the original signal.

Competitive Local Exchange Carrier (CLEC). A company that competes with the already established local telephone business by providing its own network and switching. The term distinguishes new or potential competitors from established local exchange carriers (ILECs) and arises from the Telecommunications Act of 1996, which was intended to promote competition among both long-distance and local phone service providers.

Complete Sequence Number PDU. Protocol data unit (PDU) sent by the designated router in an OSPF network to maintain database synchronization.

Compression. Reducing the size of a data set to lower the bandwidth or space required for transmission or storage.

Concatenation. Mechanism that provides continuous bandwidth through the network from end to end for transport of a payload associated with a "super-rate service" such as SONET or SDH. The set of bits in the payload is treated as a single entity and accepted, multiplexed, switched, transported, and delivered as a single piece of payload data. Applications that use concatenation include bandwidth-intensive video such as HDTV and high-speed data.

Concentrator. A device that serves as a point of consolidation of network links so that multiple circuits may share common limited network resources. Generally serves as a wiring hub in star-topology network. Sometimes refers to a device containing multiple modules of network equipment.

Conditioned Analog Line. Analog line to which devices have been added or removed to improve the electrical signal.

Congestion. Excessive network traffic.

Connection (or call) Spoofing. The concept of mimicking correct responses to keep level requests alive at the local end of a temporarily broken connection. Saves connect time charges by allowing the call to be disconnected without causing the NOS to time-out the client/host connection. It also enhances data throughput by keeping the line clear of network administration packets.

Connection Management. Process in FDDI for controlling the transition of the ring through its various operating states (off, connect, active, etc.); defined by the X3T9.5 specification.

Connection Oriented. A term applied to network architecture and services which require the establishment of an end-to-end, predefined circuit prior to the start of a communications session. Frame relay circuits are examples of connection-oriented sessions. See Connectionless.

Connectionless. A term applied to network architecture and services which do not involve the establishment of an end-to-end, predefined circuit prior to the start of a communications session. Cells or packets are sent into the connectionless network, and are sent to their destination based on addresses contained within their headers. The Internet and SMDS are two examples of connectionless networks. See Connection Oriented.

Connectionless Broadband Data Service. European high-speed, packet-switched, datagram-based WAN networking technology; SMDS alternative.

Connectionless Communications. A form of packet-switching that relies on global addresses in each packet rather than on predefined virtual circuits.

Connectionless Network Protocol. OSI protocol used by a competitive local network service for the delivery of data; uses datagrams that include addressing information for routing network messages. Used in LANs rather than WANs.

Connectionless Network Service. OSI packet-switched network in which each packet of data is independent and contains complete address and control information; can minimize the effect of individual line failures and distribute the load more efficiently across the network. Does not require a circuit to be established before data is transmitted.

Connection-Oriented Communications. A form of packet switching that requires a predefined circuit from source to destination to be established before data can be transferred.

Connection-Oriented Network Service. An OSI protocol for packet-switched networks that exchange information over a virtual circuit (a logical circuit where connection methods and protocols are pre-established); address information is exchanged only once. The CONS must detect a virtual circuit between the sending and receiving systems before it can send packets.

Connectivity System. A collection of network devices that are logically related and managed as a single entity.

Constant Bit Rate (CBR). Specifies a fixed bit rate so that data is sent in a steady stream. This is analogous to a leased line.

Consumer DSL (CDSL). A trademarked version of DSL that is somewhat slower than ADSL (1 Mbps downstream, probably less upstream) but has the advantage that a "splitter" does not need to be installed at the user's end. Rockwell, which owns the technology and makes a chipset for it, believes that phone companies should be able to deliver it in the $40-45 a month price range. CDSL uses its own carrier technology rather than DMT or CAP ADSL technology. CDSL has never really enjoyed much DSL market penetration.

Contention. Network access method where devices compete for the right to access the physical medium.

Controlled Environmental Vault (CEV). Underground outside plant network equipment hut.

Controller Access Unit. A managed concentrator on a Token Ring network; essentially, an intelligent version of an MAU; handles the ring in/ring out function.

Convergence. Convergence originally described the merging of traditional circuit-switched networks (e.g., PSTN) with modern packet networks, such as the Internet. Convergence has become more broad a concept in recent years: At some unspecified time in the future, all information will be digital, all networks will become one (no more separate TV, voice, or data networks), all user devices will just be various forms of computers.

Copper Distributed Data Interface (CDDI). Standard for FDDI over UTP or STP copper media. Provides data rates of 100 Mbps over transmission distances of about 100 meters. Uses a dual-ring architecture to provide redundancy. Compare with FDDI.

Copper To The Curb (CTTC). Also called Copper From The Curb (CFTC). This is the use of copper wire to extend from a network fiber multiplexer hub in the outside plant to the customer premises. Used for POTS and with DSL, HDSL, ADSL or UDC for transport of digital services. See FITL, FTSA and FTTC.

Core Network. Combination of switching offices and transmission plants connecting switching offices together. In the U.S., local exchange core networks are linked by several competing interexchange networks; in the rest of the world, the core network extends to national boundaries.

CoS. See Class of Service.

COT. Central Office Terminal, a universal DLC terminal located in the CO.

CP. See Call-Progress Tones.

CPE. See Customer Premises Equipment.

CP-IWF. Customer Premises Interworking Function

Craft Interface or Craft Port. An interface based upon an RS232 port, asynchronous ASCII, and a command line interface used for direct access to the element by a technician. The connection can be either direct or via a modem.

CRC. See Cyclic Redundancy Check

CRC Errors. See Cyclic Redundancy Errors.

Crosstalk. The interference caused by signals on adjacent circuits in a network. Annoying enough in the analog voice network, crosstalk is a hazard that limits distance and speed on digital networks.

CSA. See Carrier Service Area.

CSD. See Circuit-Switched Data.

CSN. See Circuit-Switched Network.

CSU. See Channel Service Unit.

CSU/DSU. See Channel Service Unit (CSU) / Data Service Unit (DSU).

CSV. See Circuit-Switched Voice.

CSV/CSD. See Alternate Circuit-Switched Voice/Circuit-Switched Data.

CTD. See Cell Transfer Delay.

Custom Local Area Signaling Services (CLASS). Number-translation services, such as call-forwarding and caller identification, available within a local exchange of a Local Access and Transport Area (LATA).

Customer Located Equipment (CLE). Equipment residing on the customer's premises, but owned, installed, provisioned, and managed by an ICP or BLEC.

Customer Premise Equipment (CPE). Service provider equipment that is located on the customer's premises (physical location) rather than on the provider's premises or in between.

Customer Premises. Typically the building owned or leased by the end user who uses services provided from the central office to enter the public-switched telephone network (PSTN).

Cyclic Code. An error correcting code implemented with a feedback shift register.

Cyclic Redundancy Check (CRC). Method for checking the accuracy of a digital transmission over a communications link. The sending computer performs a calculation on the data and attaches the resulting value; the receiving computer performs the same calculation and compare its result to the original value. If they do not match, a transmission error has occurred and the receiving computer requests retransmission of the data.

CVoDSL. See Channelized VoDSL.

D

D Channel. In an ISDN interface, the "delta", "data" or D channel is used to carry control signals and customer call data in a packet-switched mode. In the BRI (basic rate interface) the D channel operates at 16 Kbps, part of which will handle setup, teardown, and other characteristics of the call. Also, 9600 BPS will be free for a separate conversion by the user. In the PRI (Primary Rate Interface), the D channel runs at 64 Kbps.

DA. See Distribution Area.

DACS. See Digital Access & Cross-Connect System.

Data Communications Equipment. See DCE.

Data Encoding. The method by which a modem encodes digital data onto an analog signal for transmission to a remote modem to which it is connected via the PSTN.

Data Encryption Standard. See DES.

Data Exchange Interface. Allows a DTE (such as a router) and a DCE (such as an ATM DSU) to provide an ATM UNI for networks.

Data Link Connection Identifier (DLCI). An address (in a Frame Relay header) which corresponds to a particular destination value for the frame.

Data Link Connection Identifier (DLCI). The

frame relay virtual circuit number used in internetworking to denote the port to which the destination LAN is attached.

Data Link Switching. Method of encapsulating, or tunneling, Logical Link Control Type 2 (LLC2) packets from LAN-based SNA and NetBIOS applications, enabling them to traverse a non-SNA backbone. Specified in FRC 1434.

Data Over Cable Service Interface Specifications (DOCSIS). A standard used for defining the implementation of broadband cable networks. DOCSIS 2.1 defines a standard way for implementing Voice over Cable, which is a competitor to Voice over DSL (VoDSL).

Data Terminal Equipment (DTE). End-user equipment, typically a terminal or computer, that can function as the source or destination point of communication on the network.

Datagram. Logical block of data sent as a network layer unit over a transmission medium without prior establishment of a virtual circuit. Contains source and destination address information as well as the data itself. IP datagrams are the primary information units in the Internet.

dB. See Decibel.

DCE (Data Communications Equipment). A communications device that can establish, maintain, and terminate a connection (for example, a modem). A DCE may also provide signal conversion between the data terminal equipment (DTE) and the common carrier's channel.

DCS (Digital Cross-connect System). Allows flexible electronic interchange of subchannels for DS hierarchy channels.

DDS (Digital Data Service). Private line digital service that provides digital communication circuits with data rates of 56/64 Kbps.

Decibel (dB). One tenth of a Bel. A Bel is the log decimal of a power ratio.

Dedicated Line. A transmission circuit installed between two sites of a private network and "open, " or available, at all times.

Dedicated Line. A transmission circuit that is reserved by the provider for the full-time use of the subscriber. Also called a Private Line.

Delay. A contributing measure of the carrying capacity of a link, delay indicates how long it takes bits to find their way through a network, but says nothing about the bandwidth through the network. Delay can be zero, but the network can be useless if it only delivers one bit per hour. This is important for bandwidth bound applications such as bulk data transfers and VoDSL.

Delay Bound. An application, which will not necessarily benefit from more bandwidth in a network, but can only run properly with a minimum and stable delay at their disposal. A voice telephone call is a good example of a delay bound application. Adding more bandwidth beyond what it needs will not make the voice call any better.

Demarcation Point. The point at the customer premises where the line from the telephone company meets the premises wiring. From the demarcation point, the end-user is responsible for the wiring.

Demodulation. (1) Conversion of a carrier signal or waveform (analog) into an electrical signal (digital). (2) Opposite of modulation; the process of retrieving data from a modulated carrier wave.

DEMUX, demux. Demultiplexer that takes voice or data from TDM timeslots and reconstructs the original channels.

Dense Wave Division Multiplexing (DWDM). A SONET term. High-speed versions of WDM, which is a means of increasing the capacity of SONET fiber optic transmission systems through the multiplexing of multiple wavelengths of light. Each wavelength channel typically supports OC-48 transmission at 2.5 GBPS. A 32-channel system will support an aggregate 80 GBPS.

Desktop Collaboration. Using xDSL lines, linking desktop computers so teleworkers, suppliers and clients can share documents and work together no matter where they are.

Desktop Videoconferencing. By combining xDSL technology and individual PCs, people can meet "face-to-face" without leaving their offices.

It's a unique way to reduce costly and time-consuming travel. The basic desktop video conferencing system includes a video camera, a video card, and an adapter card.

DF (Distribution Frame). A cross-connect arrangement for flexible linking of cables and/or equipment. See also MDF, IDF.

DFE (Decision Feedback Equalizer). An adaptive filter used to compensate for the frequency response of the channel.

DFT. See Discrete Fourier Transform.

DHCP. See Dynamic Host Configuration Protocol.

Dial Up. The process of initiating a switched connection through the network; when used as an adjective, this is a type of communication that is established by a switched-circuit connection.

Differentiated Services (DiffServ). A protocol for specifying and controlling network traffic by class so that certain types of traffic get precedence. For example, voice traffic, which requires a relatively uninterrupted flow of data, might get precedence over other kinds of traffic. DiffServ is known as an implementation methodology for QoS service for IP networks. With DiffServ there is a cost associated with higher quality services and a risk with lower quality services.

DiffServ. See Differentiated Services.

Digital. Having only discrete values, such as 0 or 1. Opposite of analog, which is continuously varying over time. A text file on a computer is a good example of digital information and voice is the prime example of analog information. However, either can be sent over a telecommunications link with an analog or digital signal.

Digital Access & Cross-Connect System (DACS). A device that allows DS-0 channels to be individually routed and reconfigured.

Digital Hierarchy. The progression of digital transmission standards typically starting with DS-0 (64) and going up through at least DS-3. Twenty-four DS-0s make up a DS-1; 28 DS-1s make up a DS-3. There are other links (including a DS-2), but these are less common.

Digital Loop Carrier (DLC). A device the phone company uses to extend the reach of the phone service to business parks and remote locations. The network transmission equipment, consisting of a CO terminal and a remote terminal, is also used to provide a pair gain function. It concentrates many local loop pairs onto a few high-speed digital pairs or one fiber optic pair for transport back to the CO.

Digital Service Unit (DSU). Converts synchronous interface to 4 wire digital circuit.

Digital Service Unit/Channel Service Unit (DSU/CSU). The interface required to change one form of digital signal to another. Many of the devices used in xDSL technologies are basically advanced forms of DSU/CSU, such as the HTU (HTML Termination Unit). Contrast with Modem.

Digital Services Level 1 (DS-1 or DS1). The 1.44 Mbps (U.S.) or 2.108 Mbps (Europe) digital signal carried on a T-1 circuit.

Digital Services Level 3 (DS-3 or DS3). The 44 Mbps digital signal carried on a T-3 circuit.

Digital Signal. Standard specifying the electrical characteristics for data transmission over four-wire Telco circuits. DS1 is 1.544 Mbps, and DS3 is 44.736 Mbps. Also referred to as T-1 and T-3.

Digital Signal Processor (DSP). A DSP is a specialized type of processor that is optimized for performing detailed algorithmic operations on analog signals after they have been digitized. DSPs handle line signaling in modems and IADs.

Digital Subscriber Line (DSL). Point-to-point public network access technologies that allow multiple forms of data, voice, and video to be carried over twisted-pair copper wire on the local loop between a network service provider's central office and the customer site.

Digital Subscriber Line Access Multiplexer (DSLAM). The piece of equipment that resides in a central office that concentrates all remote DSL lines into a single terminating point or device. A small DSLAM in an office, apartment building, hospital or hotel is a "mini-DSLAM."

Digital Transmission. Transmission of voice,

video, and other data that has been encoded as binary values and then transmitted as electrical pulses. Analog-to-digital conversion converts continuous waveforms into digital information that can be processed and stored in a computer.

Digital-to-Analog Converter (DAC). Device that reconstructs analog voltage waveforms from an incoming sequence of binary digits; does not in itself introduce noise.

Directory. A special-purpose database that contains information about the nodes or devices attached to an enterprise network.

Discrete Fourier Transform (DFT). A signal transformation that is often implemented as a fast Fourier transformation on a digital signal processor.

Discrete MultiTone (DMT). DSL technology using digital signal processors (DSPs) to divide an ADSL circuit signal into 256 subchannels or subcarriers (in practice fewer are used) to modulate VoDSL voice and data over the local loop. To deliver extremely high bandwidth VDSL services over copper wire, DMT uses advanced digital signal processing techniques such as the Fast Fourier Transform (FFT) to modulate data on up to 4,096 subcarriers. DMT modems maximize data throughput by dynamically adapting the power level and payload size of each sub-carrier to match noise conditions. During start-up, the modems characterize the signal-to-noise ratio (SNR) on each sub-carrier and, depending on the SNR value, modulate the phase and amplitude (QAM modulation) of each sub-carrier wave so as to carry between 1 to 15 bits of data payload.

Discrete Wavelet Transform (DWT). An alternative to DCT. The sidelobes of the sub-channels are narrower than those for the wavelet transform than those of the Fourier cosine transform. See also DCT, Wavelet Transform, Fractal Transform.

Distributed Management. Approach to network management disperses data collection, monitoring, and management responsibilities among multiple consoles across the network. It provides the capability to collect data from all data points and sources on the network, regardless of network topology, giving network managers the power to manage large, geographically dispersed enterprise networks more effectively.

Distributed Queue Dual Bus. Communication protocol proposed by IEEE 802.6 committee for use in MANs.

Distributed Recovery Intelligence. The ability to track down a network problem and automatically isolate the malfunctioning node.

Distribution. Portion of the telephone cabling plant that connects subscribers to feeder cables from the CO.

Distribution Area (DA). A loop serving area for a Feeder Distribution Interface (FDI). Each CSA has several DAs. The maximum DA loop tends to be about 2/3 the length of the maximum "legal" CSA loop.

Distribution Cable. The portion of the telephone loop plant that connects the feeder cable to the drop wires.

DLC. See Digital Loop Carrier.

DLCI. See Data Link Connection Identifier.

DMS100. A digital central office switching system made by Northern Telecom.

DMT. See Discrete Multi-Tone.

DN. See Directory Number.

DNS. See Domain Name Service.

DOCSIS. See Data Over Cable Service Interface Specifications.

Domain. Part of an Internet naming hierarchy. An Internet domain name consists of a sequence of names (labels) separated by periods; for example "dsl.com".

Domain Name. A text-based alias for an IP address based on the domain name system. While an IP address is a string of digits separated by periods, as in 132.251.125.120, the domain name can even include the name of a particular organization, such as Verizon.com.

Domain Name Service (DNS). A TCP/IP protocol for discovering and maintaining network resource information distributed among different servers.

Domain Name System. Internet electronic-mail

system for translating names of network nodes into addresses.

Download. The process of transferring a file from a server to a client.

Downstream. In a communications circuit there are two circuits: one coming to you, the end user, and one going away from you. The downstream channel comes to you. Also, the digital transmission path from the central site or Internet service provider (ISP) to the end user site.

Drop Cable. A cable that connects a network device such as a computer to a physical medium such as an Ethernet network. Drop cable is also called transceiver cable because it runs from the network node to a transceiver (a transmit/receiver) attached to the trunk cable. Compare with AUI cable.

Drop Wire. The section of the local loop connecting the distribution cable to the customer premises.

Dry Copper. A term used to describe copper telephone lines that are installed but currently not used.

DS-0 (or DS0). Digital Signal Level Zero, or one 64 Kbps voice channel.

DS-1 (or DS1). Digital Signal Level 1: In the digital hierarchy, this signaling standard defines a transmission speed of 1.544 Mbps; a DS-1 can be composed of 24 DS-0 signals; this term is often used interchangeably with T-1. The unframed version, or payload, is 192 bits at a rate of 1.536 Mbps.

DS1C (or DS-1C). Digital Signal Level 1C: Has a 3.152 Mbps payload, can transport two asynchronous DS-1s via bit-stuffing.

DS-2 (or DS1). Digital Signal Level 2: Four T-1 frames packed into a higher-level frame transmitted at 6.312 Mbps.

DS-3 (or DS3)— The Digital Signal Level 3: A data communications service for data-intensive businesses requiring extremely high transmission speeds.In the digital hierarchy, this signaling standard defines a transmission speed of 44.736 Mbps; a DS-3 is composed of 28 DS-1 encapsulated into one digital "pipe;" this term is often used interchangeably with T-3.

DSL. See Digital Subscriber Line.

DSL Access Multiplexer (DSLAM). A device that terminates multiple local subscriber DSL lines and passes the data to an access switch before connection to the network. The DSLAM is analogous to the CIMS in broadband cable networks.

DSL Concentrator. A device that serves as a point of consolidation of DSL network links so that multiple circuits may share common limited network resources.

DSL Forum. Standards body with strong influence over development of DSL direction.

DSLAM. See Digital Subscriber Line Access Multiplexer.

DSP. See Digital Signal Processor.

DSU/CSU. See Digital Service Unit/Channel Service Unit.

DSX-1 - DS-1 Cross-connect. A 1.544 Mbps AMI signal used for short distances to interconnect equipment within a CO.

DTE (Data Terminal Equipment). The equipment, such as a computer or terminal, that provides data in the form of digital signals for transmission.

DTMF. See Dual Tone Multiple Frequencies.

Dual Tone Multiple Frequencies (DTMF). Also called Touchtones (AT&T trademark) and touch-tones. These are tones that are generated when a button is pressed on a telephone handset's touch-tone keypad.

Dual-Attached Concentrator. A device that is attached to and allows access to both rings in an FDDI network.

Dual-Attached Servers. Servers attached to both paths of an FDDI ring for load balancing and redundancy.

Dual-Attached Station. A station with two connections to an FDDI network, one to each logical ring. If one of the rings should fail, the network automatically reconfigures to continue normal operation. Compare with SAS.

DWDM. See Dense Wave Division Multiplexing.

DWMT (Discrete Wavelet Multi-Tone). A multicarrier modulation system pioneered by Aware Inc. that, according to the vendor, isolates its subchannels in a method that is superior to conventional DMT modulation. In the vendor's own words, "DWMT is able to maintain near optimum throughput in the narrow band noise environments typical of ADSL, VDSL, and Hybrid Fiber Coax, while DMT systems may be catastrophically impaired."

DWT. See Discrete Wavelet Transform.

Dynamic Bandwidth Allocation. A key feature of xDSL devices that allows automatic adjustment of the number of channels and the size of the "data pipe."

Dynamic Bandwidth Allocation. See Bandwidth-on-Demand.

Dynamic Host Configuration Protocol (DHCP). A service (a TCP/IP protocol) that lets clients on a LAN request configuration information, such as IP host addresses, from a server.

Dynamic IP Addressing. An IP address is assigned to the client for the current session only. After the session ends, the IP address returns to a pool of IP addresses. Contrast with Static IP Addressing.

Dynamic Routing. Routing that adjusts automatically to changes in network topology or traffic.

E

E.164. ITU-T recommendations for telecommunications numbering, including ISDN, BISDN, and SMDS; includes telephone numbers up to 15 digits long.

E-1 (or E1). European equivalent of a T-1 circuit. It is a term for a wide band digital interface used for transmitting data over a telephone network at 2.048 Mbps.

E-3. European designation for a T-3, a long-distance, point-to-point communications circuit service created by AT&T; it operates at 44 Mbps and can carry 672 channels of 64 Kbps.

E911 POTS. A minimal telephone service to provide life-line service when there has been a loss of power.

Early Packet Discard (EPD). An intelligent packet discard algorithm used to control congestion. EPD checks the first cell of a VC for buffer threshold problems. If it encounters a problem, it drops all cells (up to and including the next end-of-frame) from the particular AAL5 packet.

EC. See Echo Cancellation.

ECH. See Echo-Cancelled Hybrid.

Echo. The reflecting of a signal back to its source due to a variety of reasons. Whenever the same bandwidth is used for transmission in both directions, echo is a concern. In all cases, some form of echo control must be used to compensate for these effects, which can be annoying for voice but devastating for data. Both the voice network and simple modems employ echo cancellation techniques. Also known as "positive feedback" or "singing."

Echo Cancellation (EC). Echo describes the reflection of voice spoken into a POTs handset back to the earpiece. Echo cancellation is the process by which a transmitter/receiver cancels out the transmitted signal as to "hear" the received signal better.

Echo Canceller. Echo cancellers reduce or eliminate echo that is inherent in all voice traffic, and which is most pronounced in long-distance and wireless telephone calls. Echoes are produced and enhanced at various points in the network. Without echo cancellation, a talker would be able to hear their speech after a set delay, resulting in a very unnatural sounding conversation.

Echo-Cancelled Hybrid (ECH). A 2-to-4 wire conversion with Echo Cancellation. A hybrid transformer is often used to interface to the line.

ECSA. See Exchange Carrier Standards Association.

Edge Connection. Connection by which desktop computers are connected to one another and to local servers, usually by means of switches and hubs; any set of interconnected computers and servers form a local area network (LAN). LANs can also be connected to each other.

Edge Device. A device, such as a router or Ethernet-to-ATM switch, that is directly connected to an ATM network. The UNI defines the connection between the edge device and the ATM network switch. It is the first device a user sees when sending traffic to the ATM network. Also called an end device.

EDI. See Electronic Data Interchange.

EFS. See Error-Free Seconds.

Egress. Refers to outgoing traffic.

EIA/TIA. See Electronic Industries Association/ Telecommunications Industry Association.

EKTS. See Electronic Key Telephone Service.

Electromagnetic Compatibility (EMC). Prevents unintended radio frequency interference (see RFI).

Electronic Industries Association. Groups that together have specified data transmission standards such as EIA/TIA-232 (formerly known as RS-232).

Electronic Industries Association/Telecommunications Industry Association (EIA/TIA). This organization provides standards for the data communications industry to ensure uniformity of the interface between DTEs and DCEs.

Element Management System (EMS). A management system that provides functions at the element management layer.

Embedded Operations Channel (EOC, eoc). Communications channel for network use and separate from the customer channels in the data stream.

EMC. See Electromagnetic Compatibility.

Emergency 911 dialing. The capability to access a public emergency response system by dialing the digits 9-1-1. State and local requirements for support of Emergency 911 Dialing service by Customer Premises Equipment vary. Consult your local telecommunications service provider regarding compliance with applicable laws and regulations.

EMI (ElectroMagnetic Interference). Interference, usually manifested as a hum, static, or buzz, in audio equipment, that is produced by the equipment or cabling picking up stray electromagnetic fields.

EMS. See Element Management System.

Encapsulation. Wrapping a data set in a protocol header. For example, Ethernet data is wrapped in a specific Ethernet header before network transit. Also, a method of bridging dissimilar networks where the entire frame from one network is simply enclosed in the header used by the link-layer protocol of the other network.

End System. End-user device on a network; also, a non routing host or node in an OSI network.

End System-to-Intermediate System Protocol. The OSI protocol by which end systems such as networks personal computers announce themselves to intermediate systems such as hubs.

End User. A person who uses applications; typically with the aid of a computer system.

Enterprise Gateway. A gateway designed for use in a corporate or campus environment, typically with a lower call capacity and designed with different physical/environmental constraints than a carrier-class gateway.

Enterprise Network. A widely dispersed, multifaceted telecommunications network for a particular purpose or organization; a term for all of an organization's telecommunications networking services and equipment.

Entity. Individual, manageable device in a network. Also, OSI terminology for a layer protocol machine. An entity within a layer performs the functions of the layer within a single computer system, accessing the layer entity below and providing services to the layer entity above at local service access points.

EOC. See Embedded Operations Channel.

EPD. See Early Packet Discard.

Error Correction. Techniques used to correct errors in data transmission, typically caused by noise.

Errored Second (ES). A second of received data with one or more bit errors. A tariffed measure of circuit performance. See also EFS, SES.

Error-Free Seconds (EFS). A tariffed measure of performance. See also ES, SES.

ES. See Errored Second.

ESF. See Extended Superframe.

Ethernet. A physical medium for transmitting

local area network (LAN) traffic at speeds up to 1 Gbps and metro area network (MAN) traffic at up to 10 Gbps. Within the OSI model, Ethernet is defined at layer one (physical) and Layer Two (data link layer). Based on Carrier Sense Multiple Access/Collision Detection (CSMA/CD), Ethernet works by simply checking the wire before sending data. Sometimes two stations send at precisely the same time in which case a collision is detected and retransmission is attempted. Ethernet is a widely implemented standard for LANs. See also 10Base-T or 100Base-T.

Ethernet Phone. New type of phone that plugs directly into an Ethernet network. The benefit is that moves, adds, and changes of phones are easy. Over time enterprises will need only a single wired network for both voice and data communications. Ethernet can be carried over xDSL.

ETSI. See European Telecommunications Standardization Institute.

EU. European Union. Formerly known as EC, European Commission.

European Telecommunications Standardization Institute (ETSI). An organization that produces technical standards in the area of telecommunications.

Excessive Zeroes (EXZ). Condition of greater than 7 consecutive zeroes for T-1 B8ZS, greater than 15 consecutive zeroes for T-1 AMI, or greater than 3 consecutive zeroes for E-1 AMI or E-1 HDB3.

Exchange Area. A geographical area in which a single, uniform set of tariffs for telephone service is in place. A call between any two points in an exchange area is considered a local call. See also LATA (local access and transport area).

Exchange Carrier Standards Association (ECSA). Sponsor of T1 Standards Working Groups. Supplanted by ATIS (late 1993).

Extended Superframe (ESF). A DS-1 signal frame format: frames in a 3 ms superframe with CRC and a data link.

Exterior Gateway Protocol. Internet routing protocol by which gateways exchange information about what systems they can reach; documented in RFC 904. Generally, an exterior gateway protocol is any internetworking protocol for passing routing information between autonomous systems. Compare with BGP.

EXZ. See Excessive Zeroes.

F

Far End CrossTalk (FEXT). Leakage of one or more foreign signaling sources into the receiver of a system at the remote end of a transmission system, causing interference.

FAS. See Frame Alignment Signal.

Fast Ethernet. A LAN used to connect devices within a single building or campus at speeds up to 100 Mbps. Within the OSI model, Fast Ethernet is defined at layer one (physical) and layer two (data link). Like Ethernet, it uses CSMA/CD.

Fast Fourier Transform (FFT). An algorithm for efficiently implementing via digital signal processors the conversion from the time-domain to the frequency-domain.

Fast IP. Fast IP is 3Com's strategy for providing IP switching across all types of network backbone technologies, including Ethernet, Fast Ethernet, Gigabit Ethernet, FDDI, Token Ring, and ATM. An extension of 3Com's High-Function Switching, Fast IP is designed to scale to the forwarding requirements of today's high-speed networks and tomorrow's next-generation network technologies. Fast IP combines the control-policy function of routing with the wire-speed forwarding performance of switching.

Fault Tolerance. Generally, the ability to prevent a problem on a device from affecting other devices on the same port.

Fax Relay. Fax relay is a technique for the transmission of fax calls over packet networks. Fax relay systems demodulate (receive) the fax transmission at the near-side gateway, then transmit the fax across the packet network according to the fax relay standard, and finally modulate (transmit) the fax from the far-side gateway to the receiving fax machine.

FCC. See Federal Communications Commission.

FDDI (Fiber Distributed Data Interface). LAN technology that permits data transfer on fiber-optic cable speeds up to 100 Mbps or more over a dual, counter-rotating token ring; ANSI standard X3T9.5.

FDI. See Feeder Distribution Interfaces.

FDM. See Frequency Division Multiplexing.

FEC. See Forward Error Control.

Federal Communications Commission (FCC). The U.S. federal regulatory agency responsible for regulating interstate and international communications.

Feeder. That portion of the telephone cable plant that extends from the CO to distribution frames where distribution cables deliver traffic to subscribers.

Feeder Distribution Interfaces (FDI). Points where cable bundles from the telephony switch use drop lines extended out to the customer premises.

Feeder Network. That part of a public switched network which connects access nodes to the core network.

FEXT. See Far End CrossTalk.

FFT. See Fast Fourier Transform.

Fiber Distributed Data Interface. See FDDI.

Fiber In The Loop (FITL). General term for application of optical fiber and associated remote electronics in the outside plant. Defined in TA-NWT-000909. See also FTTH and FTTC.

Fiber Optic Cable. A transmission medium composed of glass or plastic fibers; pulses of light are emitted from a LED or laser-type source. Fiber optic cabling is the present cabling of choice for all interexchange networks, and increasingly for the local exchange loops as well; it is high security, high bandwidth, and takes little conduit space. Considered the physical medium of all future, land-based communications.

Fiber Optic Inter-Repeater Link (FOIRL). Fiber-optic signaling methodology based on the IEEE 802.3 fiber-optic specification.

Fiber Optics. Communications technology that uses thin filaments of glass or other transparent materials. Fiber optic technology offers extremely high transmission speeds, and in the future, will allow for services such as "video on demand."

Fiber to Service Area (FTSA). A CATV term.

Fiber to the Cabinet. Generic term for Fiber To the Curb. Network architecture where an optical fiber connects the telephone switch to a streetside cabinet where the signal is converted to feed the subscriber over a twisted copper pair.

Fiber To The Curb (FTTC). A telephone company service delivery system that delivers voice and video programming to small clusters of residences using fiber optics as the feeder and either twisted pairs or coax cable as the distribution plant to each home.

Fiber To The Home (FTTH). Network where an optical fiber runs from telephone switch to the subscriber's premises or home.

File Transfer Protocol (FTP). A file transfer protocol typically used for uploading and downloading of files and operational code. A common use of FTP on the Internet is to download software programs.

Firewall. A system or group of systems that enforces an access control policy between an organization's network and the Internet for purposes of security.

FITL. See Fiber In The Loop.

Flat Pair. Transmission media in which the two wires are parallel with each other. Often the pair is covered with an insulator for mechanical structure and control of the electrical properties. See also Balanced Line, Drop Wire, Twisted Pair.

FM. See Frequency Modulation.

FoIP (Fax-over-IP). Two FoIP protocols exist T.37 (store and forwarded) and T.38 (real-time).

Forward Error Control (FEC). Errors are corrected by the receiver using redundant information sent by the transmitter.

Fourier Transform. Uses a sinusoidal expansion to represent a signal.

FPS. See Frames Per Second.

FR. See Frame Relay.

Fractional T-1. A WAN communications service that allows a customer to subscribe to some portion of a T-1 (1.544 Mbps) circuit. Originally these were divided into 24 separate 64 Kbps channels, but modern technology allows the sale of bandwidth in increments from 2.4 Kbps to 56/64 Kbps. In Europe, they deal in Fractional E-1s.

Fractionally Spaced Equalizer. An equalizer using multiples of the symbol rate.

FRAD. See Frame Relay Assembler/Dissembler.

Frame. (1) A variable length unit of information. Frames contain packets and are subject to varying delays as they make their way through a network. Nevertheless, frames are the most popular way of transporting packets. Contrast with Cell. (2) A fixed length unit used for the transport of bits over a physical link. This is technically a transmission frame and forms part of a framed transport. All xDSL technologies are framed transports.

Frame Relay. A high-speed connection-oriented packet switching WAN protocol using variable-length frames, similar to X.25. Frame Relay is a leading contender for LAN-to-LAN interconnect services, and is well suited to the bursty demands of LAN environments. See also Permanent Virtual Circuit and Switched Virtual Circuit.

Frame Relay Assembler/Dissembler (FRAD). Also called a Frame Relay Access Device. The FRAD connects non-frame relay devices to the frame relay network.

Frame-based ATM User Network Interface (FUNI). Standard protocol stack for supporting both ATM and Frame Relay in the same device.

Frames Per Second (FPS). The number of frames per second of video images displayed on the screen. Term most often used when talking about the speed of video capture and playback. The higher the frame rate, the more fluid the motion appears. The highest, or best, quality frame rate available is 30 fps. Lower frame rates (below 10) still appear as motion but are noticeably "jerky," and zero fps corresponds to a still frame (no motion).

Frequency Division Duplex (FDD). Two-way transmission via Frequency Division Multiplexing (FDM).

Frequency Division Multiplexing (FDM). A multiplexing technique that uses different frequencies to combine multiple streams of data for transmission over a communications medium. FDM assigns a discrete carrier frequency to each data stream and then combines many modulated carrier frequencies for transmission.

Frequency Modulation (FM). FM uses changes in frequency as a carrier signal to represent information.

FRSP. Frame Relay Service Provider.

FTP. See File Transfer Protocol.

FTSA. See Fiber to Service Area.

FTTC. See Fiber To The Curb.

FTTCab. See Fiber To The Cabinet.

FTTH. See Fiber To The Home.

Full Duplex. Refers to the ability of a device or line to transmit data in two directions simultaneously.

FUNI. See Frame-based ATM User Network Interface.

G

G.168. The ITU's most recent and complete standard for line echo cancellation. G.168 does not define how to implement line echo cancellation, but rather defines an extensive test suite for the echo canceller. G.168 (2000) is the most recent version of the G.168 standard.

G.711. The ITU's standard for transmitting pulse-code modulated voice on the PSTN. G.711 is a noncompressed voice standard, with each voice channel occupying a bandwidth of 64 Kbps.

G.dmt. Another term for full-rate Asymmetric Digital Subscriber Line, which allows the ADSL line to support up to 8 Mbps downstream and 1 Mbps upstream and requires that a device called a POTS splitter be installed at the subscriber's premises.

G.Lite. G.lite ADSL (also known as universal ADSL) is a standard for DSL service that became available in mid-to-late 1999. The cost for equipment and service is less than other varieties. It's all easier to install than other varieties. you can do it yourself. It is based on ADSL, and offers downstream speeds up to 1.5 megabits per second and a maximum upstream data rate of 384 kilobits per second. The major downside to G.lite ADSL is that it may have too little bandwidth to be acceptable for voice and entertainment applications.

G.SHDSL. See Single pair High-bit-rate Digital Subscriber Line.

Gateway. A computer system that transfers data between applications or networks that use different protocols. A gateway reformats the data to make it acceptable for the new application or network before passing the data on. There are many kinds of gateways. Gateways provide address translation services, but don't necessarily translate data. In next generation networks, a gateway commonly refers to a box which transcodes voice traffic between the PSTN and modern packet networks. The gateway performs a layer-7 protocol-conversion to translate one set of protocols to another (for example, from TCP/IP to SNA or from TCP/IP to X.25). A gateway operates at Open Systems Interconnection (OSI) layers up through the Session Layer. Contrast it with the Bridge and the Router.

GBPS. See Gigabits per second.

GCRA. See Generic Cell Rate Algorithm.

Generic Cell Rate Algorithm (GCRA). An algorithm set forth by the ATM Forum to measure conformance for cells in relation to the traffic contract of the associated connection.

Gigabit Ethernet. A 1000-Mbps technology based on the 10BASE-T Ethernet CSMA/CD network access method to accommodate the operation of local area networks.

Gigabits per second (Gbps). A measure of bandwidth or throughput based on 230 (1,073,741,824) bits per second (slightly over a billion).

GR-303 (Generic Requirement 303). Formerly known also as TR-303. A specification issued by Bellcore (now Telcordia) that governs the interconnection of loop carrier equipment with Class 5 digital switches. Although often called an "interface" it is really an access protocol that runs between the Class 5 Local Digital Switch (LDS), also called the Integrated Digital Terminal (IDT), and the Access Equipment (the Remote Digital Terminals or RDTs) which provides network access for the subscribers. Because of its flexibility in concentrating bandwidth analog and digital (e.g. ISDN) services bandwidth into the switch, GR-303 enables service provider to scale up to accommodate additional subscribers who may be using both digital equipment and analog media. GR-303 runs on dedicated redundant control channels also known as Timeslot Management Channels (TMCs) and Embedded Operation Channels (EOCs). TMCs are used for timeslot allocation and deallocation during the assignment and management operations on the LDS interface; EOCs are used for remote management operations such as maintenance and alarm surveillance. As for its physical location, the GR-303 interface protocol can reside on T-1 cards or other line trunk cards, so that additional hardware need not be installed. Although a relic of the PSTN, the GR-303 standard has actually become critical in helping service providers inexpensively migrate to VoATM-based VoDSL as well as ultimately migrate from legacy Time Division Multiplexed (TDM) networks to the new packet-based technologies. GR-303 enables gateways in the packet network to provide POTS, private line, coin, and ISDN services through the installed base of DLC, WLL, and Fiber In The Loop (FITL) systems. Despite the fact that GR-303 is accepted as an open standard, there are several variations on how it is implemented.

Graphical User Interface (GUI). A graphical and mouse-oriented interface used for configuring devices. (Contrasted to a Command Line Interface).

Group 3 Fax. Currently, the most widely used facsimile protocol, which operates over analog telephone lines or with a terminal adapter over DSL.

Group 4 Fax. A facsimile protocol that allows high-speed, digital fax machines to operate over DSL.

GSTN (General Switched Telephone Network). European term for the public telephone network. See Public Switched Telephone Network (PSTN).

GUI. See Graphical User Interface.

H

H (MDF). Horizontal side of MDF: Switch and equipment connection side. See also V (MDF).

H.248. An ITU-T recommendation, also called Megaco by the IETF. Like MGCP, Megaco/H.248 is a control protocol used by Media Gateway Controllers to control Media Gateways. Megaco/H.248 is essentially an extension of MGCP, to handle various forms of communications including voice, video and data.

H.261, px64. ITU-T (CCITT) compression standard for digitally encoded visual telephony. For video including motion such as teleconferences. Data rate = px64 where p = 1 to 30.

H.263. ITU-T standard for video telephony that advances the techniques in H.261. See H.324.

H.320. ITU-T standard for video conferencing for 64 Kbps to 1.544 Mbps in the ISDN environment.

H.324. ITU-T standard for video telephony below 64 Kbps. The umbrella standard for audio and H.263 video.

Half Duplex. Data transmission that can occur in two directions over a single line, but only one direction at a time.

Hamming Distance. The number of bits having a different value for a pair of codewords.

HCDS (High Capacity Digital Service). Service based on DS-1 (1.544 Mbps) and higher rates.

HDB3. See High Density Bipolar Three Zeros Substitution.

HDLC. See High-level Data Link Control.

HDSL. See High Bit Rate Digital Subscriber Line.

HDSL Terminal Unit-Central Office (HTU-C). The HDSL modem or line card that physically terminates an HDSL connection at the telephone service provider's serving central office. Also known as a Line Termination Unit (LTU).

HDSL Terminal Unit-Remote (HTU-R). The HDSL modem or PC card that physically terminates an HDSL connection at the end-user's location. Also known as a Network Termination Unit (NTU).

HDSL-2 (2nd generation HDSL). This variant delivers 1.5 Mbps service each way, supporting voice, data, and video using either ATM, private-line service or frame relay. This ANSI standard for this symmetric service which gives the same fixed rate both up and downstream (up to 1.5 Mbps speed). HDSL2 does not provide standard voice telephone service on the same wire pair. HSDL2 differs from HDSL in two ways: 1) HDSL2 normally uses one pair of wires to convey 1.5 Mbps whereas HDSL uses two wire pairs, and 2) HDSL2 is standardized.

HEC (Header Error Check). Used in ATM cells.

Hertz (Hz). Basic unit of frequency measurement. 1 Hertz = 1 cycle per second.

Heterogeneous LAN Management. Management of LANs that contain dissimilar devices running different protocols and different applications.

HFC. See Hybrid Fibre Coax.

Hierarchical Routing. Routing based on a hierarchical addressing system. IP routing algorithms use IP addresses, for example, which contain network numbers, host numbers and, frequently, subnet numbers.

High Bit Rate Digital Subscriber Line (HDSL). A mature, medium-speed, symmetric technology. It's often used to implement T-1 data circuits over phone lines. HDSL requires two pairs of wire for transmitting and receiving. Downstream and upstream bandwidth is about 1.5 Mbps using two telephone lines with DSL service. See also HDSL2.

High Capacity Digital Service. See HCDS.

High Density Bipolar Three Zeros Substitution (HDB3). A bipolar coding method that does not allow more than three consecutive zeros. Used to accommodate the ones density requirements of E1 lines.

High Pass Filter. A signal filter, which would be installed in a customer premises ADSL modem

(ATU-R), which only allows higher frequency data to be delivered to the modem. See Low Pass Filter.

High-level Data Link Control (HDLC). HDLC is a group of protocols or rules for transmitting data between network points or nodes, it specifies an encapsulation method for data on synchronous serial data links. Various manufacturers have proprietary versions of HDLC, including IBM's SDLC. HDLC data traffic is organized into a unit, called a frame, and sent across a network to a destination that verifies its successful arrival. The HDLC protocol also manages the flow or pacing at which data is sent. HDLC is one of the most commonly-used protocols in Layer 2 of the industry communication reference model, Open Systems Interconnection (OSI). It's used in X.25 communicatoins.

Holding Time. An amount of time that users interact with or over a network. Holding times vary widely, from hours on a cable TV network to minutes for a voice telephone call. Data sessions involving clients and servers on the Internet and Web typically have holding times of about one hour. The voice network becomes stressed because long holding time data sessions are run over voice links designed for voice calls lasting a few minutes.

Host. A single, addressable device on a network. Computers, networked printers, and routers are hosts.

Hot-Swapping. The ability to replace a card or other hardware part in a hardware device without turning it off or losing functionality.

HRU (HDSL Repeater Unit). Used for local loops longer than normally allowed for by CSA. See also HTU-C, HTU-R.

HTML. See Hypertext Markup Language.

HTTP. See Hypertext Transport Protocol.

HTU-C. See HDSL Terminal Unit-Central Office.

HTU-R. See HDSL Terminal Unit-Remote.

Hub. A LAN device that serves as a central "meeting place" for cables from computers, servers, and peripherals. Hubs typically "repeat" signals from one computer to the others on the LAN.

Hybrid Fibre Coax (HFC). A system (usually CATV) where fibre is run to a distribution point close to the subscriber and then the signal is converted to run to the subscriber's premises over coaxial cable.

Hypertext. The usual method of presenting information on the World Wide Web. Information is linked by a series of jumps from place to place. Today the term is misleading because much of the content is audio and view as well as text. However, the term hypermedia is slow to catch on. See also HTML (Hypertext Markup Language).

Hypertext Markup Language (HTML). An authoring software used on the Internet's World Wide Web. HTML is basically ASCII text with HTML commands ("tags") and is used to convert plain text files into the multimedia pages that make up the Web. HTML is a subset of SGML, a more powerful editing language.

Hypertext Transfer Protocol (HTTP). The set of rules for exchanging text, graphic images, sound, video, and other multimedia files on the World Wide Web.

Hz (hertz). A unit of measure for indicating frequency in cycles per second.

I

IAB (Internet Activities Board). The technical body that oversees the development of the Internet suite of protocols.

IAD. See Integrated Access Device.

ICD. See Integrated Concentration Device.

ICP. See IMA Control Protocol (ICP); Integrated Communications Provider.

IDC. See Insulation Displacement Connection.

IDF. See Intermediate Distribution Frame.

IDFT. See Inverse Discrete Fourier Transform.

IDLC. See Integrated Digital Loop Carrier.

IDSL. See Integrated Services Digital Network DSL.

IDSL. See ISDN Digital Subscriber Line.

IDT. See Integrated Digital Terminal.

IEC. Interexchange Carrier (see IXC).

IEEE. See Institute of Electrical and Electronic Engineers.

IEEE 802. IEEE committee to develop local area network (LAN) standards.

IEEE 802.6. The protocol that details MANs that rely on DQDB, a connectionless packet-switched protocol.

IEEE 803.2. The protocol that defines an Ethernet network at the physical layer of network signaling and cabling.

IETF. See Internet Engineering Task Force.

IFFT. See Inverse Fast Fourier Transform.

ILEC. See Incumbent Local Exchange Carrier.

ILMI. See Interim Local Management Interface.

IMA. See Inverse Multiplexing over ATM.

IMA Control Protocol (ICP). Cells that are sent down each ATM link to test availability and to measure delay. Individual links are added to or removed from the group based on the results of the control cells.

IME. See Interface Management Entity.

Impedance. The resistance and reactance a wire offers to a change in current as the current runs down the length of the wire, measured in ohms.

Impulse Noise. An unwanted signal of short duration, often resulting from the coupling of energy from an electrical transient from a nearby source.

Impulse Response. The time-domain response of a network to an input impulse.

In-band. Tones that pass within the voice frequency band and are carried along the same circuit as the talk path.

In-band Signaling. Network signaling carried in the same channel as the bearer traffic. In analog telephone communications, the same circuits used to carry voice are used to transmit the signal for the telephone network. Touch-tone signals are an example of in-band signaling.

Incumbent Local Exchange Carrier (ILEC). A local exchange carrier (LEC), such as Verizon or Bell South that was established before the Telecommunications Reform Act of 1996. All RBOCs are ILECs but not all ILECs are RBOCs. ILECs are in competition with competitive local exchange carriers (CLECs).

Independent Telephone Company (ITC). In the U.S., a telephone company that was not owned by AT&T before divestiture.

Ingress. Refers to incoming traffic.

Institute of Electrical and Electronic Engineers (IEEE). A professional group that designs and defines network standards, LAN standards in particular. Their committees actually develop and propose computer standards, which define the physical and data link protocols of communication networks. Members represent an international cross section of users, vendors and engineering professionals.

Insulation Displacement Connection (IDC). A type of wire connection device in which a wire is punched down into a metal holder that strips away the insulation to achieve electrical connection.

Integrated Access Device (IAD). Enables CLECs, ILECs, ICPs, etc., to deliver voice, video, and data services over packet-based networks using a single platform.

Integrated Concentration Device (ICD). A new category of multiservice CPE platforms introduced and promoted by Avail Networks. The ICD combines the features of IADs, ATM switches, and compact DSLAMs, providing options for deploying broadband multiservice offerings in multi-subscriber environments.

Integrated Digital Loop Carrier (IDLC). A DLC system with IDT and RDT and no separate COT. COT functions are integrated directly into the switch as the IDT. See also DLC, IDT.

Integrated Digital Terminal (IDT). Provides a Digital Loop Carrier (DLC) interface on the switch. The "switch side" of the GR-303 interface. An IDT can send provisioning and status information to a DLC or a Remote Digital Terminal (RDT) using the Embedded Operations Channel (EOC) as defined in the Telcordia GR-303 Specification. See also DLC.

Integrated Services Digital Network (ISDN). An early dial-up predecessor of the DSL family that can support up 128 Kbps symmetrical service. ISDN is a usage-based service offered by the telephone companies that is being rapidly eclipsed by DSL. ISDN comes in two varieties: 1) Basic Rate Interface which uses two 64 Kbps circuit switched bearer or "B" channels and one 16 Kbps packet switched delta or "D" channels for a total bandwidth of 144 Kbps, and 2) Primary Rate Interface which runs at 1.544 Mbps. See also Basic Rate Interface, Broadband ISDN, and Pimary Rate Interface.

Integrated Services Digital Network Digital Subscriber Line (IDSL). A form of DSL that supports symmetric data rates of up to 144 Kbps using existing phone lines. It is unique in that it has the ability to deliver services through a DLC. While DLCs provide a means of simplifying the delivery of traditional voice services to newer neighborhoods, they also provide a unique challenge in delivering DSL into those same neighborhoods. IDSL differs from its relative ISDN in that it is an "always-on" service, but capable of using the same terminal adapter, or modem, used for ISDN.

Interconnection Agreement. A negotiated agreement between a CLEC and an ILEC that sets the terms and conditions under which the CLEC purchases services from the ILEC.

Inter-eXchange Carrier (IXC). Also known as IEC. A U.S. term for a long-distance communications carrier that provides telecommunications services between exchanges (or central offices). IXCs often rely on local exchange carriers (LECs) or competitive access providers for the local origination and termination of their traffic. In Europe, Asia, and other nations around the world, the local telco also serves as the major IXC in the country.

Interface. A point of connection between two systems, networks, or devices.

Interface Management Entity (IME). Software components that execute the ILMI protocol.

Interim Local Management Interface (ILMI). An interim requirements definition in ATM Forum UNI 3.1. It supports bidirectional exchange of management information between UNI management entities related to the ATM layer and physical layer parameters.

Interior Gateway Protocol. Protocols used in TCP/IP networks to exchange routing information between collaborating routers in the Internet. ART, RIP, and OSPF are examples of IGPs.

InterLATA. Connections between local access companies. For example, long-distance connections and services that originate in one LATA and terminate in another.

Intermediate Distribution Frame (IDF). An IDF usually has no protection from over-voltage or over-current as does an MDF. See also MDF, DF.

Intermediate System. A bridge, router, gateway, or hub that interconnects network segments.

Intermediate System-to-Intermediate System. Link-state OSI protocol by which packets are dynamically routed between routers or intermediate systems; provides routing services for both TCP/IP and OSI.

International Organization for Standardization (ISO). International body that is responsible for establishing standards for communications and information exchange; developed the OSI Reference Model. ISO is not an acronym, but the Greek word for "equal." The U.S. representative to ISO is ANSI.

International Telecommunication Union (ITU). The international standards body that helps and defines emerging standards. Charter organization of the United Nations. The ITU is the parent organization for ITU-T (formerly CCITT).

Internet. A global TCP/IP network linking millions of computers for communications purposes. The Internet Engineering Task Force (IETF) is the committee that defines standard Internet operating protocols such as TCP/IP. The IETF is supervised by the Internet Society's Internet Architecture Board (IAB). Internet-standard Network Management Framework Device configuration and monitoring via SNMP. The Internet is not just a network for a huge network of networks, or internetwork, comprising large backbone nets (MILNET, NSFNET, etc) and an array of regional and local campus networks worldwide.

Internet Access Node. The Internet access provider's facility for receiving communications from subscribers and prepping it for transmission into the Internet.

Internet Activities Board. See IAB.

Internet Address. The 32-bit address assigned to hosts using TCP/IP.

Internet Control Message Protocol (ICMP). In TCP/IP, the collection of messages exchanged by IP modules in both hosts and gateways to report errors, problems and operating information.

Internet Engineering Task Force (IETF). The primary working body developing TCP/IP standards for the Internet.

Internet Group Management Protocol (IGMP). Protocol that runs between hosts and their immediately neighboring multicast routers; the mechanisms of the protocol allow a host to inform its local router that it wishes to receive transmissions addressed to a specific multicast group.

Internet Packet Exchange (IPX). Part of Novell's NetWare protocol stack, used to transfer data between the server and workstations on the network.

Internet Protocol (IP). An open networking protocol used for Internet packet delivery. It keeps track of the Internet's addresses for different nodes, routes outgoing messages, recognizes incoming messages, and allows a packet to traverse multiple networks on the way to its final destination.

Internet Protocol Address. A unique identifier that allows communication over the Internet to be directed to the appropriate destination. Every computer on the Internet must have a unique IP address. IP addresses are allocated by an ISP in following format: nnn.nnn.nnn.nnn, where nnn is a numeric value from 0 to 255. IP addressing might be referred to as being static (fixed) or dynamic.

Internet Security Scanner. Publicly available program that scans an entire domain or subnetwork, looking for security holes; can be used to locate network security vulnerabilities, or by hackers to breach network security.

Internet Service Provider (ISP). ISPs provide access to the Internet. The ISPs that offer DSL usually don't own the equipment that makes the service possible. Instead, they buy the service from a traditional phone company or one of the newer competitive ones. The distinctions between telephone companies and Internet service providers are already blurred because ISPs can also be telephone companies. Also, many telephone companies sell Internet access. The terms NSP (network service provider) and USP (universal service provider) are coming into use to describe these companies that sell many different communication services.

Internet Telephony Service Provider (ITSP). New class of carriers that provide packet-based WAN networks to carry voice traffic.

Internetwork. Series of networks interconnected by routers or other devices that function as a single network. Sometimes called an internet, which is not the same thing as The Internet.

Internetwork Packet Exchange (IPX). A LAN communications protocol for Novell networks used to move data between server and workstation programs running on different network nodes.

Inter-Networking. Data communications across different network operating systems.

Interoperability. The ability of equipment from multiple vendors to communicate using standardized protocols.

Interworking Function (IWF). The ability to communicate between devices supporting dissimilar protocols (such as Frame Relay and ATM).

Intranet. A term that describes an internal TCP/IP-based network that mimics the protocols and functionality of the Internet. It's designed to be used within the confines of a company, university or organization.

Inverse Discrete Fourier Transform (IDFT). A discrete form of the IFFT.

Inverse Fast Fourier Transform (IFFT). An algorithm for efficiently implementing via digital signal processors the conversion from the frequency-domain to the time-domain.

Inverse Multiplexing. The logical aggregation of multiple switched circuits to achieve a higher effective transmission speed.

Inverse Multiplexing of ATM (IMA). A UNI standard that enables "right-sizing" and "right-pricing" of ATM solutions for organizations with low- to mid-range WAN traffic requirements. IMA takes a stream of ATM cells and, on a cell-by-cell basis, divides it across multiple T-1/E1 WAN circuits. The aggregate bandwidth of any number of these T-1/E1 lines determines the rate of the ATM connection.

IP. See Internet Protocol.

IP Address. See Internet Protocol Address.

IP Security. IETF security protocol for virtual private networks. It provides cryptographic security services supporting a combination of authentication, integrity, access control, and confidentiality. These services are in the network layer of the protocol stack.

IP Spoofing. The use of a forged IP source address to circumvent a firewall. The packet appears to have come from inside the protected network and to be eligible for forwarding into the network.

IP Switching. IP switching is a technology designed to address the need for faster throughput between separate subnets or Virtual LANs (VLANs). IP switching retains the control functions of routing while leveraging the wire-speed forwarding capabilities of switching. This approach offers the ability to scale to gigabit network requirements.

IP telephony gateway (ITG). Also called a Voice over IP (VoIP) Gateway. A bridge between traditional circuit-switched telephony and the Internet that extends the advantages of IP telephony to the standard telephone by digitizing the standard telephone signal (if it isn't already digital), significantly compressing it, packetizing it for the Internet using Internet Protocol (IP), and routing it to a destination over the Internet.

IPX. See Internetwork Packet Exchange.

IPX Control Protocol. See IPXCP.

IPXCP (IPX Control Protocol). Protocol for transporting IPX traffic over a PPP connection.

IRR (Internal Rate of Return). The rate of return associated with a stream of future cash flows. Typically, the early cash flows are negative (i.e., an investment).

ISDN. See Integrated Services Digital Network.

ISO. See International Standards Organization (also called International Organization for Standardization)

ISP. See Internet Service Provider.

ITC. See Independent Telephone Company.

ITG. See IP telephony gateway.

ITSP. See Internet Telephony Service Provider.

ITU. See International Telecommunications Union.

ITU-T. International Telecommunications Union Telecommunications Standardization Sector (See International Telecommunications Union).

IWF. See Interworking Function.

IXC. See Inter-Exchange Carrier.

J

Jabber. The uncontrolled transmission of oversized frames to the network by a faulty device.

Jitter. An expression of the end-to-end delay variations during the course of a transmission. Jitter and wander are defined as the short-term and long-term variations of the significant instants of a digital signal from their ideal positions in time. When a digital signal is compared to a primary reference signal (PRS), the digital signal's transitions occur either before or after the point where the PRS would expect them to occur. See wander.

Jitter-buffer. A jitter-buffer is a signal-processing function implemented at the receiving end of a packet-based network to buffer incoming packets, rearrange them into correct order and output them at a constant rate to ensure smooth playback.

K

K56Flex. Proprietary 56K technology developed by Rockwell and Lucent Technologies. A predecessor of modern 56 Kbps modem technology.

Kbps. See Kilobits per second.

Key Systems. Telephone equipment with extra buttons that provide users with more functionality than regular telephones. ISDN phones and NT1 Plus devices that support analog telephones include key systems. A key system is a protocol invoked when you press a sequence of keys on the analog or ISDN telephone's dialing pad. Key systems can be plugged into an IAD and phone calls can therefore be made over VoDSL.

Kilobits per second (Kbps). A measure of bandwidth or throughput, usually taken to be 210 (1,024) bits per second.

Kilobytes. A measure, representing 1,024 bytes, generally used to express the storage capacity of digital components. A byte represents a single character, or a group of eight bits.

KiloHertz (kHz). Frequency measurement, of 1,000 cycles per second.

L

L2TP. See Layer Two Tunneling Protocol.

LAN. See Local Area Network.

LAN Emulation (LANE). A standard paradigm for integrating legacy LANs and applications transparently with ATM networks. As specified by the ATM Forum, it is a standard implementation for making edge devices appear as though they were attached to a LAN. An emulated LAN has all of the benefits (and weaknesses) of the traditional LANs they are emulating. Currently only Ethernet and Token Ring LAN Emulation are specified. 3Com, and most other vendors, are implementing Ethernet LAN Emulation first.

LANE. See LAN Emulation.

LAP. See Link Access Procedure.

Last Mile. Refers to the local loop and is the difference between a local telephone company office and the customer premises; a distance of about 3 miles or 4 kilometers.

LATA. See Local Access and Transport Area.

Latency. A measure of the temporal delay. Typically, in xDSL, latency refers to the delay in time between the sending of a unit of data at one end of a connection until the receipt of that unit at the destination.

Layer. OSI reference model; each layer performs certain tasks to move the information from the sender to the receiver. Protocols within the layers define the tasks for the networks but not how the tasks are accomplished.

Layer Two Tunneling Protocol (L2TP). An extension to the PPP protocol that enables ISPs to operate VPNs. L2TP merges the best features of two other tunneling protocols: PPTP and L2F.

LDS (Local Digital Switch). A digital network Class 5 telephone switch as compared to older analog switching systems.

Leaky-Bucket-Algorithm. An algorithm designed to monitor the flow of cells to verify they conform to the stated traffic contract for the associated connection.

Leased Line. A telecommunications transmission circuit that is reserved by a communications provider for the private use of a customer. Also called a private line or nailed up circuit.

LEC. See Local Exchange Carrier.

LES. See Loop Emulation Services.

Ligne d'abonné Numérique. French name for DSL.

Line Termination (LT). Defines the local loop at the telephone company side of a DSL connection. In case of ISDN, matches NT1 function at the customer end of the local loop.

Line Unit "T" Interface (LUTI). A digital carrier channel unit version of an ISDN Basic Access NT as used in UDLC systems to interface with the ET. The LUNT presents an ANSI T1.605 "S/T" interface toward the customer. See also LULT and LUNT.

Line Unit LT (LULT). A digital carrier channel unit version of an ISDN Basic Access LT as used in DLC systems, either IDLC or UDLC. The LULT presents an ANSI T1.601 "U" interface toward the customer. See also LUNT and LUTI.

Line Unit NT (LUNT). A digital carrier channel unit version of an ISDN Basic Access NT as used in UDLC systems to interface with the ET. The LUNT presents an ANSI T1.601 "U" interface toward the switch. See also LULT and LUTI.

Línea de Abonado Digital. Spanish name for DSL.

Line-Side T-1. A T-1 that undergoes at least one analog-to-digital conversion in the path between the V.90 digital modem and the PSTN.

Link. A certain kind of connection between communicating entities, such as an xDSL link.

Link Access Procedure (LAP). Link-level protocol specified by the CCITT X.1 recommendation used for communications between data communications equipment and data terminal equipment.

LMI. See Local Management Interface.

Loaded Cable. Twisted wire pair into which inductors have been inserted at periodic intervals to approximate the optimum ratios of the primary cable constants for minimum loss. A loaded cable acts like a lowpass filter. Transmission loss below the cutoff frequency is reduced below that for the nonleaded cable and is nearly flat with ripples. Above the cutoff frequency, loss increases very rapidly.

Loaded Loop. Also called a Loaded Pair (loaded twisted pair); a loop that contains series inductors, typically spaced every 6000 feet for the purpose of improving the voice-band performance of long loops. However, high bandwidth DSL operation over loaded loops is not possible because of excessive loss at higher frequencies.

Loading Coil. A device used to extend the range of a local loop for voice grade communications. They are wire coil inductors added in series with the phone line which compensate for the parallel capacitance of the line. They benefit the frequencies in the high end of the voice spectrum at the expense of the frequencies above 3.4 KHz. Thus, loading coils prevent DSL connections and must be removed.

Local Access and Transport Area (LATA). An "Exchange Area." The U.S. term LATA arose out of the post-divestiture fight between the local telephone companies and AT&T over who could carry which traffic as AT&T split itself up. Roughly, a LATA may be geographically defined as larger than a local calling area and smaller than a whole state.

Local Area Network (LAN). A data communications network covering a small area, usually within the confines of a building or floors within a building; a relatively high-speed computer communications network for in-building data transfer and applications. Common LAN protocols are Ethernet and Token Ring. See also CAN (Campus Area Network), MAN (Metropolitan Area Network), and WAN (Wide Area Network).

Local Exchange Carrier (LEC). A company that provides intra-LATA (local access transport area) telecommunications services. LECs appeared as a result of U.S. deregulation of telecommunications. See also RBOC.

Local Loop. A generic term for the physical connection (typically, twisted pair of copper wire) connecting the end user (subscriber) to the provider's central office. Also colloquially referred to as "the last mile" (even though the actual distance can vary).

Local Management Interface (LMI). Set of enhancements to basic Frame Relay; includes methods of exchanging status information between devices.

Logical Ring. Network that is treated logically as a ring, even though it may be cabled as a physical star topology.

Long Distance. Representing the communications of information over a distance other than the local calling area. Also called "long haul" traffic.

Long Haul. A term for long distance.

Loop. Portion of the telephone network that connects the subscriber to the CO. See Local Loop.

Loop Emulation Services (LES). The standards for delivering VoDSL as defined by the ATM Forum. A term that tends to be used interchangeably with Broadband Loop Emulation Services (BLES).

Loop Qualification. The process of determining if a line (or loop) will support a specific type of DSL transmission at a given rate.

Loop Signaling. POTS signaling: Nominal 48 Volt DC battery used to signal on-hook and off-hook and to power the telephone station set. Pulse or touchtone dialing is used.

Loopback Tests. Any tests in which a test signal is injected at one end of a circuit, is looped back at the other end, and monitored at the originating end.

Loss of Frame (LOF). The inability of the framer to find (and lock into) the framing bits.

Loss of Signal (LOS). The inability of the framer to detect a signal on a DS1 line.

Low Pass Filter. A signal filter installed in a customer premises ADSL modem (ATU-R), which would not modify the low frequencies present in its input signal (the POTS transmission is sent unmodified to a phone), but would prevent the high-frequency components (data) from reaching a customer's telephone. See High Pass Filter.

LT. See Line Termination.

LULT. See Line Unit LT.

LUNT. See Line Unit NT.

LUTI. See Line Unit "T" Interface.

M

M13. A piece of telecommunications equipment which multiplexes (combines) 28 DS-1 signals into a single DS-3 signal, commonly used for concentrating traffic for economies of transmission. See also Multiplexer.

MAC. See Media-specific Access Control.

MAE. See Metropolitan Area Ethernet.

Main Distribution Frame (MDF). Central point where all local loops terminate in the CO.

Maintenance Termination Unit (MTU). Used to support POTS service. POTS MTU has no specifications above 4 kHz. May be incompatible with ADSL transmission if it were in direct path. See also ACTS and ARTS.

MAN. See Metropolitan Area Network.

Management Information Base (MIB). Database of network management information used and maintained by a network management protocol such as SNMP. The value of a MIB object can be changed or retrieved using SNMP commands. MIB objects are organized in a tree structure that includes public (standard) and private (proprietary) branches.

Mask. A mask provides a way to subdivide networks using address modification. Written in dotted-decimal notation, it specifies which bits of a destination address are significant (see subnet mask).

MAU. See Multistation Access Unit.

Maximum Burst Size (MBS). The maximum number of cells that can be sent at the Peak Cell Rate.

Maximum Transmission Unit (MTU). The largest possible unit of data that can be sent on a given physical medium; for example, the MTU of Ethernet is 1,500 bytes.

MBS. See Maximum Burst Size.

MCR. See Minimum Cell Rate.

MDF. See Main Distribution Frame.

MDSL. See Multi-rate Digital Subscriber Line.

MDU. See Multiple Dwelling Unit.

Mean Time Between Failures (MTBF). Measure of continuous performance that can be expected of a hardware device; expressed in thousands or tens of thousands of hours.

Mean Time To Repair (MTTR). Measure of the average time required to repair a failed hardware device.

Mechanized Loop Testing (MLT). POTS testing system. See LMOS/MLT. Capable of "multimeter" metallic loop testing for voltage, current, resistance, capacitance. Multi-point Microwave.

Media Access Control (MAC). This is a physical address that is the portion of the data-link layer in networks that controls addressing information of the packet and enables data to be sent and received across a local area network.

Media Controller. See Media Gateway Controller.

Media Gateway. A media gateway handles the actual transport of voice in a distributed Softswitch architecture, performing the transcoding of the voice traffic from the PSTN to and from packet networks.

Media Gateway Controller. In a distributed Softswitch architecture, the media gateway controller (also known as a "Call Agent", "Session Agent," or Media Controller) is located centrally in a network, and takes SS7 or other signaling information from a signaling gateway, provides call routing control and billing information to multiple media gateways and other elements of the network.

Media Gateway Controller Protocol (MGCP). A standardized protocol used for sending call-setup, tear-down, and related commands from the Call Agent to the Media Gateway. A new standard called H.248 by the ITU and Megaco by the IETF may supersede MGCP. Media Gateways receive commands from Call Agents, usually according to the MGCP or Megaco/H.248 standard. A media gateway is sometimes referred to as a Trunking Gateway.

Megabits per second (Mbps). Megabits per second means millions of bits per second.

Megaco. An IETF recommendation, also called H.248 by the ITU-T. Like MGCP, Megaco/H.248 is a control protocol used by Media Gateway Controllers to control Media Gateways. Megaco/H.248 is an extension of MGCP, to cover various forms of communications including voice, video and data.

Message Telephone Service (MTS). Term used by ANSI instead of POTS. See also PSTN.

Metropolitan Area Ethernet (MAE). A MAE is a NAP where ISPs can connect with each other. The original MAE was set up by a company called MFS and is based in Washington, DC. Later, MFS built another one in Silicon Valley, dubbed MAE West. In addition to the MAEs from MFS, there are many other NAPs. Although MAE refers really only to the NAPs from MFS, the two terms are often used interchangeably.

Metropolitan Area Network (MAN). A data communication network covering the geographic area of a city; generally larger than a LAN but smaller than a WAN.

Metropolitan Serving Area. A regional area served by a provider which is classified based on the metropolitan coverage area.

MGCP. See Media Gateway Control Protocol.

MHz. Abbreviation for megahertz. This is a unit of measure indicating frequency in millions of cycles per second.

MIB. See Management Information Base.

Million Instructions Per Second (MIPS). MIPS refers to the number of instructions a processor can perform in one second. MIPS is commonly used to describe the performance of a fixed-point DSP. MIPS consumption is also used to describe the amount of processing required to perform a given algorithm on a processor.

Minimum Cell Rate (MCR). An ATM traffic parameter used by the ABR service category that specifies the guaranteed minimum rate for a virtual channel.

Minimum Point of Entry. The closest practical point to where the carrier facilities cross the property line or the closest practical point to where the carrier cabling enters a multiunit building or buildings.

MIPS. See Million Instructions Per Second.

ML-PPP. See Multilink PPP.

MLT. See Mechanized Loop Testing.

MMDS. See Multichannel Multipoint Distributed Service.

Modem. Short for "Modulator/Demodulator." Data communications equipment that connects a computer to the telephone network. Technically, a modem converts a computer's digital signals to analog signals that can be transmitted over standard telephone lines. Digital, ISDN or xDSL modems use more complicated modulation techniques to send digital data over analog lines. See also Demodulation and Modulation.

Modem Pool. Today, the term typically refers to large banks of modems that enhance translation of cellular protocols to landline protocols.

Modulation. Process by which signal characteristics of an electrical carrier wave is altered to represent information and thus be used to transmit a signal from a source to a destination.

Monitoring Capabilities. The capabilities of network management software for monitoring LAN traffic to ensure that it is distributed according to the policy implemented by the call center server and that LAN resources continue to be over provisioned as the call center grows. Both RMON-2 and distributed RMON can be used for this purpose.

Motion Picture Experts Group (MPEG). This is an industry organization whose goal is to develop standards and specifications for the encoding, compression, transmission, decompression and unencoding of video information over various media and network technologies.

Movies on Demand. See Video on Demand.

MPEG. See Motion Picture Experts Group.

MPLS. See Multi-Protocol Label Switching.

MPoA. See Multi-Protocol Over ATM.

MPT. See Multipoint-to-Point Tunneling.

MSDN. See Multiservice Data Network.

MTBF. See Mean Time Between Failures.

MTS. See Message Telephone Service.

MTTR. See Mean Time To Repair.

MTU. See Maintenance Termination Unit; Maximum Transmission Unit; Multiple Tenant Unit.

mu-law. A standard analog signal compression algorithm, used in digital communications systems of the North American digital hierarchy, to optimize the dynamic range of an analog signal prior to digitizing.

Multichannel Multipoint Distributed Service (MMDS). Also known as wireless cable. MMDS is a pay television delivery system that delivers up to 33 channels of video programming via microwave transmission. MMDS systems consist of four parts (1) a head end, located atop a tall building, where broadcast and cable TV signals are received and converted to Microwave radio signals for retransmission. They are sent using (2) an omnidirectional transmit antenna to subscribers who are equipped with (3) receiving antennas which convert the microwave frequencies to lower frequencies, and send them to (4) a TV or VCR. MMDS is line of sight transmission; the receiving antenna must have unobstructed view of the transmitting antenna. MMDS systems operate a band of radio spectrum that ranges from 2.5 billion cycles a second to 7 billion. That band can only be used for broadcast (one-way) communications.

Multilink PPP (ML-PPP). A standard for aggregating/combining multiple ISDN B-channels using synchronous PPP framing. MLPPP allows ISDN equipment to combine two 64 Kbps B-channels for a total bandwidth of 128 Kbps.

Multimedia. Audio, video, image, and other types of data presentation in the same session.

Multimedia Exchange. An advanced Ethernet switch with integrated call processing for voice/video real-time communication. This product combines the virtues of current LAN switches with those of PBX, and can effectively handle both multimedia PCs and Ethernet phones.

Multimode Fiber Cabling. Fiber cable with a wide core. Light is reflected along the core at multiple angles, and is propagated along multiple paths, each path with a different length and hence a different time to traverse the fiber. These multiple angles or modes cause the signal elements to spread out in time, so that distortions occur that limit the distance over which the integrity of the light signal can be maintained. Multimode fiber is the predominant type of LAN fiber installed within buildings and is less expensive than single-mode fiber.

Multinetting. The ability to support multiple TCP/IP subnets on a single physical LAN interface.

Multiple Dwelling Unit (MDU). A building with multiple residences such as an apartment building.

Multiple Level Protocol Switching (MLPS). See Multi-Protocol Label Switching.

Multiple Tenant Unit (MTU). A building with multiple tenants such as an office building.

Multiple Virtual Lines (MVL). New local loop access technology developed by Paradyne. Designed and optimized for multiple concurrent services for residential, SOHO and small business markets. MVL transforms a single copper wire loop into multiple virtual lines supporting multiple and independent services simultaneously.

Multiple-Layer Interface Driver. Device driver

for the network interface card that is attached to the MLI layer; manages the sending and receiving of packets to and from the network.

Multiplexer. An electronic device for combining multiple data or voice signals into one signal group for transmission over a high-speed trunk. Several communications paths or channels may be either permanently or dynamically established over the medium to accomplish this. Also known as a "mux." See also Channel; Multiplexing; Time Division Multiplexing.

Multiplexing. A technique that enables several data streams to be sent over a single physical link; also, in the OSI Reference Model, a function by which one connection from a layer is used to support more than one connection to the next higher layer. In protocol multiplexing, multiple protocol stacks can be used at the same time in the same computer. See Multiplexer.

Multipoint-to-Point Tunneling (MPT). Methodology for reducing the number of virtual circuits; an MPT acts as a reverse-path forwarding tree for data destined to the specific switch located at the tree's root; unlike a multicast delivery tree, traffic flows from the leaves of the tree toward the root.

Multi-Protocol Label Switching (MPLS). An IETF standard intended for Internet applications, MPLS is a widely supported method of speeding up IP-based data communication over ATM networks and providing Quality of Service (QoS). As Frame Relay, IP and ATM converge, the MPLS promotes the concept of "route at the edge and switch in the core." Under such a model, routers are used at the ingress and egress edges of the network, where their high levels of intelligence can be best used. The MPLS model is based on using very small standardized headers which include routing information to specify the route of the packets from source to destination.

Multi-Protocol Over ATM (MPoA). Enables an ATM device or application to add a standard protocol identifier to the LAN data which allows higher-layer protocols, such as IP, to be routed over ATM.

Multi-rate Digital Subscriber Line (MDSL). An interesting, low-bandwidth form of DSL designed for the evolving networks which require a "mix" of digital and traditional analog services to be delivered to the customer premises on already existing copper cable plants. MDSL is capable of transmitting 288 Kbps (using 2B1Q line encoding) over 18,000 feet (3.5 miles or 5.45 km) on any unconditioned, twisted 0.4 mm diameter (26 AWG) copper pair. It can even reach 31,200 feet (9.5 km) if the wires are 0.6 mm in diameter (22 AWG). Also, a copper pair carrying MDSL may "run" in the same binder along with other pairs carrying traditional analog POTS or other digital HDSL services. An example of MDSL technology is the VCL-288 from Valiant Communications, Ltd. (New Delhi, India — (+ 91-11) 542-2215, www.valiantcom.com) which delivers the following pair gain subscriber services:

- Four 64 Kbps, POTS lines based on Multi-Rate Digital Subscriber Line technology. on a single twisted copper pair to the user premises. All POTS (voice) lines operate simultaneously at 64 Kbps, without any compression or line concentration.
- Two 64 Kbps, POTS Lines and one Basic Rate, "U" Interface, ISDN subscriber connection.
- Two 64 Kbps, POTS and digital data at 64 or 128 Kbps for a leased (dedicated) data line.
- Frame Relay access.

MultiService Data Network (MSDN). A network that can handle various protocols, such as ATM, IP, Frame Relay, etc. See also Voice over Multiservice Broadband Networks.

Multistation Access Unit (MAU). A hub in a Token Ring network; each MAU supports up to eight nodes and servers and can be connected to other units to create large networks.

MUX, mux. Multiplexer. See Multiplexer.

MVL. See Multiple Virtual Lines.

N

Nail Up. The process of dedicating a telecommunications circuit for a particular use; the physical or logical dedication of a line for a particular use. See also Leased Line.

NANP. See North American Numbering Plan.

NAP. See Network Access Point or Network Access Provider.

NarrowBand. A term used to describe services with up to and including T-1 or 1.544 Mbps.

NAT. See Network Address Translation.

National ISDN1. The ISDN standard in the U.S. It is the first successful attempt to standardize at a level allowing the same end-user equipment to connect transparently to different switch vendors' equipment. Prior to this standard, all end-user devices had to understand the particulars of the switch to which they were connected.

NDIS. See Network Driver Interface Specification.

NE. Network Element: Switch or transmission terminal.

Near-End Crosstalk (NEXT). Any interference that occurs near a connector at either end of a cable; usually measured near the source of the test signal.

NEBS. See Network Equipment Building System.

Net. A nickname for the Internet, and generally refers to the World Wide Web.

NetBIOS. Developed by IBM and used as the basis for DOS and NT networks.

Network Access Point (NAP). A public network exchange facility where ISPs can connect with one another in peering arrangements. The NAPs are a key component of the Internet backbone because the connections within them determine how traffic is routed. They are also the points of most Internet congestion.

Network Access Provider (NAP). Another name for the provider of networked telephone and associated services, usually in the U.S.

Network Address Translation (NAT). The translation of an Internet Protocol address (IP address) used within one network to a different IP address known within another network. One network is designated the inside network and the other is the outside. Typically, a company maps its local inside network addresses to one or more global outside IP addresses and unmaps the global IP addresses on incoming packets back into local IP addresses. This helps ensure security since each outgoing or incoming request must go through a translation process that also offers the opportunity to qualify or authenticate the request or match it to a previous request. NAT also conserves on the number of global IP addresses that a company needs and it lets the company use a single IP address in its communication with the world.

Network Computer (NC). A sealed workstation that has no local disk drive; NC users run applications and access files entirely from servers.

Network Driver Interface Specification (NDIS). Used for all communication with network adapters. NDIS works primarily with LAN manager and allows multiple protocol stacks to share a single NIC. Also known as Network Design Interface Specification.

Network Equipment Building System (NEBS). A set of requirements for the reliability and usability of equipment, established by Bellcore (now Telcordia) in the 1980s.

Network Interface (NI). The FCC definition: "The point of demarcation between the network and the CI." This refers to a network-customer interface, instead of a network-network interface. See also Demarcation Point.

Network Interface Box (or Network Interface Device). The physical device that provides the means to connect the telephone company's wire to the premises wiring. In the U.S., in cases where the service provider wishes to provide equipment beyond the network demarcation, the provider must first have permission from any regulators involved and the customer must agree. Outside the U.S., users have less control over their wiring and equipment. Also called NTU (Network termination unit) or "demarc." See NT1, NT1 Plus device, and NT2.

Network Interface Card (NIC). The circuit board or other form of computer hardware, which serves as the interface between a computer, or other form of communicating DTE, and the communications network; in ADSL, a common NIC card is an Ethernet NIC card, which serves as the interface to the ADSL modem from the computer. See also Adapter.

Network Interface Device (NID). The (typically) gray box attached to the side of your home or office that marks the point of demarcation between the service provider and your business or home. Also known as NIU (Network Interface Unit).

Network Management System (NMS). A Windows-based system that is responsible for managing a network. In xDSL, network management systems allow providers to control and monitor those services based on the ADSL streams, at both the physical and logical layers of the services.

Network Node. The heart of any network. The network node hooks the users together. In the voice network, the links between CO (or LEC) network nodes are called trunks and the links to the users are called access lines. On the Internet and Web, the network node is a router (or sometimes an ATM switch). DSL technologies in one sense try to get Internet traffic office the CO network node. See also Access line, AN (Access node), and Router.

Network Operating System (NOS). System software running on the network's file server, with a smaller component on each device attached to the network. Examples of client/server NOSs include Novell NetWare, Banyan VINES, Microsoft Windows.

Network Operations Center (NOC). A site that controls and monitors a LAN or WAN network and that communicates with other networks on the Internet to improve services and solve problems.

Network Service Node (NSN). See Network Node.

Network Service Provider (NSP). A vendor, such as an ISP, local telephone company, CLEC or corporate LAN that provides network services to subscribers.

Network Termination (NT). The DSL Transceiver at the network end of the line.

Network Termination, Type 1 (NT1). The equipment required to convert from the two-wire U interface to the four-wire S/T interface. This equipment is not required outside of North America.

Network-Node Interface (NNI). ATM Forum standard that defines the interface between two ATM switches that are both located in a private network or are both located in a public network. The interface between a public switch and private one is defined by the UNI standard.

NEXT. See Near End CrossTalk.

NI. See Network Interface.

NI-1 (National ISDN-1). A specification for a "standard ISDN" phone line.

NIC. See Network Interface Card.

NID. See Network Interface Device.

NMS. See Network Management System.

NNI. change to Network-Network Interface. Also see Network-Node Interface.

NOC. See Network Operations Center.

Nonblocking Network. Network in which terminals and nodes are interconnected in such a way that any unused input/output pair can be connected by a path through unused nodes, no matter what other paths exist at the time.

Nonloaded Cable. Twisted pair cable with no loading coils. Transmission loss increases approximately at the square root of the frequency with no hard cutoff as with Loaded Cable. For the outside loop plant, about 80% is nonloaded cable and 20% has been loaded. See Loaded Cable.

Non Real Time Variable Bit Rate (nrt-VBR or VBR-nrt). This ATM service category is used for connections that transport variable bit rate traffic for which there is no inherent reliance on time synchronization between the traffic source and destination, but there is a need for an attempt at a guaranteed bandwidth or latency. An application that might require an nrt-VBR service category is Frame Relay interworking, where the Frame Relay CIR (Committed Information Rate) is mapped to a bandwidth guarantee in the ATM network. No delay bounds are associated with nrt-VBR service.

North American Numbering Plan (NANP). The familiar ten-digit numbering system used today in the U.S., Canada, and Mexico, which includes the three-digit area code followed by the seven-digit local telephone number.

NOS. See Network Operating System.

nrt-VBR. See Non Real Time Variable Bit Rate.

NSGR (Network Standards Generic Requirements). Bellcore (Telcordia) Technical Reference TR-TSV-000800. The umbrella document for network standards requirements. See also OTGR and TSGR.

NSN. See Network Service Node.

NSP. See Network Service Provider.

NT. See Network Termination.

NT1 Plus Device. Device that includes a built-in NT1 as well as ports to connect other devices (analog or digital) to an ISDN line.

Nx64. Describes a contiguous bit stream to an application at the Nx64 rate. Examples are LAN interconnect and point-to-point videoconferencing.

O

OAM(&P). See Operations, Administration, Maintenance (and Provisioning).

OC. See Optical Carrier.

OC-3. See Optical Carrier Level 3.

OC-n. See Optical Carrier Level n Signal.

ODI. See Open Datalink Interface.

Off-Hook. Signal to the network that the POTS station set is ready for a call by removal of the handset from the hook (rest). The active condition of switched access or a telephone exchange service line, which permits direct current flow to station set. See also on-hook.

Office Repeater Bay (ORB). Mounting and powering arrangement for digital regenerators, such as for T-1 lines.

On-Hook. Signal to the network that the POTS station set is terminating a call by placement of the handset on the hook (rest). Interrupts direct current flow to station set.

ONU. See Optical Network Unit.

Open Datalink Interface (ODI). The specification developed by Novell for supporting different adapters and network operating systems (NOS).

Open Systems Interconnection (OSI). A seven-layer architecture model for data communications systems; the OSI Reference Model was created by the ISO and the ITU-T. Each layer specifies particular network functions: Layer 7, the application layer, the highest layer of the model, defines the way applications interact with the network. Layer 6, the presentation layer, includes protocols that are part of the operating system, and defines how information is formatted for display or printing and how data is encrypted, and translation of other character sets. Layer 5, the session layer, coordinates communication between systems, maintaining sessions for as long as needed and performing security, logging, and administrative functions. Layer 4, the transport layer, controls the movement of data between systems, defines protocols for structuring messages, and supervises the validity of transmissions by performing error checking. Layer 3, the network layer, defines protocols for routing data by opening and maintaining a path on the network between systems to ensure that data arrives at the correct destination node. Layer 2, the data-link layer, defines the rules for sending and receiving information from one node to another between systems. Layer 1, the physical layer, governs hardware connections and byte-stream encoding for transmission. It is the only layer that involves a physical transfer of information between network nodes.

Operations, Administration, Maintenance (and Provisioning). A group of network management functions that proved network fault notification, performance information, and diagnosis functions.

Optical Carrier. Refers to a Synchronous Optical Network (SONET) signal. An OC level is the optical equivalent of an STS signal. Transmission rates are based on 51.84 Mbps (OC-1). A "c" following an OC level identifies concatenation of payload (for example, OC-3c).

Optical Carrier 3 (OC3, or OC-3). OC-3 is a widely used level of SONET which defines an STS-

3c signaling rate (155.52 Mbps) rate over of data over fiber at 155.52 Mbps. A channelized OC-3 line carries 2016 simultaneous uncompressed phone calls, whereas an unchannelized OC-3 line transmits data at the full 155.52 Mbps.

Optical Carrier Level n Signal (OC-n or OCn). Refers to the Optical Carrier interface, which is designed to work with STS signaling rates in a SONET. Transmission rates begin at 51.84 Mbps (STS-1), and each "n" represents increments of 51.84 Mbps (so, OC-1 is 51.84 Mbps, OC-3 is 155 Mbps, and OC-12 is 622 Mbps). The physical portion of an OC interface consists of two optical fibers (one transmit and one receive), which form a point-to-point connection between two devices.

Optical Network Unit (ONU). A form of access node that converts optical signals transmitted via fiber to electrical signals that can be transmitted via coaxial cable or twisted pair copper writing to individual subscribers.

ORB. See Office Repeater Bay.

OS. See Operations System or Operating System.

OS/NE (Operations System/Network Element). Interface between OS and network element.

OSI. See Open Systems Interconnection.

OTGR (Operations Technology Generic Requirements). Bellcore Technical Reference TR-TSV-000439. See also NSGR and TSGR.

Out-of-Band Signaling. Allows telephone network management signaling functions and other services to be sent over a separate channel rather than the bearer channel. ISDN uses out-of-band signaling via the D channel. Out-of-band signaling in DSL, including ISDN, consists of digital messages rather than audio signals, as is the case with the touch-tone analog telephone system.

P

PABX. Private Automatic Branch Exchange, the European term for a PBX (See Private Branch Exchange).

Packet. A subunit of a data stream; a logical grouping of information that is sent over a network; it includes a header (containing control information such as sender, receiver, and error-control data), as well as the message ("payload") itself. May be of fixed or variable length.

Packet-Switched Network. A network in which data is transmitted in units called packets. Unlike a circuit-switched network, a packet-switched network does not establish a dedicated path through the network for the duration of a session, opting instead to segment and transmit data in units in a connectionless manner. The packets can be routed individually over the best available network connection and reassembled in the original order at the destination. Also called a Switched Packet Network (SPN).

Packetized Data. Packetized data refers to a data-stream that is broken into many discrete units, called packets. Each packet begins with a header that describes the contents, destination, and other information about the packet, and the payload, which contains the actual data being transported. A packet with a small payload is "padded" with zeroes.

Pair Gain or Pairgain. The multiplexing of a number of signals over fewer facilities.

PAP. See Password Authentication Protocol.

Partial Packet Discard (PPD). An intelligent packet discard algorithm used to control congestion. PPD checks for any of the following conditions: a policing violation, a CLP threshold violation, or no free buffer space available. If it encounters one of these conditions, it discards all remaining cells from point of encounter (including violating cells) up to but not including the next end-of-frame cell.

Passband. A range of frequencies that has a non-zero lower limit and some upper limit. For example, local loops (or access lines) will limit the range to the voice passband (around 300 to 3300 Hz). Any frequency outside this range is not carried on the link. This passband is adequate for voice but severely limits the speed of digital information flow on the link.

Passive Optical Network (PON). A fiber-based transmission network containing no active electronics.

Password Authentication Protocol (PAP). A security protocol that establishes a simple PPP authentication method using a two-way handshake to verify the identity of the two computers or communicating devices. PAP sends passwords as text, which makes it vulnerable to hackers.

Payload. That portion of a frame or cell that carries user traffic. It is effectively what remains in the frame or cell if you take out all headers or trailers.

Payload Type Identifier (PTI). A 3-bit descriptor in the cell header which indicates the type of payload the cell contains (for example: user or management payload; or last cell in frame information).

PBX. See Private Branch Exchange.

PC. Personal Computer.

PC Card. A credit-card sized removable module that contains memory, I/O, or a hard disk. The term may refer to a variety of proprietary card-sized products; however, the term "PC Card" is PCMCIA's trademark for its PC Card standard.

PCI. See Peripheral Component Interconnect PC Bus Interface.

PCM. See Pulse Code Modulation.

PCR. See Peak Cell Rate.

PDH. See Plesiosynchronous Digital Hierarchy.

PDU. See Protocol Data Unit.

Peak Cell Rate (PCR). An ATM traffic parameter used in ATM service categories to specify the maximum rate at which a source may transmit on a given virtual channel.

Peer-to-Peer Communications. A type of communications and data exchange between peer entities on two or more networks.

Performance Monitoring (PM). In-service surveillance of error performance, contrasted to most testing functions which are conducted out-of-service.

Peripheral Component Interconnect PC Bus Interface (PCI). The connection interface in a Personal Computer for addition of 3rd party devices such as modems or NICs.

Permanent Virtual Circuit (PVC). A software-defined virtual (logical) connection between fixed endpoints established through a network. Called a virtual circuit (in X.25), virtual connection (in Frame Relay), or virtual channel connection (in ATM). The connection is established by administrative means, provisioned much like a leased or dedicated real circuit. A PVC can be a point-to-point, point-to-multipoint, or multipoint-to-multipoint connection. See also Switched Virtual Circuit.

Permanent Virtual Path (PVP). A virtual path made up of a number of PVCs.

Permanent Virtual Path Tunneling (PVP Tunneling). A means of linking two private ATM networks across a public network using a virtual path.

Personal Office Internetworking. Network services that provide individual remote users with access to corporate LAN resources. A different term for what is now known as a virtual private network (VPN).

Phantom Power. The capability of the NT1 to provide power to the TE1 or terminal adapters via two wires in an eight-wire cable.

Phase Locked Loop (PLL). A feedback loop with narrow bandwidth used to recover signal timing.

Phase Modulation (PM). Uses changes in the phase of a carrier signal to represent information.

PIC. See Polyethylene Insulated Cable.

Plain Old Telephone Service (POTS). Refers to ordinary, analog telephone service over the PSTN, with an analog bandwidth of less than 4 kHz traveling over copper wire. Any service sharing a line with POTS (such as ADSL) must either use frequencies above POTS or convert POTS to digital and interleave it with other data signals.

Plesiosynchronous Digital Hierarchy (PDH). Term for the current pre-SONET and pre-SDH digital hierarchies, DS-1, DS-3, etc.

PLL. See Phase Locked Loop.

Plug-and-Play. The ability of a computer system

to configure expansion boards and other devices automatically without demanding that the user turn off the system during installation.

PM. See Performance Monitoring; Phase Modulation.

PMD. See Physical Media Dependent.

PNNI. See Private Network Node Interface.

Podule. An external cable combined with a functional device to connect a PC Card to a telephone line or ISDN BRI line. For example, an external cable/device that converts a two-wire U interface PC Card connection into a four-wire S/T interface.

Point of Presence (POP). Physical connection to the telephone company or other long-distance carriers. In an IXC, the POP is the place within a Local Access and Transport Area (LATA) where your long-distance carrier terminates long-distance lines just before they are connected to the phone lines of the local telephone company servicing the LATA. Each IXC can have multiple POPs within one LATA. Also refers to a node of an ISP or other NSP, usually a network node.

Point of Sale (POS). Any device used for handling transactions, such as card readers for credit card or debit card transactions.

Point-to-Multipoint Configuration. A physical connection between one device on one end and more than one device on the other end.

Point-to-Point Connection. A connection established between two devices.

Point-to-Point Protocol (PPP). A common, Layer 2 protocol used with Internet protocols and services. It sets up the exchange of information between a personal computer or data terminal "client" to a remote "server" via a dial-up telephone line. It provides router-to-router and host-to-network connections over both synchronous and asynchronous circuits The successor to the less popular SLIP (the Serial Line Internet Protocol), which achieves a similar result.

Point-to-Point Tunneling Protocol (PPTP). A technology for creating Virtual Private Networks (VPNs) over the Internet, developed jointly by Microsoft Corporation, U.S. Robotics, and several remote access vendor companies, known collectively as the PPTP Forum. Because the Internet is basically an open network, PPTP ensures that messages transmitted from one VPN node to another are secure. With PPTP, users can access their corporate network via the Internet.

Policing. The process of measuring traffic flow, comparing it to predefined limitations, and then labeling packets for potential discard.

Policy. A set of rules governing the allocation of networking resources.

Policy-Powered Networking. Networking model in which transactions are evaluated for completion based on policies that specify who can use particular resources, and at what times.

Polling. Method of controlling the sequence of transmission by devices on a multipoint line by requiring each device to wait until the controlling processor requests it to transmit.

Polyethylene Insulated Cable (PIC). In contrast to paper or pulp insulated cable. Sometimes also called "plastic insulated cable." Excellent dielectric. See also PVC.

Polyvinyl Chloride (PVC). A tough insulation for some types of drop wire. Dielectric properties vary with frequency.

PON. See Passive Optical Network.

POP. See Point-of-Presence.

Port. A location for passing data in and out of a device, and, in some cases, also for attaching other devices or cables.

Port Density. The number of ports, physical or logical, per network device.

POS. See Point Of Sale.

Post, Telephone and Telegraph administration (PTT). Generic non-U.S. term for a provider of access services. A PTT is often state-owned, existing as a governmental agency in many countries in Latin America, Europe and Asia. One example of a PTT is Germany's Deutsche Bundespost.

POTS. See Plain Old Telephone Service.

Power Spectral Density (PSD). Signal power

at a function of frequency.

Powering. The powering of NT1 and CPE equipment. The NT1 and any CPE connected to it must be powered locally. Usually, these powering capabilities are built into the NT1 or NT1 Plus device.

PPD. See Partial Packet Discard.

PPP. See Point-to-Point Protocol.

PPTP. See Point-to-Point Tunneling Protocol.

PRA. Primary Rate Access. Same as Primary Rate Interface.

PRBS. See Pseudo-Random Binary Sequence.

PRI. See Primary Rate Interface.

Primary Rate Interface (PRI). This is an ISDN interface typically used by larger customers. In the U.S., the Primary Rate Interface is split into 23 B channels and one 64 Kbps D channel. PRI is delivered over the same physical link as a T-1, which is a 1.55 Mbps link. In the U.S. the 1.544 Mbps service is referred to as H1 and may or not be channelized as 23B+D. There is also a 384 Mbps service referred to as H0. In Europe, PRI is split into 30 B channels and one 64 Kbps D channel and is delivered over a single E-1 link (2.08 Mbps). Also called Primary Rate Access.

Primary Reference Signal (PRS). A highly accurate clock (for example, a cesium-based clock).

Private Branch Exchange (PBX). A PBX is a small telephone switch within an enterprise or campus that switches calls between enterprise users on local lines while allowing all users to share a certain number of external phone lines. Called a Private Automatic Branch Exchange (PABX) in Europe.

Private Line. See Dedicated Line.

Private Network Node Interface (PNNI). Interface between two network nodes in a private ATM network. Allows multivendor switch interoperability for SVC setup. It will eventually allow dynamic ATM networks to be constructed with heterogeneous (multivendor) components.

Protector. Line protector protects CPE from overvoltage (>1000V).

Protector Block. The point where the lines from the telephone company meet the lines from premises wiring before the network interface box. It can be found in a "house protector" or "station protector" at the subscriber network interface's demarcation point.

Protocol. A formal, standardized description of message formats and the rules that defines how those messages can be exchanged among two or more systems. Protocol definitions range from how bits are placed on a wire to the format of an e-mail message. Standard protocols allow different manufacturers' computers to communicate.

Protocol Converter. Device for translating the data transmission code and/or protocol of one network or device to the corresponding code or protocol of another network or device, enabling equipment with different conventions to communicate with one another.

Protocol Data Unit (PDU). OSI term for packet. A segment of data generated by a specific layer in a protocol stack; usually consists of a block of data from a higher layer (the Service Data Unit or SDU) encapsulated by the next lower layer with a header and trailer

Protocol Translator. Network device or software that converts one protocol into another, similar, protocol. See Transcoder.

Proxy Service. Special-purpose, application-level code installed on an Internet firewall gateway. The proxy service allows the network administrator to permit or deny specific applications or specific features of an application.

PRS. See Primary Reference Signal.

PSC. See Public Service Commission.

PSD. See Power Spectral Density.

Pseudo-Random Binary Sequence (PRBS). A sequence of generated random numbers that are used for transmission system testing.

PSTN. See Public Switched Telephone Network.

PTI. See Payload Type Identifier.

PTT. See Post, Telephone, and Telegraph administration.

Public Service Commission (PSC). U.S.

State-level regulators of the local phone companies. PSCs define how you are charged for telephone service (charges are called tariffs). The FCC (Federal Communications Commission) deals with some similar functions at the federal level. Also known as PUC (Public Utilities Commission). See Federal Communications Commission.

Public Switched Telephone Network (PSTN). A generic term used to refer to the circuit switched, TDM-based network currently used to deliver most voice and dial-up data to subscribers to the network. This network basically composed of Local Exchange Carriers (LECs) and Inter-eXchange Carriers (IXCs) who transmit telephone calls digitally (see Pulse Code Modulation), although local loops and older areas of the network transmit analog signals. Also known as the dial network.

PUC. Public Utility Commission: State or local regulatory authority. See also Public Service Commission, Federal Communications Commission, NTIA.

Pulse Code Modulation (PCM). Technique for converting an analog signal with an infinite number of possible values into discrete binary digital words that have a finite number of values. The waveform is sampled, and then the sample is quantized into PCM codes. A commonly used encoding method for converting an analog voice signal into a digital bit stream.

PVC. See Permanent Virtual Circuit; Polyvinyl cloride.

PVP. See Permanent Virtual Path.

PVP tunneling. See Permanent Virtual Path tunneling.

Px64. See H.261.

Q

QAM. See Quadrature Amplitude Modulation.
QoS. See Quality of Service.
QPSK. See Quadrature Phase Shift Keying.

Quad Cable. Cables where four wires are twisted as a unit. High crosstalk is usually encountered among the wires within a quad unit.

Quadrature Amplitude Modulation (QAM). A modulation technique used by modems and DSL equipment in which a carrier's amplitude and phase are simultaneously modulated by the digital traffic.

Quadrature Phase Shift Keying (QPSK). A passband modulation method.

Quality of Service (QoS). For the analog PSTN, this is the measure of telephone service as specified by the PSC. On the Internet and in other "next generation" networks, QoS is the idea that the service quality provided to a user (transmission rates, error rates, and other characteristics) can be measured, improved, and, to some extent, guaranteed in advance. QoS is of particular concern for the continuous transmission of voice, high-bandwidth video and multimedia information through virtual circuits. To most users in simplest terms, QoS refers to the audible quality of a voice connection, particularly a Voice Over IP (VoIP) connection. Local telephone calls on the PSTN are often used as the benchmark for determining minimal acceptable QoS (the quality of such calls is referred to as toll quality).

Quantization. Process of representing a voltage with a discrete binary digital number. The process is used in analog to digital conversion to "digitize" the successive changes in the height (voltage) of an analog waveform, transforming it into a series of successive binary numbers. However, approximating a smooth, infinite valued signal with a finite number system introduces an error called quantization error, which lowers to quality of the waveform when it is reconstructed from the binary numbers.

Quantization Noise. The noise resulting from analog to digital conversion.

Quat. A quaternary symbol representing two bits with four-level symbol.

Queue. An ordered list of packets waiting to be forwarded over a routing interface.

R

Radio Frequency Interference (RFI). The electromagnetic spectrum from 3 kHz to 300 GHz.

All computer equipment and most electronic devices generate radio waves. Levels are regulated by the FCC.

RADSL. See Rate Adaptive Digital Subscriber Line.

RAI. See Remote Alarm Indication.

Rate Adaptive Digital Subscriber Line (RADSL). A non-standard version of ADSL where the modems test the line at start up and adapt their operating speed to whatever the fastest bit rate is that can be handled by the line at that moment. The modems are therefore able to adjust to varying line lengths and qualities of lines. Unlike fixed rate ADSL modems, these modems will connect over varying lines at varying speeds, making them a good choice for service providers attempting to deploy ADSL past 18,000 feet. Modems can be designed to select their connection speed at train up, during a connection, or upon signal from the CO. Allows adaptive data rates up to 7 Mbps. Many RADSL modems use CAP modulation. Note, however, that standard DMT-encoded ADSL also has some ability for the ADSL modem to adapt its speeds of data transfer.

RBOC. See Regional Bell Operating Company.

RD. See Resistance Design.

RDT. See Remote Digital Terminal.

Real-Time Control Protocol (RTCP). Protocol that accompanies RTP, allowing receiving end station to provide feedback to senders. Applications can use information from RTP and RTCP to detect network congestion and implement basic flow control. RTP and RTCP make networked multimedia applications more robust against network congestion and errors, and facilitate realtime applications such as Voice over IP (VoIP).

Real-Time Protocol (RTP). See Real-Time Transfer Protocol.

Real-Time Streaming Protocol (RTSP). Proposed IETF protocol for standardizing streaming media on the Internet; it works on top of RTP to provide control mechanisms and take care of high-level issues such as session establishment and licensing.

Real-time Transfer Protocol (RTP). RTP is a protocol that is commonly used to provide end-to-end services to support real-time multimedia applications. In particular it addresses the problem of transmitting delay-sensitive voice traffic across a packet-based network. It is commonly used in conjunction with UDP/IP. An RTP header incorporates a time-stamp that is used to indicate time-of-arrival of packet information. This IETF protocol also provides payload type identification, and packet sequence numbering for each data packet. RTP usually works on top of UDP and thus provides no delivery guarantee.

Real-Time Variable Bit Rate (rt-VBR or VBR-rt). An ATM service category used with VoDSL for sending multimedia applications with "lossy" properties, applications that can tolerate a small amount of packet loss without noticeably degrading the quality of the transmission. One traffic type requires this type of service category is variable rate, compressed video streams. Sources that use rt-VBR connections are expected to transmit at a rate that varies with time (e.g., traffic that can be considered bursty). Real-time VBR connections can be characterized by a Peak Cell Rate (PCR), Sustained Cell Rate (SCR), and Maximum Burst Size (MBS). Cells delayed beyond the value specified by the maximum CTD (Cell Transfer Delay) are considered to be of almost no value to the application, thus, delay is considered in the rt-VBR service category.

Redirector. Software that intercepts requests for resources within a computer and analyzes them for remote access requirements.

Reed Solomon. A forward error correcting code that is used to offset the effects of bit error bursts in the receive-bit stream. Used by DSLs such as ADSL.

Regional Bell Operating Company (RBOC). A "Baby Bell" company resulting from the divestiture of AT&T in 1984. There were seven original Baby Bells: Ameritec, Bell Atlantic, Bell Northern, BellSouth, Pacific Telesis, U.S. West, and Nynex. Since 1984, mergers among RBOCs have reduced their number.

Glossary

Remote Alarm Indication (RAI). Also known as a yellow alarm. RAI is carried in the Facilities Data Link for T-1. RAI is carried in Timeslot 0 for E-1.

Remote Bridge. Bridge that connects physically dissimilar network segments across WAN links.

Remote Digital Terminal (RT). Digital Loop Carrier (DLC) or Integrated Digital Loop Carrier (IDLC) terminal located in the outside plant nearer to the customer than the central office terminal. See also COT, Digital Loop Carrier.

Remote Networking. Networking that extends the logical boundaries of a corporate LAN over wide-area links to give remote offices, teleworkers, and mobile users access to critical information and resources. Modern incarnation of this is referred to as a Virtual Private Network (VPN). See Virtual Private Network.

Remote Office Internetworking. Connecting LAN-based remote offices to the corporate LAN. See Virtual Private Network (VPN).

Remote Terminal (RT). If a phone subscriber is too far from a central office, the subscriber's local loop terminates at a remote terminal situated at an intermediate point between the subscriber and the central office.

Remote Termination Unit (RTU). A device (such as an IAD) installed at the customer premises that connects to the local loop to provide high-speed connectivity. Also referred to as the ATU-R.

Remote User. A teleworker, individual contractor, business traveler, or "Road Warrior" type of nomadic user who needs client access to a corporate enterprise LAN over dial-up WAN links.

REN. See Ringer Equivalency Number.

Repeater. A device that amplifies or regenerates a signal, and is usually embedded in a transmission line. See also Loop Qualification and Local Loop.

Resistance Design (RD). A loop design plan in which the maximum DC loop resistance on a copper pair is limited to about 1300 to 1500 ohms (the actual value is dependent on the given switch used). Loops longer than 18,000 feet in working length are inductively loaded. Nonloaded loops may have up to 6000 feet of bridged tap. Drop wire allowances are 700 feet or 25 ohms. A 9 dB maximum signal loss over the loop at 1 KHz is acceptable. Under the more modern and accepted Carrier Serving Area (CSA) rules, the average loop has 600 ohms DC resistance and a 4 dB loss at 1 KHz. See also Carrier Serving Area, Distribution Area and Revised Resistance Design.

Resource Reservation Setup Protocol (RSVP). Provides priority data transmissions based on reservation protocol.

Reverse ADSL. A term for a unique DSL stream that is asymmetrical in the upstream direction; that is, a reverse ADSL link has a small downstream and large upstream communications path.

Revised Resistance Design (RRD). Revised loop design plan in which bridged tap length is included in the 18,000 foot maximum length for nonleaded loops. See also Carrier Serving Area, Distribution Area, and Resistance Design.

RFI. See Radio Frequency Interference.

Ring Topology. Network topology in which a series of repeaters are connected to one another by unidirectional transmission links to form a single closed loop. Each station on the network connects to the network at a repeater.

Ring Trip. The process of stopping the AC ringing signal at the CO when the telephone being rung goes off-hook.

Ringer Equivalency Number (REN). A number representing the ringer loading effect on a line. A ringer equivalency number of 1 represents the loading effect of a single traditional telephone set ringing circuit.

Ringing Voltage. The voltage of the signal used to ring a subscriber

RJ-11. A six-conductor modular jack that is typically wired for four conductors and has four pins. The RJ-11 jack is the most common telephone jack in use worldwide.

RJ-12. Six-conductor modular jack used with four-wire cabling. Most common phone jacks in the world and is used commonly on phones, modems, and fax machines.

RJ-22. A four-position modular jack that is typically used for connecting telephone handsets to telephone instruments; it is always wired with four conductors (also called wires).

RJ-45. Eight-pin connector used to attach data transmission devices to standard telephone wiring. Commonly used in LAN connections.

RLCG. Acronym representing Resistance (R, ohms/km), Inductance (L mH/km), Capacitance (C, nF/km), and Conductance (G, mho/km). These primary constants, as distributed values per distance along the line, are used to model most transmission lines. Secondary parameters such as characteristic impedance, propagation constant (attenuation and phase) or the chain parameters ABCD may be calculated from the primary constants. These "constants" actually vary in value with frequency, temperature and humidity.

Router. A intelligent, protocol-dependent device in a packet network which receives, then re-transmits ("routes") packets to other routers in the network (or to a client) based on addressing information contained in the header of each incoming packet and subject to prevailing network traffic conditions (degree of congestion, outages, etc.). Routers near the edge of the network can connect LANs by dynamically routing data on the most expedient route through the networks. Since they can connect subnetworks together, routers are useful in dividing what would be a large, complex, congestive network into smaller subnetworks. However, routers introduce longer delays and typically have much lower throughput rates than bridges. Routers operate at the Layer 3 Network Layer of the OSI Reference Model.

RRD. Revised Resistant Design.

RS-232. An industry standard for serial communications connections. The current version of this standard is RS-232E. Most PCs include one or more RS-232 ports for connecting devices such as a modem and a mouse.

RSVP. See Resource Reservation Setup Protocol.

RT. See Remote Terminal.
RTCP. See Real-Time Control Protocol.
RTP. See Real-time Transfer Protocol.
RTSP. Real-Time Streaming Protocol.
RTU. See Remote Termination Unit.
rt-VBR. See Real Time Variable Bit Rate.

S

Satellite Access. A competitor to xDSL, satellites can deliver high-bandwidth Internet access to a huge geographic area. This involves the use of technology/service known as Direct Broadcast Satellite (DBS), also known as Digital Satellite Service (DSS). If your home or business has a clear line of sight to the southern sky, you can order a special type of Internet access through your DBS service. This requires the installation of a small dish (usually 18 to 21 inches across), mounted outdoors to receive data sent from a stationary satellite. Tests have shown that DBS service providers can theoretically deliver download speeds of about 350 Kbps.

Scalable. Capable of being scaled; the ability to pattern, make, set, regulate, or estimate according to some rate or standard. The best telecom systems are modular and can be easily scaled up or down in size, depending on the number of users served.

SCR. See Sustainable Cell Rate.
SDH. See Synchronous Digital Hierarchy.
SDSL. See Symmetric Digital Subscriber Line.
SEF. Severely Errored Frame.

Semiconductor. Typically made from either silicon or germanium, semiconductors make it possible to miniaturize electronic components, such as transistors, into integrated circuits which little space, are fast and extremely energy-efficient.

Serial Communication. The transmission of data one bit at a time over a single line. Serial communications can be synchronous or asynchronous.

Serial Interface. Interface that requires serial transmission, or the transfer of information in which the bits composing a character are sent sequentially. Implies only a single transmission channel.

Serial Line Internet Protocol (SLIP). An older Layer 2 protocol used to run IP packets over serial lines such as telephone circuits or RS-232 cables interconnecting two systems Internet traffic; less sophisticated than PPP, which pretty much replaced it for use in establishing dialup connections to the Internet.

Server. A device or system (usually housed in a networked computer) which has been specifically configured to provide a service, usually to a group of clients. "Client-server" computing is the driving force behind the construction of many corporate local area networks and the World Wide Web.

Server Clustering. Network design in which all the servers on one or more rings are placed in a central location.

Server Farm. Network design in which a cluster of servers in a centralized location serve a wide user population.

Service Level Agreement (SLA). A contract between a network service provider and a customer that specifies, usually in measurable terms, what services the network service provider will furnish to the customer/subscriber. Often an SLA outlines the minimum acceptable performance parameters (such as delay, throughput, percent availability, etc.) for public data services such as ATM, Frame Relay, VPNs, and Internet access. A related ATM term is "Traffic Contract" which is an agreement between the user and the network regarding the expected QoS that the ATM network is to provide that user and the set of traffic parameters with which the ATM network user is expected to comply.

Service User (SU). The end-user at the customer premises.

Serving Central Office. Another name for the central office in the local communications network that is directly connected to the end-user location in question. See Central Office, Service Wire Center.

Service Wire Center (SWC). Also called a Serving Wire Center. The wire center from which the customer's premises or CO would normally obtain dial tone from the Local Exchange Company.

Serving Wire Center Area. The territory encompassed by a Service Wire Center.

Session Initiation Protocol (SIP). Is a text-based protocol used by media gateway controllers to communicate with each other and provide call control to users making voice calls in packet networks. SIP competes to some degree with H.323 and will become that standard signaling protocol to support multimedia sessions in next-generation networks.

Set Top Box. A transmission/reception device that acts as an interface typically to a television or other video output display device. In addition to ADSL, SDSL, HDSL, and VDSL interfaces, set top units are increasingly modular, and other interfaces can include Ethernet, MMDS, coaxial cable, V.34 modem, and ISDN, among others.

SF. In T-carrier parlance, a DS-1 Superframe, which consists of 12 frames in a 1.5 ms superframe. This has been mostly superseded by the Extended SuperFrame (ESF).

SHDSL. See Single High-bit-rate Digital Subscriber Line.

Shielded Twisted Pair. Transmission medium that consists of a Receive (RX) and a Transmit (TX) wire twisted together to reduce crosstalk. The twisted pair is shielded from RFI by a braided outer metallic sheath; used for rates of around 16 Mbps. Compare with Twisted Pair.

SID. See Silence Insertion Descriptor.

Signaling. The process of sending a transmission over a physical medium for purposes of communication.

Signaling Gateway. In a distributed Softswitch architecture, a signaling gateway provides the connection to the SS7 network, and then provides signaling control information to the Call Agent.

Signaling System 7 (SS7). The SS7 network carries the signaling information regarding the set up, teardown, billing, and various other telephone services for calls on the PSTN. The SS7 network is a completely and physically separate network from the PSTN, and is thus based upon "out-of-band" signaling.

Signal-to-Noise Ratio (SNR). A signal quality measure. Measure of link performance arrived at by dividing signal power by noise power. Typically measured in decibels. The higher the ratio, the clearer the connection.

Silence Insertion Descriptor (SID). One method used to improve VoDSL and packet transmissions is "silence suppression" which prevents bandwidth from being wasted on the transmission of silent pauses between words or sentences. Silence suppression involves the use of sending and receiving vocoders, which have the ability to exchange details on the length of silence, overhead, etc. A "silence insertion descriptor" is used to carry this information between vocoders to ensure consistent voice quality. See Silence Suppression.

Silence Suppression. Refers to removal of the data that represents the periods of silence in a conversation from a digitized voice stream, to reduce bandwidth consumption of the voice stream.

Simple Network Management Protocol (SNMP). Popular network management protocol used to manage and monitor network devices on TCP/IP-based networks such as the Internet.

Simplex transmission. Data transmission that can occur in only one direction on a given line. Compare with half duplex and full duplex.

Single-Mode Fiber. Fiber with a relatively narrow diameter, through which only one mode will propagate. Carries higher bandwidth than multimode fiber, but requires a light source with a narrow spectral width.

Single Pair High-bit-rate Digital Subscriber Line (G.SHDSL). Also known as SHDSL or Symmetric High-bit-rate DSL, G.SHDSL was the first true international DSL standard, having been ratified by the ITU-T as G.991.2 in April 2001. A hybrid of HDSL2 and SDSL, that offers businesses high-speed local loop access to the Internet with simultaneous telephone network access over a single twisted copper pair. G.SHDSL supports bit rates from 192 Kbps to 2.312 Mbps (2.304 Mbps with some equipment) on one wire pair and 384 kbps to 4.624 Mbps on two pairs. G.SHDSL and can support T-1, E-1, ISDN, ATM, and IP signals.

SIP. See Session Initiation Protocol

SLA. See Service Level Agreement.

SLC. See Subscriber-Loop Carrier.

SLIP. See Serial Line Internet Protocol.

Small Office/Home Office (SOHO). A term used to a 1- or 2-person office or an office that someone has set up in his or her home.

SMDS. See Switched Multimegabit Data Device.

SNA. See Systems Network Architecture.

SNMP. See Simple Network Management Protocol.

SNR. See Signal-to-Noise Ratio.

Softswitch. The term refers to modern software-based telephony switches which replace the traditional Class 4 and Class 5 switches of the PSTN and which can deal with packet-based next generation networks. Usage of the term Softswitch has varied and as a result the idea of a softswitch is somewhat amorphous, but it is generally said to contain at least a media gateway controller and a signaling gateway.

SOHO. See Small Office/Home Office.

SONET. See Synchronous Optical Network.

Source-Route Transparent Bridge. Proposed IEEE 802.1 bridge to combine source routing (in which the source end system provides routing information) with transparent bridging (in which the bridge makes independent message-handling choices and therefore is transparent to the message source and destination).

Special Services. Any telephone customer service that is neither POTS nor an interoffice message trunk.

Splitter. A filtering device used in ADSL to allow users to continue to use their analog telephones while at the same time accessing the Internet and

Glossary

web for digital information. Backward compatibility is the key idea here. The analog voice goes into the voice switch at the CO while data packets can be sent onto the Internet and Web through a router. Also known as a POTS splitter.

Spoofing. The use of a forged IP source address to circumvent a firewall; the packet appears to have come from inside the protected network, and to be eligible for forwarding into the network.

Spread Spectrum. A technique of sharing a media by greatly expanding the bandwidth of each channel by coordinated frequency hops, carrier sweeping or by code division multiplexing (CDM). See also FDM, TDM.

SS#7. Signaling System Number 7. See Signaling System 7.

SS7. See Signaling System 7.

Stack. A group of network devices that are logically integrated into a single system.

Standard. A set of technical specifications used to establish uniformity in hardware and in software.

Static IP Addressing. An assigned IP address used to connect to a TCP/IP network. The same IP number is used every time the connection is made. Contrast with Dynamic IP Addressing.

STM. See Synchronous Transfer Mode; Synchronous Transport Module.

STP. See Shielded Twisted Pair.

Streaming Video. Streaming video is a sequence of images sent in compressed form over the Internet and displayed by the viewer as they arrive. Streaming media consists of streaming video with sound. With streaming video or streaming media, you do not have to wait to download a large file before seeing the video or hearing the sound. Instead, the media is sent in a continuous stream and is played as it arrives.

Structured Service. Refers to a type of Circuit Emulation (i.e., CES) service in which DS-1 channels on a physical interface may be split off and transported across different ATM PVCs.

STS. See Synchronous Transport Signal.

STS-1 (Synchronous Transport Signal 1). The fundamental SONET standard for transmission over optical fiber at 51.84 Mbps.

STS-3 (Synchronous Transport Signal 3). The fundamental SONET standard for transmission over optical fiber at 155.52 Mbps.

STS-n (Synchronous Transport Signal-n). Electrical counterpart to optical OC-n. See also SONET, OC-n, VT.

SU. See Service User.

Subnet mask. A 32-bit number that determines how an IP address is split into network and device portions on a bit-by-bit basis. For example, the subnet mask of 255.255.000.000 identifies that the first two bytes of an IP address belong to the network, and the last two bytes refer to the device (i.e., representing a standard class B subnet mask). One subnet mask applies to all IP devices on an individual IP network.

Subscriber Loop. The pair of copper wires that connects the end-user to the telephone network. Another name for the Local Loop.

Subscriber Network Interface. SMDS term describing generic access to an SMDS network over a dedicated circuit, which can be DS-0, DS-1, or DS-3.

Subscriber-Loop Carrier (SLC). AT&T term for computerized aggregating substation for voice traffic used outside the 3.4 mile range of the central office switch. Aggregates local loops from customers into T-1 lines leading back to CO. Also used instead of a repeater. See Digital Loop Carrier.

Sustainable Cell Rate (SCR). The maximum throughput bursty traffic can achieve within a given virtual circuit without cell loss.

SVC. See Switched Virtual Circuit.

SWC. See Service Wire Center.

Switch. The equipment that connects users of the telecommunications network. In the PSTN world, each subscriber has a dedicated loop to the nearest telephone switch. All of these switches have access to trunk lines for making calls beyond the local exchange area. A call from one user to another consists of a loop at each end of the connection, with switches and trunk lines used to route the connection between them. In the packet-switched world, a data switch connects computing devices to host computers, allowing a large number of devices to share a limited number of ports. ATM switches are multiport devices that perform cell switching.

Switched 56. Switched data transmission service at 56 Kbps (as opposed to service on dedicated leased lines).

Switched Ethernet. Ethernet hub with integrated MAC-layer bridging or switching capability to provide each port with 10 Mbps or more of bandwidth; separate transmissions can occur on each port of the switching hub, and the switch filters traffic based on destination MAC address.

Switched Multimegabit Data Device (SMDS). A packet switching, connectionless data service based on global addressing that enables communications between LANs, typically at speeds between 1.5 Mbps and 34 Mbps.

Switched Virtual Circuit (SVC). A term found in ATM and Frame Relay networking (and used with some VoDSL equipment) in which a virtual connection, with variable endpoints, is established through an ATM network at the time the call is begun; the SVC is de-established at the conclusion of the call. See also Permanent Virtual Circuit (PVC).

Switching Hubs. Hubs that use intelligent Ethernet switching technology to interconnect multiple Ethernet LANs and higher-speed LANs such as FDDI.

Symmetric Digital Subscriber Line (SDSL). A "flavor" of DSL that offers a symmetric transmission of data at the same speed as HDSL, with two important differences: it can be done using only one phone line and the user must be no more than 10,000 feet from the phone company's central office. It also supports POTS service "under" DSL. SDSL is the forerunner to HDSL2. Asymmetric data flow is acceptable for most residential or very small applications, but some applications (especially for businesses) require more upstream capacity than ADSL offers.

SDSL is called symmetric because it supports the same data rates for upstream and downstream traffic. The equal speeds make this good for local area network (LAN) access, distributed applications, video-conferencing, and for sites hosting their own Web sites. SDSL also requires a special modem. This is an umbrella term for which a number of supplier-specific implementations over a single copper pair providing variable rates of symmetric service exist, with or without ordinary telephone service. SDSL is moving towards standardization within ETSI (European Telecommunications Standards Institute). SDSL tops out at 768 Kbps both upstream and downstream, and is one of the more popular "flavors" of DSL used in VoDSL deployment.

Symmetric Flavors of DSL. Symmetrical variations include SDSL, HDSL, HDSL-2, IDSL and G.SHDSL

Synchronous. Data communications in which transmissions are sent at a fixed rate, with the sending and receiving devices transmitting at the same rate.

Synchronous Clocking. An ATM network transmit clock that is produced by an external source. This clock (or primary reference signal or PRS) is distributed throughout the ATM network so that other devices can synchronize to the PRS.

Synchronous Digital Hierarchy (SDH). An ITU-T optical-transmission standard, similar to SONET, for synchronous digital transmission rates used outside North America. The basic SDH rate (STM-1) is 155.52 Mbps.

Synchronous Optical Network (SONET). An ANSI standard for high capacity transmissions of digital information over optical networks. Fiber optic transmission rates range from 51.84 Mbps to 13.22 Gbps. It defines a physical interface, optical line rates known as Optical Carrier (OC) signals, frame formats, and an OAM(&P) protocol. The base rate is known as OC-1 and runs at 51.84 Mbps. Higher rates are a multiple of this such that OC-12 is equal to 622 Mbps (12 times 51.84 Mbps).

Synchronous Transfer Mode (STM). Newer

Glossary

term for traditional TDM data switching and transmission to distinguish it from ATM. See also ATM.

Synchronous Transmission. Data transmission using synchronization bytes, instead of start or stop bits, to control the transmission. More complex than asynchronous transmission. In xDSL, video streams are considered to be synchronous in nature. See Asynchronous Transmission.

Synchronous Transport Module (STM). The basic electrical signal of SDH. STM-1 is a single payload STM signal running at 155.52 Mbps.

Synchronous Transport Signal (STS). The basic electrical signal of SONET. Examples are as follows: STS-1 is an STS signal running at 51.84 Mbps; STS-3 represents three multiplexed STS-1 signals (155.52 Mbps); STS-3c represents a Concatenated (single payload) STS running at 155.52 Mbps.

Systems Network Architecture (SNA). A description of the logical structure and protocols that transmit information and control the operation on an IBM network.

T

T-1 (also T1). Old Bell System term for a point-to-point carrier facility that contains two-sets of twisted-pair cables for transmission of digital data at 1.544 megabits per second in each direction. A T-1 can carry 24 simultaneous uncompressed phone calls (each at 64 Kbps) or a single 1.544 Kbps "pipe."

T-1 Robbed Bit. Type of signaling used during T-1 transmissions in which the control and signaling information is buried with the voice channels, using the least-significant bits. T-1 Robbed Bit circuits provide a simple signaling mechanism between the customer premise equipment and the central office.

T-3 (also T3). A digital circuit connection having a transmission capacity of 44.736 Mbps. A channelized T-3 carries 672 simultaneous uncompressed phone calls (each at 64 Kbps), while an unchannelized T-3 can transmit a single 44.736 mbps datastream "pipe". A channelized T-3 is also referred to as a Digital Services Level 3 or DS-3.

TA. See Terminal Adapter.

TAPI. See Telephony Application Programming Interface.

Tariff. A rate and availability schedule for telecommunications services that is filed with and approved by a regulatory body to become effective. Tariffs also include general terms and conditions of service.

TC. Traffic Contract (See Service Level Agreement).

TCM. See Time Compression Multiplexing; Trellis Coded Modulation.

TCP. See Transmission Control Protocol.

TCP/IP. See Transmission Control Protocol/Internet Protocol.

TCP/IP Stack. The software that allows a computer to communicate via TCP/IP.

TDM. See Time Division Multiplexing.

TDMA (Time Division Multiple Access). A shared channel access mechanism based on time division multiplexing. Often used in wireless transmissions.

Telco. A popular generic term for a local telephone company (including RBOCs, LECs and PTTs) In the U.S., the major telcos are the seven regional Bell operating companies and the leading independent telcos, GTE, SNET, and Sprint; in Europe, Asia, and elsewhere, the term "telco" generally refers to the incumbent monopoly, but increasingly refers to competing local providers as well.

Telcordia. Originally Bellcore (Bell Communications Research). Former research arm of the regional telephone companies. Bellcore was part of Bell Laboratories before the breakup of AT&T. Name changed to Telecordia Technologies in 1998.

Telecommunications Act of 1996. Statute passed by the U.S. Congress to open up competition in the telecommunications and cable industries.

Telecommunications Industries Association (TIA). Industry standards group, accredited by ANSI, that deals with telecommunications equipment and systems. See also EIA.

Telephony. The marriage of computers and

telecommunications.

Telephony Application Programming Interface (TAPI). A Windows standard for controlling any kind of telephone interaction. TAPI also arbitrates conflicts between applications requesting use of communications ports, modems, and so on.

Telephony over Passive Optical Network (TPON). Telephony using a PON as all or part of the transmission system between telephone switch and subscriber.

Telework. The combination of computer and telecommunications technology which enables office workers to work at home or away from the main office on a part-time or full-time basis.

Telnet. The virtual terminal protocol in the Internet suite of protocols. Allows users of one host to log into a remote host and interact as normal terminal users of that host.

Terminal Emulation. Network software application that makes a PC or workstation appear to a network host to be a directly attached dumb terminal.

Terminal Server. Communications processor that connects asynchronous devices to a LAN or WAN through network and terminal emulation software.

TFTP. See Trivial File Transfer Protocol.

Throughput. The effective rate of data flow you an transfer through a voice or data network.

TIA. See Telecommunications Industries Association.

Time Compression Multiplexing (TCM). Permits two-way transmission by the use of alternating short one-way transmission bursts. Also known as Ping-Pong.

Time Division Multiple Access. See TDMA.

Time Division Multiplexing (TDM). A way of combining multiple data streams into a single large data stream by allotting the large "pipe" to different channels, one at a time at regular intervals (called "time slots"). TDM systems allow for the synchronous transmission of data over a fixed bandwidth channel. The popular TDM transports include T-1 (1.5 Mbps) transport and a T-3 (45 Mbps) transport. This transmission medium is ideal for traffic requiring a CBR transport, such as voice.

TIRKS. See Trunk Integrated Records Keeping System.

TM. See Traffic Management.

Token. Control information frame, possession of which grants a network device the right to transmit.

Token Ring. As defined in IEEE 802.5, a communications method that uses a token to control access to the LAN. The difference between a token bus and a Token Ring is that a Token Ring LAN does not use a master controller to control the token. Instead, each computer knows the address of the computer that should receive the token next. When a computer with the token has nothing to transmit, it passes the token to the next computer in line.

Topology. A network's topology is a logical characterization of how the devices on the network are connected and the distances between them. The most common network devices include hubs, switches, routers, and gateways. Most large networks contain several levels of interconnection, the most important of which include edge connections, backbone connections, and wide-area connections.

TP. See Tunneling Protocol.

TPON. See Telephony over Passive Optical Network.

Traffic Management (TM). The set of algorithms and techniques used to manage the flow of traffic through the ATM network to avoid congestion and deliver the contracted QoS for each connection.

Traffic Shaping. A method for adapting the flow of traffic entering the ATM network by temporarily buffering it and transmitting those buffered cells at a rate conforming to the stated traffic contract.

Transcoder. A device that converts a data stream encoded according to a set of protocols, to a different set of protocols.

Transmission Control Protocol (TCP). Connection-oriented protocol that provides a reliable byte stream over IP. A reliable connection means that each end of the session is guaranteed to receive all of the data transmitted by the other end of the connection, in the same order that it was originally transmitted, without receiving duplicates.

Transmission Control Protocol/Internet Protocol (TCP/IP). A packet-based communication protocol that is the dominant protocol suite in the worldwide Internet, TCP is Layer 4, the transport layer. IP is Layer 3, the network layer. The protocols are the result of the Defense Advanced Research Projects Agency (DARPA) project to interconnect disparate computer networks of the 1970s.

Transmission Technologies. Fundamental mechanisms for network communications. Transmission technologies include Ethernet, Frame Relay, ISDN, and Asynchronous Transfer Mode (ATM).

Transparent Bridge. Acts like a station on two or more LANs; it listens to all packets on the wire, stores each received packet, and forwards it on to the other LANs when it has permission to send. To cut down on unnecessary forwarded packets, the bridge learns the location of stations and does not forward packets that it knows it does not need to. Bridging scheme preferred by Ethernet and IEEE 802.3 networks in which bridges pass frames along one hop at a time based on tables associating end nodes with bridge ports.

Transport Systems Generic Requirements (TSGR). Bellcore Technical Reference TR-TSV-000499. The umbrella document for network transmission requirements. A module of Network Systems Generic Requirements (NSGR), Technical Reference TR-TSV-000440. See also NSGR and OTGR.

Trellis Coded Modulation (TCM). Provides error protection without adding bandwidth, but does add delay (latency).

Trellis Coding. A form of error correction found in many modems that allows for forward error correcting to account for bit errors from various interference on communications lines, such as crosstalk and background noise.

Trivial. See File Transfer Protocol (TFTP).

Trojan Horse. A packet sniffer that hides its sniffing activity. These packet sniffers can collect account names and passwords for Internet services, allowing a hacker to gain unauthorized access to other machines.

Trunk. A link between network nodes. Most commonly used to refer to links between voice CO (or LEC) switches, though the term has been frequently applied to links between a telephone service provider and an Internet service provider, although technically these are just local loops (or access lines). However, modern usage employs the term trunk to indicate any links that are not specifically between a user and a network (e.g., voice switch network node to Internet router network node).

Trunk Integrated Records Keeping System (TIRKS). OS for special service and interoffice message trunk design and records of equipment, facilities and circuits. (TIRKS is a registered Bellcore trademark).

Trunk-side T-1. A T-1 line that has a direct digital connection to the phone network, and therefore undergoes no analog conversions in the path between the V.90 digital modem and the PSTN.

TSGR. See Transport Systems Generic requirements.

Tunnel Switching. The ability to terminate a tunnel and initiate a new tunnel to one of a number of subsequent tunnel terminators; tunnel switching extends the PPP connection to a further end point.

Tunneling. Encapsulation of one protocol within another for security purposes and for transporting a LAN protocol across a backbone that does not support that protocol.

Tunneling Protocol (TP). A technology that enables one network to send its data via another network's connections. Tunneling works by encapsulating a network protocol within packets carried by the second network. For example, Microsoft's PPTP technology enables organizations to use the Internet to transmit data across a VPN. It does this by embedding its own network protocol within the TCP/IP packets carried by the Internet.

Twinaxial. Cable containing two coaxial cable runs; typically used in a IBM AS/400 environment to connect IBM 5250s to the host.

Twisted Pair. A common form of copper cabling

used for telephony and data communications. It consists of two copper lines (named tip and ring) twisted around each other at the six twists per inch; the twisting protects the communications from electromagnetic frequency and radio frequency interference. Cable may be shielded (STP) or unshielded (UTP). See also Unshielded Twisted Pair.

U

U Interface. Two-wire interface presented to the customer by the Telco in the U.S. market. The customer is responsible for converting this signal to the four-wire S/T interface in order to make a connection, usually by adding an external NT-1 device. Also called the U Reference Point.

UADSL. See Universal ADSL.

UART. See Universal Asynchronous Receiver/Transmitter.

UBR. See Unspecified Bit Rate.

UDC. See Universal Digital Channel.

UI (Unit Interval). The unit of measure for jitter. One UI of jitter is equal to the width of 1 data bit, regardless of the data rate. Thus, at the T-1 rate of 1544 kbs, one UI is equal to 647 ns. At the E1 data rate of 2048 kbs, one UI is equal to 488 ns.

Unbalanced Line. A transmission line in which the two conductors have unequal electromagnetic characteristics with respect to ground and the external environment including other conductors. Coaxial cable is a common example. See also Balanced Line, Balun.

Unbundled Elements (UNE). The Federal Telecom Act of 1996 and the accompanying FCC order requires ILECs to unbundle telecommunications services into their component elements. The act further requires that the ILECs sell these parts to competitors at cost plus a reasonable rate of return. One of these unbundled elements is the local loop.

UNE. See Unbundled Elements.

UNI. See User Network Interface.

Uniform Resource Locator (URL). Internet address that provides the information needed to access a particular page.

Uninterruptible Power Supply (UPS). A device that ensures a backup power supply for electrical devices in the event of a power outage. NT1 Plus devices can include UPS for maintaining power for analog voice communications during a power outage.

Universal ADSL (UADSL). Also referred to as DSL lite or G.lite, UADSL is focused on providing a mass-market version of ADSL, which is interoperable with full rate ADSL, but with fewer complexities and less overall requirements at a tradeoff for speed. The solution is intended to reduce the need for a "splitter" box installed outside the home or new wiring in the home. UADSL enables plug-and-play and PC-integrated solutions.

Universal Asynchronous Receiver/Transmitter (UART). UART chips are the part of your PC's COM port that handles communications between the CPU and the device attached to the COM port.

Universal Digital Channel (UDC). A small, single-loop, pairgain system using a DSL system as the transport technique.

Universal Serial Bus (USB). A serial bus standard developed in the mid-1990s. USB connections attach personal computers to keyboards, printers, and other peripherals, delivering power to devices on the bus and eliminating separate power cords. The USB 1.0 standard provides a data rate of 12 megabits per second and supports up to 127 devices. It's slower than its rival FireWire (IEEE 1394), but is less expensive to implement. USB delivers complete plug-and-play capabilities to devices as well as hot-swap capability. And because of the small size of the USB connector, it can be used on notebook and handheld computers. Devices adopting the newer USB 2.0 standard can transmit at a more impressive 480 Mbps.

Unshielded Twisted Pair (UTP). A cable with one or more twisted copper wires bound in a plastic sheath. Preferred method to transport data and voice to business workstations and telephones. Unshielded wire is preferred for transporting high-speed data because at higher speeds electromagnetic radiation is created. If shielded cabling is used, the radiation is not released and creates interference. UTP cabling does not need the fixed spacing between connections that is necessary with coaxial-type connections. See Category 5 cabling.

Unspecified Bit Rate (UBR). An ATM service category which does not specify traffic related service guarantees and does does not include a per-connection negotiated bandwidth. No numerical commitments are made with respect to the cell loss ratio experienced by a UBR connection, or as to the cell transfer delay experienced by cells on the connection. The UBR service category is thus somewhat similar to the nrt-VBR category, because it is used for connections that transport variable bit rate traffic for which there is no reliance on time synchronization between the traffic source and destination. However, unlike the ABR service category, there are no flow-control mechanisms to dynamically adjust the amount of bandwidth available to the user. UBR generally is used for applications that are very tolerant of delay and cell loss, and has been used in Internet, LAN and WAN environments for store-and-forward traffic, such as file transfers and e-mail.

Unstructured Service. Refers to a type of Circuit Emulation (i.e., CES) service in which all DS1 channels on a physical interface are transported over the same ATM PVC.

UPS. See Uninterruptible Power Supply.

Upstream. Typically refers to the transmission direction from the customer premises toward the DSLAM and the CO or ISP, such as sending e-mail or uploading files.

URL. See Uniform Resource Locator.

Usage Sensitive. The cost of a service, such as ISDN or analog telephone service, that is based on the time you actually use the service.

USB. See Universal Serial Bus.

User Network Interface (UNI). ATM Forum specification that defines an interoperability standard for the interface between ATM-based products (a router or an ATM switch) located in a private network and the ATM switches located within the public carrier networks.

USOC. Universal Service Order Code: Identification code for a tariffed network service.

UTP. See Unshielded Twisted Pair.

V (MDF). Vertical side of MDF: Loop connections and primary over-current protection. See also H (MDF) and MDF.

VAD. See Voice Activity Detection.

VADSL. See Very High Speed ADSL.

Variable Bit Rate (VBR). Service in which the payload bit rate varies with time and information content. See also CBR.

VBR. See Variable Bit Rate.

VBR-nrt. See Non Real Time Variable Bit Rate

VBR-rt. See Real Time Variable Bit Rate.

VC. See Virtual Channel; Virtual Circuit.

VCC. See Virtual Channel Connection.

VCI. See Virtual Channel Identifier.

VCL. See Virtual Channel Link.

VDSL. Also called VHDSL. See Very High bit rate Digital Subscriber Line.

VDSL Transmission Unit-Central Office (VTU-C). The VDSL modem or line card that physically terminates a VDSL connection at the telephone service provider's serving central office

Very High Speed ADSL (VADSL). Same as VDSL (or a subset of VDSL, if VDSL includes symmetric mode transmission).

Very High Speed Digital Subscriber Line (VHDSL, or VDSL). See Very High-bit-rate Digital Subscriber Line

Very High-bit-rate Digital Subscriber Line (VHDSL, or VDSL). Also called Very High Speed Digital Subscriber Line. The fastest xDSL technology, using a modem for twisted pair access operating at data rates from 12.9 to 52.8 Mbps downstream and 1.5 to 2.3 megabits (or more) upstream, with corresponding maximum reach ranging from 4500 feet to 1000 feet of 24-gauge twisted pair. In most cases,

Glossary

VDSL lines will be served from neighborhood cabinets that link to a Central Office via optical fiber. It is well-suited for "campus" environments such as universities, business parks and resorts. VDSL will make high-quality video on demand possible over existing phone lines.

VHDSL. See Very High-bit-rate Digital Subscriber Line.

Video Information Provider (VIP). Source of video programming material for distribution over high-bandwidth xDSL or other broadband delivery services.

Video On Demand (VOD). The ability to activate a stored or live motion picture stream; in xDSL the application that allows subscribers to view movies or other video programming on request, similar to cable television's Pay-Per-View. See Pay-Per-View.

Video Telephony. The ability to view real-time video communications on a two-way or multipoint basis. Also called videoconferencing.

VIP. See Video Information Provider.

Virtual Channel (VC). As an ATM term, it means, according to the ITU-T: "a unidirectional communication capability for the transport of ATM cells." As a SONET term, it represents a SONET emulation of a traditional TDM channel.

Virtual Channel Connection (VCC). The set of virtual channel links that connect a source and destination across an intervening ATM network.

Virtual Channel Identifier (VCI). The address or label of a virtual circuit. As an ATM term, it is a unique numerical tag as defined by a 16-bit field in the ATM cell header that identifies a virtual channel, over which the cell is to travel.

Virtual Channel Link (VCL). The segment of a VCC over which a VCI applies; the VC connecting two ATM nodes.

Virtual Circuit (VC). A logical connection, or private packet-switching mechanism created between two devices at the beginning of a transmission over a public facility, such as an ATM or IP network.

Virtual LAN (VLAN). Workstations connected to an intelligent device which provides capabilities to define LAN membership.

Virtual Path Identifier (VPI). As an ATM term, it is an eight-bit field in the ATM cell header which indicates the virtual path over which the cell should be routed.

Virtual Private Network (VPN). A way to deliver private data safely over a public network, such as the Internet. The data traveling between two hosts are encrypted for privacy using both hardware and software solutions (such a tunneling protocol). VPNs give companies the same capabilities provided by owned or leased lines at much lower cost by using the shared public infrastructure rather than a private one.

Virtual Tributary (VT). SONET channels defined to transport traditional digital hierarchy signals. VT1.5, VT2, VT3, and VT6 can carry payloads of DS-1 (1.544 Mbps), E-1 (2.048 Mbps), DS1C (3.152 Mbps) and DS-2 (6.312 Mbps) respectively.

Viterbi Algorithm. An algorithm used for reception of trellis coded modulation.

VLAN. See Virtual LAN.

VMOA. See Voice and Media over ATM.

VoB. Voice over Broadband. Umbrella term covering almost any means of delivering packetized voice (as opposed to TDM) over broadband.

Vocoder. A vocoder is another name for voice codec, which is used for the compression/decompression of voice signals in order to reduce and thus conserve the bandwidth or bit-rate of the voice stream.

VOD. See Video On Demand.

VoDSL. See Voice over Digital Subscriber Line.

VoFR. Voice Over Frame Relay.

Voice Activity Detection (VAD). VAD is used to detect speech on a line in order to facilitate silence suppression.

Voice and Multimedia over ATM (VMOA). Generic set of standards defined by the ATM Forum by a working group of the same name.

Voice and Telephone over ATM (VToA). Former name of standards governing VoDSL. For

ATM-based VoDSL this evolved into Broadband Loop Emulation Service (BLES). Not to be confused with the more general term Voice Traffic over ATM.

Voice Frequency. In telephony, typically the range is from zero to 3.4 kHz.

Voice Gateway. An IP or ATM-based network element that converts digital voice so it can be sent to the analog-based telephone network.

Voice over Cable (VoCable). The transmission of a packet-voice stream over a broadband data cable network. A competitor to VoDSL.

Voice over Digital Subscriber Line (VoDSL). The transmission of a channelized or packet-voice stream over a DSL network.

Voice over Ethernet. Using Ethernet instead of ATM to send voice packets over xDSL services.

Voice over IP (VoIP). A term used for a set of facilities for managing the delivery of voice information using Internet Protocol (IP).

Voice over Multiservice Broadband Networks (VoMBN). The delivery of voice over one or more networks where the main transport is a packet network, not TDM. Call control is decentralized and separated from the actual call being controlled.

Voice over Packet (VoP). The transmission of a packet-voice stream over a packet data network.

Voice over Wireless Local Loop (VoWLL). The transmission of a packet-voice stream over a broadband wireless network to a fixed customer premise.

VoIP. See Voice over IP.

VoIP Gateway. A Gateway that is used for transcoding between an IP-based packet network and a switched voice network (commonly the PSTN).

VoMBN. See Voice over Multiservice Broadband Networks.

VoMSDN. Voice over Multiservice Data Network. See Voice over Multiservice Broadband Networks.

VoP. See Voice over Packet.

VoWLL. See Voice over Wireless Local Loop.

VP. Virtual Path. A set of VCs bundled together between two ATM nodes.

VPC. See Virtual Path Connection. The set of virtual path links that connect a source and destination across an intervening ATM network.

VPI. See Virtual Path Identifier.

VPL. See Virtual Path Link.

VPN. See Virtual Private Network.

VT. See Virtual Tributary.

VToA. See Voice and Telephone over ATM.

VTU-C. See VDSL Transmission Unit-Central office

W

Wall Jack. A small hardware component used to tap into telephone wall cable. An RJ-11 wall jack usually has four pins; an RJ-45 wall jack usually has eight pins.

WAN. See Wide Area Network.

Wander. Low frequency jitter. ITU-T G.810 places the limit between jitter and wander at 10 Hz. Wander applies to frequency below 10 Hz; jitter above 10 Hz.

WC. See Wire Center.

White Noise. Noise with equal power at all frequencies.

Whiteboard. Collaboration software typically bundled with desktop video-conferencing systems. It allows two users to share a computer screen just as people share a whiteboard in a meeting room.

Wide Area Network (WAN). Private telecommunications network facilities that link geographically dispersed facilities (branch offices, for example) that reside in different regions or cities. Typically offered as a service by telephone companies but also available from alternative access providers such as CLECs or CAPs.

WinSock. A program that conforms to a set of standards called the Windows Socket API (Application Programming Interface). A WinSock program controls the link between Microsoft Windows software and a TCP/IP program.

Wire Center (WC). Telephone building that is the origin of the outside loop plant. Usually has one or more Central Offices.

Wireless Local Loop (WLL). A wireless connection between the telephone companies and service providers to a fixed-location customer premise. New wireless local loop services offer broadband data services, including Voice over Wireless Local Loop (VoWLL).

World Wide Web (WWW). The Internet-based hypertext multimedia-based system for accessing Internet resources. Commonly referred to as the Web (or web), it lets users download files, listen to audio, and view images and videos. Users can jump around the Web using hyperlinks embedded in documents.

XYZ

X.25. The protocol for packet-mode services as defined by CCITT. A CCITT interface standard that lets computing devices communicate via wide area packet-switched data networks.

xDSL. A generic term for the suite of DSL services, where the "x" can be replaced with any of a number of letters. (ADSL, HDSL, HDSL2, SDSL, SHDSL, VDSL, etc.)

Xmodem. Xmodem is an error-correcting protocol for analog modems. Modems that agree on using the Xmodem protocol send data in 128-byte blocks. If a block is received successfully, a positive (ACK) acknowledgement is returned. If an error is detected, a negative (NAK) acknowledgement is returned and the block is resent. Xmodem uses the checksum method of error checking.

ZBTSI. Zero Byte Time Slot Interchange: DS1 ESF alternative to B8ZS. Uses 2 Mbps of the 4 Mbps ESF bandwidth. Usable on any type of facility not just a T-1 Line.

Index of Figures

1.1. Magneto set telephone from the late 1890s to early 1900s. 7
1.2. 1890s Western Electric Common Battery Telephone. 7
1.3. Wall Jack. 8
1.4. Type 66 punchdown block. 9
1.5. Holtzer-Cabot ringing voltage generator. 16
1.6. Schematic of Tielines. 18
1.7. Replica of the world's first switchboard. 19
1.8. Photo of building that housed world's first telephone exchange. 19
1.9. Schematic of centralized switch configuration. 20
1.10. Major components of the Outside Plant. 21
1.11. Station Protector. 22
1.12. Photos of early telephone poles. 22
1.13. Toroidal loading coil. 25
1.14. Photo of cable vault. 31
1.15. A Main Distribution Frame at the central office. 32
1.16. The Smart-MDF from OKI. 33
1.17. A switchboard (and central office) from the late 1870s. 34
1.18. A central office in 1884. 36
1.19. A switchboard from circa 1930. 36
1.20. How telephone operators performed an exchange trunk connection
       prior to 1915. 38
2.1. TDM versus packet-switched access networks. 49
2.2. A central office-based FDDI network. 53
2.3. ATM Layers. 60
2.4. ATM Adaptation Layer. 61
2.5. ATM cell formats of the ATM Layer. 64
2.6. ATM bandwidth utilization. 66

Index of Figures

2.7. Vinton G. Cerf and Robert E. Kahn, who developed TCP/IP. 68
3.1. Alcatel / Thomson multimedia ADSL modems. 81
3.2. ADSL Lite. 87
3.3. Full Service Network Architecture using DSLAM and ATM. 102
3.4. Typical DLC deployment. 109
3.5. Cross-Connects before and after DSL deployment. 109
3.6. Typical RAM deployment. 112
3.7. Types of Crosstalk. 114
4.1. Typical VoDSL solution. 131
5.1. Derived voice taxonomy. 137
5.2. Protocol stacks for VoATM AAL2. 155
5.3. AAL2 CPS packet format. 157
5.4. AAL2 protocol in action. 157
5.5. I.366.2 packet formats. 160
5.6. CPS Lite Option in the Loop Emulation Service. 167
5.7. Broadband Loop Emulation ATM VoDSL. 168
5.8. Inefficient xDSL data packet aggregation and transport. 169
5.9. DSLAM aggregation with ATM switch. 170
5.10. Integrated voice and aggregation using General Bandwidth G6. 171
5.11. Transporting voice packet via IP. 175
5.12. Bandwidth efficiency of Voice-over-IP and AAL2 on ATM-based DSL. . . 176
5.13. Mapping of AAL2 Packet Voice to Voice over IP. 179
5.14. UDP segment. 183
5.15. RTP Segment Structure. 185
5.16. Comparison between Local Exchange Switch and Softswitch. 187
5.17. Integrated PSTN and Internet Telephony. 189
5.18. Media Gateway for packet-to-circuit access
     with integrated signaling gateway. 198
5.19. Sun Microsystems design for a complete softswitch. 198
5.20. Examples of the Gluon Networks CLX. 199
5.21. Basic steps needed to convert
     analog voice into an Ethernet packet. 200

Index of Figures

5.22. CVoDSL's less complicated architecture
compared to VoATM-based VoDSL. 203
5.23. G.SHDSL used in three kinds of Cisco access network solutions. 204
5.24. Voice entering the stack. 204
6.1. Coppercom IADs. 206
7.1. Rapidly declining port costs for packet equipment. 236
7.2. VoDSL provider outsourcing tradeoffs. 253
8.1. Avail Networks' system based on
Integrated Concentration Device (ICD). 269
8.2. ARESCOM's CDS 6000 platform for multi-tenant units. 272
8.3. Global VoDSL demand. 274
8.4. Extreme Networks architecture for Ethernet delivered over VDSL. 279
8.5. Nortel's 9110 and 9115 work with ordinary Nortel Median phones. . . . 287
9.1. Bandwidth and "reaches" of various DSLs. 309
A.1. OSI Model. 337

From CMP Books

www.cmpbooks.com

Disaster Survival Guide
For Business Communications Networks
Strategies for Planning, Response and Recovery in Data and Telecom Systems
by Richard Grigonis

Order Now! $49.95

DISASTER RECOVERY GUIDE

This book is written specifically for IT system administrators and telecom managers who want to protect their organization's increasingly complicated and diverse telecom and datacom networks, as well as the telephone systems, Internet sites, computers and data-laden storage devices attached to them. This unique guide says that managers must take a new approach in protecting their assets. Instead of just implementing monitoring and detection measures with immediate intervention to combat disasters and ensure business continuity, managers should extend and intensify their security planning phase, reducing the possible magnitude of catastrophic incidents in advance by re-engineering their business to achieve a distributed, decentralized and hardened organization. This can be done using the latest in communications technology (virtual private networks, conferencing technology, the Internet) as well as networking and managing a new generation of the traditional "uptime" technologies (fault tolerant computer telephony, convergence and power supply systems)

ISBN 1-57820-117-9

Find CMP Books in your local bookstore, or go to www.cmpbooks.com
Tel: 1-800-500-6875 • Fax 1-408-848-5784 • Email: cmp@rushorder.com

CMP Books

From CMP Books
www.cmpbooks.com

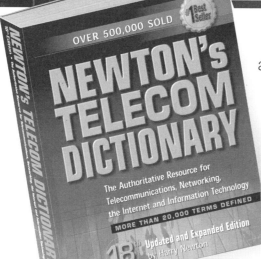

18th Updated and Expanded Edition

Order Now!

$34.95

NEWTON'S TELECOM DICTIONARY

Long considered required reading for anyone involved in the telecommunications industry, Newton's Telecom Dictionary is more essential than ever. New telecom technologies and terms are being created every day. In this classic telecom text, telecom terms and concepts are explained in easy to understand definitions and mini essays written in business (not technical) language.
ISBN 1-57820-104-7

Find CMP Books in your local bookstore, or go to www.cmpbooks.com
Tel: 1-800-500-6875 • Fax 1-408-848-5784 • Email: cmp@rushorder.com

CMPBooks

From CMP Books

www.cmpbooks.com

Order Now!

$39.95

A PRACTICAL GUIDE TO DSL

A Practical Guide to DSL is a must for anyone looking to purchase DSL for high-speed Internet service over existing telephone lines.

The book provides an in-depth discussion of what DSL is, what you can do with it and how to purchase it, including a DSL shopping checklist. But it goes beyond the basics and explains all the technological issues you must consider before venturing into the world of high-speed Internet access.

The author also addresses the problems typically associated with installing DSL service and provides helpful hints regarding what you should look for in a service provider.

The Practical Guide to DSL doesn't stop there — it deftly explains how to set up a home or small office network, the equipment involved, the installation process and gives a solid overview of the different kinds of networks available for the home and small business environments.

ISBN 1-57820-060-1

Find CMP Books in your local bookstore, or go to www.cmpbooks.com
Tel: 1-800-500-6875 • Fax 1-408-848-5784 • Email: cmp@rushorder.com

CMP Books